Communications
in Computer and Information Science   454

More information about this series at http://www.springer.com/series/7899

Ana Fred · Jan L.G. Dietz
Kecheng Liu · Joaquim Filipe (Eds.)

Knowledge Discovery, Knowledge Engineering and Knowledge Management

5th International Joint Conference, IC3K 2013
Vilamoura, Portugal, September 19–22, 2013
Revised Selected Papers

 Springer

Editors

Ana Fred
Technical University of Lisbon
Lisbon
Portugal

Jan L.G. Dietz
Faculty of Electrical Engineering,
   Mathematics and Computer Science
Delft University of Technology
Delft
The Netherlands

Kecheng Liu
Henley Business School
University of Reading
Reading
UK

Joaquim Filipe
INSTICC
Setubal
Portugal

ISSN 1865-0929   ISSN 1865-0937 (electronic)
Communications in Computer and Information Science
ISBN 978-3-662-46548-6 ·   ISBN 978-3-662-46549-3 (eBook)
DOI 10.1007/978-3-662-46549-3

Library of Congress Control Number: 2015935419

Springer Heidelberg New York Dordrecht London

Printed on acid-free paper

Springer-Verlag GmbH Berlin Heidelberg is part of Springer Science+Business Media
(www.springer.com)

Preface

The present book includes extended and revised versions of a set of selected papers from the Fifth International Joint Conference on Knowledge Discovery, Knowledge Engineering, and Knowledge Management (IC3K 2013), held in Vilamoura, Algarve, Portugal, during September 19–22, 2013. IC3K was sponsored by the Institute for Systems and Technologies of Information, Control and Communication (INSTICC) and held in Vilamoura, Algarve, Portugal and was organized in cooperation with the AAAI - Association for the Advancement of Artificial Intelligence, DICODE - Mastering Data-Intensive Collaboration and Decision Making Project, TIMBUS - Digital Preservation for Timeless Business Processes and Services, iKMS - Information & Knowledge Management Society, ACM SIGMIS - ACM Special Interest Group on Management Information Systems, ACM SIGIR - ACM Special Interest Group on Information Retrieval, ACM SIGART - ACM Special Interest Group on Artificial Intelligence and ERCIM - European Research Consortium for Informatics and Mathematics. IC3K is also technically co-sponsored by APRP - Associação Portuguesa de Reconhecimento de Padrões.

The main objective of IC3K is to provide a point of contact for scientists, engineers, and practitioners interested in the areas of Knowledge Discovery, Knowledge Engineering, and Knowledge Management.

IC3K is composed of three colocated complementary conferences, each specialized in one of the aforementioned main knowledge areas. Namely:

- International Conference on Knowledge Discovery and Information Retrieval (KDIR)
- International Conference on Knowledge Engineering and Ontology Development (KEOD)
- International Conference on Knowledge Management and Information Sharing (KMIS)

The International Conference on Knowledge Discovery and Information Retrieval (KDIR) aims to provide a major forum for the scientific and technical advancement of knowledge discovery and information retrieval. Knowledge Discovery is an interdisciplinary area focusing upon methodologies for identifying valid, novel, potentially useful and meaningful patterns from data, often based on underlying large datasets. A major aspect of Knowledge Discovery is data mining, i.e., applying data analysis and discovery algorithms that produce a particular enumeration of patterns (or models) over the data. Knowledge Discovery also includes the evaluation of patterns and identification of which add to knowledge. Information retrieval (IR) is concerned with gathering relevant information from unstructured and semantically fuzzy data in texts and other media, searching for information within documents and for metadata about documents, as well as searching relational databases and the Web. Information retrieval can be combined with knowledge discovery to create software tools that empower users of decision support systems to better understand and use the knowledge underlying large datasets.

The purpose of the International Conference on Knowledge Engineering and Ontology Development (KEOD) is to provide a major meeting point for researchers and practitioners interested in the study and development of methodologies and technologies for Knowledge Engineering and Ontology Development. Knowledge Engineering (KE) refers to all technical, scientific, and social aspects involved in building, maintaining, and using knowledge-based systems. KE is a multidisciplinary field, bringing in concepts and methods from several computer science domains such as artificial intelligence, databases, expert systems, decision support systems, and geographic information systems. Currently, KE is strongly related to the construction of shared knowledge bases or conceptual frameworks, often designated as ontologies. Ontology Development aims at building reusable semantic structures that can be informal vocabularies, catalogs, glossaries as well as more complex finite formal structures representing the entities within a domain and the relationships between those entities. A wide range of applications is emerging, especially given the current web emphasis, including library science, ontology-enhanced search, e-commerce, and business process management.

The goal of the International Conference on Knowledge Management and Information Sharing (KMIS) is to provide a major meeting point for researchers and practitioners interested in the study and application of all perspectives of Knowledge Management and Information Sharing. Knowledge Management (KM) is a discipline concerned with the analysis and technical support of practices used in an organization to identify, create, represent, distribute, and enable the adoption and leveraging of good practices embedded in collaborative settings and, in particular, in organizational processes. Effective knowledge management is an increasingly important source of competitive advantage, and a key to the success of contemporary organizations, bolstering the collective expertise of its employees and partners. Information Sharing (IS) is a term used for a long time in the information technology (IT) lexicon, related to data exchange, communication protocols, and technological infrastructures.

The joint conference, IC3K received 239 paper submissions from 57 countries, which demonstrates the success and global dimension of this conference. From these, 23 papers were published as full papers, 60 were accepted for short presentation, and another 47 for poster presentation. These numbers, leading to a "full-paper" acceptance ratio of about 10 % and an oral paper acceptance ratio close to 35 %, show the intention of preserving a high-quality forum for the next editions of this conference.

On behalf of the conference Organizing Committee, we would like to thank all participants. First of all to the authors, whose quality work is the essence of the conference and to the members of the Program Committee, who helped us with their expertise and diligence in reviewing the papers. As we all know, producing a conference requires the effort of many individuals. We wish to thank also all the members of our Organizing Committee, whose work and commitment were invaluable.

October 2014

Ana Fred
Jan L.G. Dietz
Kecheng Liu
Joaquim Filipe

Organization

Conference Chair

Joaquim Filipe   Polytechnic Institute of Setúbal/INSTICC, Portugal

Program Co-chairs

KDIR

Ana Fred   Technical University of Lisbon, Portugal

KEOD

Jan L.G. Dietz   Delft University of Technology,
The Netherlands

KMIS

Kecheng Liu   University of Reading, UK

Organizing Committee

Marina Carvalho   INSTICC, Portugal
Helder Coelhas   INSTICC, Portugal
Vera Coelho   INSTICC, Portugal
Bruno Encarnação   INSTICC, Portugal
Ana Guerreiro   INSTICC, Portugal
André Lista   INSTICC, Portugal
Andreia Moita   INSTICC, Portugal
Raquel Pedrosa   INSTICC, Portugal
Vitor Pedrosa   INSTICC, Portugal
Susana Ribeiro   INSTICC, Portugal
Sara Santiago   INSTICC, Portugal
Mara Silva   INSTICC, Portugal
José Varela   INSTICC, Portugal
Pedro Varela   INSTICC, Portugal

KDIR Program Committee

Sherief Abdallah   British University in Dubai, UAE
Muhammad Abulaish   Jamia Millia Islamia, India
Samad Ahmadi   De Montfort University, UK
Francisco Martínez Álvarez   Pablo de Olavide University of Seville, Spain

Eva Armengol	IIIA CSIC, Spain
Zeyar Aung	Masdar Institute of Science and Technology, UAE
Márcio Basgalupp	Universidade Federal de São Paulo, Brazil
Isabelle Bichindaritz	State University of New York at Oswego, USA
Marc Boullé	Orange Labs, France
Emanuele Di Buccio	University of Padua, Italy
Maria Jose Aramburu Cabo	Jaume I University, Spain
Luis M. de Campos	University of Granada, Spain
Keith C.C. Chan	The Hong Kong Polytechnic University, Hong Kong
Meng Chang Chen	Academia Sinica, Taiwan
Shu-Ching Chen	Florida International University, USA
Philipp Cimiano	University of Bielefeld, Germany
Juan Manuel Corchado	University of Salamanca, Spain
Jerome Darmont	Université de Lyon, France
Marcos Aurélio Domingues	University of São Paulo, Brazil
Dejing Dou	University of Oregon, USA
Antoine Doucet	University of Caen, France
Habiba Drias	USTHB, LRIA, Algeria
Tapio Elomaa	Tampere University of Technology, Finland
Iaakov Exman	The Jerusalem College of Engineering, Israel
Philippe Fournier-Viger	University of Moncton, Canada
Fabrício Olivetti de França	Universidade Federal do ABC, Brazil
Ana Fred	Technical University of Lisbon, Portugal
Susan Gauch	University of Arkansas, USA
Rosalba Giugno	University of Catania, Italy
Manuel Montes y Gómez	INAOE, Mexico
Nuno Pina Gonçalves	EST-Setúbal/IPS, Portugal
Daniel Gruhl	IBM Almaden Research Center, USA
Antonella Guzzo	University of Calabria, Italy
Yaakov Hacohen-Kerner	Jerusalem College of Technology (Machon Lev), Israel
Greg Hamerly	Baylor University, USA
José Hernández-Orallo	Universitat Politècnica de València, Spain
Enrique Herrera-Viedma	University of Granada, Spain
Beatriz de la Iglesia	University of East Anglia, UK
Szymon Jaroszewicz	Polish Academy of Sciences, Poland
Mouna Kamel	IRIT, France
Ron Kenett	KPA Ltd., Israel
Hagen Langer	University of Bremen, Germany
Anne Laurent	LIRMM, Montpellier 2 University, France
Carson K. Leung	University of Manitoba, Canada
Jiexun Li	Drexel University, USA
Xiaoli Li	Nanyang Technological University, Singapore

Xia Lin   Drexel University, USA
Berenike Litz   Attensity Corporation, USA
Rafael Berlanga Llavori   Jaume I University, Spain
Alicia Troncoso Lora   Pablo de Olavide University of Seville, Spain
Edson T. Matsubara   UFMS, Brazil
Misael Mongiovi   University of Catania, Italy
Stefania Montani   University of Piemonte Orientale, Italy
Henning Müller   University of Applied Sciences Western
                                       Switzerland, Switzerland
Engelbert Mephu Nguifo   LIMOS, Université Blaise-Pascal, France
Giorgio Maria Di Nunzio   Università degli Studi di Padova, Italy
Mitsunori Ogihara   University of Miami, USA
Krzysztof Pancerz   University of Information Technology and
                                       Management in Rzeszow, Poland
Luigi Pontieri   National Research Council of Italy (CNR), Italy
François Poulet   University of Rennes 1 - IRISA, France
Ronaldo Prati   Universidade Federal do ABC, Brazil
Alfredo Pulvirenti   University of Catania, Italy
Marcos Gonçalves Quiles   Federal University of São Paulo - UNIFESP, Brazil
Bijan Raahemi   University of Ottawa, Canada
Zbigniew W. Ras   University of North Carolina at Charlotte, USA
Fabio Rinaldi   University of Zurich, Switzerland
Ovidio Salvetti   National Research Council of Italy (CNR), Italy
Filippo Sciarrone   Open Informatica srl, Italy
Fabricio Silva   Brazilian Army Technological Center, Brazil
Dominik Slezak   Infobright, Poland
Marcin Sydow   IPI PAN (and PJIIT), Warsaw, Poland
Andrea Tagarelli   University of Calabria, Italy
Moritz Tenorth   University of Bremen, Germany
Andrew Beng Jin Teoh   Yonsei University, Korea
Ulrich Thiel   Fraunhofer Gesellschaft, Germany
Kar Ann Toh   Yonsei University, Korea
Yannick Toussaint   Inria, France
Domenico Ursino   Mediterranea University of Reggio Calabria, Italy
Sebastian Wandelt   Humboldt-Universität zu Berlin, Germany
Yang Xiang   The Ohio State University, USA
Jierui Xie   Samsung R&D Center, USA
JingTao Yao   University of Regina, Canada

KDIR Auxiliary Reviewers

Helena Moniz   INESC-ID, Portugal
Ricardo Ribeiro   ISCTE-IUL and INESC-ID Lisboa, Portugal

KEOD Program Committee

Alia Abdelmoty	Cardiff University, UK
Ajith Abraham	Machine Intelligence Research Labs (MIR Labs), USA
Alessandro Agostini	University of Trento, Italy
Salah Ait-Mokhtar	Xerox Research Centre Europe, France
Masanori Akiyoshi	Hiroshima Institute of Technology, Japan
Raian Ali	Bournemouth University, UK
Carlo Allocca	FORTH Research Institute, University of Crete, Greece
Yuan An	Drexel University, USA
Francisco Antunes	Institute of Computer and Systems Engineering of Coimbra and Beira Interior University, Portugal
Marie-aude Aufaure	École Centrale Paris, France
Hilton de Azevedo	Universidade Tecnológica Federal do Paraná, Brazil
Costin Badica	University of Craiova, Romania
Claudio de Souza Baptista	Universidade Federal de Campina Grande, Brazil
Jean-Paul Barthes	Université de Technologie de Compiègne, France
Teresa M.A. Basile	Università degli Studi di Bari, Italy
Sonia Bergamaschi	University of Modena and Reggio Emilia, Italy
Damir Boras	Faculty of Humanities and Social Sciences, University of Zagreb, Croatia
Patrick Brezillon	LIP6 - Pierre-and-Marie-Curie University (Paris 6), France
Giacomo Bucci	Università degli Studi di Firenze, Italy
Vladimír Bureš	University of Hradec Králové, Czech Republic
Doina Caragea	Kansas State University, USA
Jin Chen	Michigan State University, USA
Ruth Cobos	Universidad Autónoma de Madrid, Spain
James Crawford	Google, USA
Sally Jo Cunningham	University of Waikato, New Zealand
Fabiano Dalpiaz	University of Toronto, Canada
Jan L.G. Dietz	Delft University of Technology, The Netherlands
John Edwards	Aston University, UK
Magdalini Eirinaki	San José State University, USA
Anna Fensel	STI Innsbruck, University of Innsbruck, Austria
Dieter A. Fensel	University of Innsbruck, Austria
Raúl Garcia-Castro	Universidad Politécnica de Madrid, Spain
Faiez Gargouri	ISIMS, Tunisia
Dimitrios Georgakopoulos	CSIRO, Australia
Manolis Gergatsoulis	Ionian University, Greece
George Giannakopoulos	SKEL Laboratory – NCSR Demokritos, Greece
Rosario Girardi	UFMA, Brazil

Matteo Golfarelli	University of Bologna, Italy
Sven Groppe	University of Lübeck, Germany
Ourania Hatzi	Harokopio University of Athens, Greece
Christopher Hogger	Imperial College London, UK
Angus F.M. Huang	Academia Sinica, Taiwan
Kulkarni Anand Jayant	OAT Research Lab, Maharashtra Institute of Technology, India
John Josephson	The Ohio State University, USA
Achilles Kameas	Hellenic Open University, Greece
Dimitris Kanellopoulos	University of Patras, Greece
Nikos Karacapilidis	University of Patras and CTI, Greece
Pinar Karagoz	METU, Turkey
Katia Lida Kermanidis	Ionian University, Greece
Patrick Lambrix	Linköping University, Sweden
Antoni Ligeza	AGH University of Science and Technology, Poland
Elena Lloret	University of Alicante, Spain
Adolfo Lozano-Tello	Universidad de Extremadura, Spain
Xudong Luo	Sun Yat-Sen University, China
Paolo Manghi	Institute of Information Science and Technologies, Italy
Rocio Abascal Mena	Universidad Autónoma Metropolitana – Cuajimalpa, Mexico
Andres Montoyo	University of Alicante, Spain
Claude Moulin	JRU CNRS Heudiasyc, Université de Technologie de Compiègne, France
Ana Maria Moura	National Laboratory for Scientific Computing - LNCC, Brazil
Phivos Mylonas	National Technical University of Athens, Greece
Kazumi Nakamatsu	University of Hyogo, Japan
William Nelson	Devry University, USA
Erich Neuhold	University of Vienna, Austria
Jørgen Fischer Nilsson	Technical University of Denmark, Denmark
Laura Papaleo	Province of Genoa, Italy
Irina Perfilieva	University of Ostrava, Czech Republic
Mihail Popescu	University of Missourii – Columbia, USA
Nives Mikelic Preradovic	Faculty of Humanities and Social Sciences, University of Zagreb, Croatia
Violaine Prince	LIRMM-CNRS, France
Juha Puustjärvi	University of Helsinki, Finland
Amar Ramdane-Cherif	Versailles Saint-Quentin-en-Yvelines University, France
M. Teresa Romá-Ferri	University of Alicante, Spain
Inès Saad	France Business School, France
Masaki Samejima	Osaka University, Japan
Martín Serrano	National University of Ireland, Galway, Ireland

Nuno Silva	Polytechnic of Porto, Portugal
Kiril Simov	Bulgarian Academy of Sciences, Bulgaria
Anna Stavrianou	Xerox Research Centre Europe, France
Mari Carmen Suárez-Figueroa	Ontology Engineering Group, Universidad Politécnica de Madrid, Spain
Domenico Talia	ICAR-CNR and University of Calabria, Italy
Jiao Tao	Oracle, USA
Gheorghe Tecuci	George Mason University, USA
Annette Ten Teije	VU University Amsterdam, The Netherlands
Dhavalkumar Thakker	University of Leeds, UK
Orazio Tomarchio	University of Catania, Italy
Shengru Tu	University of New Orleans, USA
Manolis Tzagarakis	Department of Economics, University of Patras, Greece
Rafael Valencia-Garcia	Universidad de Murcia, Spain
Iraklis Varlamis	Harokopio University of Athens, Greece
Cristina Vicente-Chicote	Universidad de Extremadura, Spain
Bruno Volckaert	Ghent University, Belgium
Sebastian Wandelt	Humboldt-Universität zu Berlin, Germany
Gian Piero Zarri	Sorbonne University, France
Jinglan Zhang	Queensland University of Technology, Australia
Catherine Faron Zucker	I3S, Université Nice Sophia Antipolis, CNRS, France

KEOD Auxiliary Reviewers

José María García	University of Innsbruck, Austria
Nikolaos Lagos	Xerox Research Centre Europe, France

KMIS Program Committee

Marie-Helene Abel	HEUDIASYC CNRS UMR, Université de Technologie de Compiègne, France
Adriano Albuquerque	University of Fortaleza - UNIFOR, Brazil
Miriam C. Bergue Alves	Institute of Aeronautics and Space, Brazil
Rangachari Anand	IBM T.J. Watson Research Center, USA
Rajeev K. Bali	Coventry University, UK
Alessio Bechini	University of Pisa, Italy
Ralph Bergmann	University of Trier, Germany
Elsa Cardoso	ISCTE-IUL, Portugal
Marcello Castellano	Politecnico di Bari, Italy
Xiaoyu Chen	SKLSDE, Beihang University, China
Dickson K.W. Chiu	Dickson Computer Systems, Hong Kong
Byron Choi	Hong Kong Baptist University, Hong Kong
Giulio Concas	Università di Cagliari, Italy

KMIS Auxiliary Reviewers

Nicola Epicoco	Politecnico di Bari, Italy
Shixiong Liu	University of Reading, UK
Diego Magro	University of Turin, Italy
Nada Nadhrah	Henley Business School, University of Reading, UK
Shen Xu	University of Reading, UK

Invited Speakers

Alan Eardley	Staffordshire University, UK
David Aveiro	University of Madeira, Portugal
Belur V. Dasarathy	Information Fusion, USA
Nikos Karacapilidis	University of Patras and CTI, Greece
Alexandre Castro-Caldas	Portuguese Catholic University, Portugal

Contents

Knowledge Management and Information Sharing

Invited Papers

Enterprise Ontology and DEMO Benefits, Core Concepts and a Case Study

David Aveiro[1,2,3(✉)]

[1] Exact Sciences and Engineering Centre, University of Madeira,
Caminho da Penteada, 9020-105 Funchal, Portugal
[2] Center for Organizational Design and Engineering, INESC-INOV,
Rua Alves Redol 9, 1000-029 Lisbon, Portugal
[3] Madeira Interactive Technologies Institute, Caminho da Penteada,
9020-105 Funchal, Portugal
daveiro@uma.pt

Abstract. This paper aims to introduce the field of enterprise ontology and enterprise engineering, as well as the related Design and Engineering Methodology for Organizations (DEMO). Several core concepts and benefits are introduced, followed by the presentation of a case study confirming some of these benefits. In this case study of a practical enterprise change project DEMO was used in the initial stage as to give a neutral and concise but comprehensive view of the organization of a local government administration having the purpose to implement an e-government project. On how applying DEMO we were able to confirm, in practice, its qualities of conciseness and comprehensiveness. Namely, we present in this paper a generic pattern that reflects the functioning of a local government Center of Veterinary Care (CVC) service and its integration with other local and regional government entities. The DEMO based specification of this CVC gave important insights to (1) perceive current operational constraints and (2) devise a strategical roadmap for the implementation phase of the e-government project.

Keywords: Enterprise ontology · Enterprise engineering · E-Government · Enterprise change · DEMO · Case study · Method · Tool

1 Introduction

The complexity of both organizations and information and communication technology (ICT) are becoming unmanageable. The maintenance costs of ICT applications are ever increasing and a positive correlation between the investments and the revenues cannot still be found. Information systems professionals fall short in assisting enterprises to implement change initiatives effectively with ICT projects, failing to meet final user's expectations. The vast majority of strategic initiatives fail, meaning that enterprises are unable to gain success from their strategy. From [1], where some case studies were made, a recent survey with 800 IT managers [2, 3], found that 63 % of software development projects failed, 49 % suffered budget overruns, 47 % had higher than expected maintenance costs and 41 % failed to deliver the expected business value and

© Springer-Verlag Berlin Heidelberg 2015
A. Fred et al. (Eds.): IC3K 2013, CCIS 454, pp. 3–20, 2015.
DOI: 10.1007/978-3-662-46549-3_1

user's expectations. From these case studies, it was found that some of the common causes of software failures are: the lack of clear, well-thought-out goals and specifications, poor management and poor communication among costumers, designers and programmers [4], unrealistically low budget requests, and underestimates of time requirements, use of new technologies maybe for which the software developers don't have adequate experience and expertise and refusal to recognize or admit that a project is in trouble [1].

The organizational professionals also fall short in assisting enterprises to implement change initiatives effectively and the key reason is their predominant managerial approach (function-orientation, black-box thinking), whereas changing an enterprise requires an engineering approach (construction-orientation, white-box thinking). Research indicates that the key reason for strategic failures is the lack of coherence and consistency among the various components of an enterprise [5].

To address this challenges a paradigm shift is required as proposed by the discipline of Enterprise Engineering (EE) [6]. The mission of the EE discipline is to develop new and appropriate theories, models, methods and other artifacts for the analysis, design, implementation, and governance of enterprises by combining (relevant parts of) management and organization science, information systems science, and computer science.

In this paper we introduce the field of enterprise ontology and enterprise engineering, as well as the related Design and Engineering Methodology for Organizations (DEMO), where several core concepts and benefits are presented.

This paper also presents a case study reflecting some useful outcomes of a practical enterprise change project where the Design and Engineering Methodology for Organizations (DEMO) was used in the initial stage as to give a neutral and concise but comprehensive view of the organization of a local government administration having the purpose to implement an e-government project.

In Sect. 2, we do a brief introduction on Enterprise Engineering and DEMO. In Sect. 3 we present our Case Study. Section 4 explores ways of Improving DEMO's Way of Working based on our experience. Section 5 wraps it up with a Results Analysis and Evaluation and finally, in Sect. 6, we present our Conclusions.

2   Enterprise Engineering and DEMO

2.1   Enterprise Engineering

Premises of Enterprise Engineering proposed in [7] are: (1) the enterprise can be viewed as a complex system, (2) the enterprise is to be viewed as a system of processes that can be engineered both individually and holistically, (3) the use of engineering rigor in transforming the enterprise. The enterprise is viewed as a complex system of processes that can be engineered to accomplish specific organizational objectives. In [8] it is explained that by enterprise engineering it is meant the whole body of knowledge regarding the development, implementation, and operational use of enterprises, as well as its practical application in engineering projects. In [9] it is stated that the current situation in the organizational sciences resembles very much the one that existed in the information system sciences around 1970. At that time, a revolution took place in the way people conceived information technology and its applications. Since then, people are aware of the distinction between the form and the content of information. This revolution marks the transition from the era of data systems engineering to the era of information systems engineering. The comparison drawn with the information sciences is not an arbitrary one. On the one hand, the key enabling technology for shaping future enterprises is the modern Information and Computer Technology (ICT). On the other hand, there is a growing insight within the information system sciences that the central notion for understanding profoundly the relationship between organization and ICT is the entering into and complying with commitments between social individuals. These commitments are raised in communication, through the so-called intention of communicative acts. Examples of intentions are requesting, promising, stating, and accepting. Therefore, just like the content of communication was put on top of its form in the 1970 s, the intention of communication is now put on top of its content. It explains and clarifies the organizational notions of collaboration and cooperation, as well as notions like authority and responsibility.

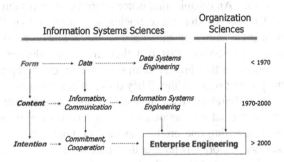

Fig. 1. The roots of enterprise engineering.

This current revolution in the information systems sciences marks the transition from the era of information systems engineering to the era of enterprise engineering. At the same time, it enables information systems engineering to converge with the traditional organizational sciences, as illustrated in Fig. 1 [9].

The mission in current efforts of developing the EE discipline is: to develop new and appropriate theories, models, methods and other artifacts for the analysis, design, implementation, and governance of enterprises by combining (relevant parts of) management and organization science, information systems science, and computer science [6]. The goals of the EE discipline are summarized in the following points:

Intellectual Manageability – Define proper theories about the construction and operation of enterprises, in order to get and keep insight and overview concerning enterprises and enterprise changes, and to master their complexities, and define ideas of enterprise evolvability for making changes expeditious and manageable;

Organizational Concinnity – Define a viable theory and methodology for enterprise engineering, able to address all relevant aspects, even those that cannot be foreseen presently, in a properly integrated way, so that the operational enterprise is always a coherent and consistent whole thus designing organizational concinnity and not hopping for it to emerge in a natural way;

Social Devotion – Taking a human centered view i.e. to consider the importance of employee involvement and participation for enterprise productivity, product and service quality, customer orientation, learning and innovation (and subsequent enterprise change), as well as for coping with enterprise dynamics, complexity, and uncertainty leading to emerging enterprise developments. All employees are fully empowered and competent for the tasks they have to perform and endorsed with transparent authority and have access to all information they need in order to perform their tasks in a responsible way;

With these goals it is expected that, the results of the efforts should be theoretically rigorous and practically adequate [6].

2.2   The Ψ – Theory

The Ψ-theory provides an explanation of the construction and the operation of organizations, regardless of their particular kind or branch (like industry or government, or manufacturing or service). An organization is considered to be a system in the category of social systems. This means that the elements are social individuals, i.e. human beings or subjects in their ability of entering into and complying with commitments regarding the products or services that that they create and deliver. The operating principle of organizations is that human beings enter into and comply with commitments regarding the production of things. They do so in communication, and against a shared background of cultural norms and values. These commitments occur in processes that follow the universal transaction pattern, a structure of coordination acts/facts between two actors concerning one production fact/act. One initiator (consumer) and one executor (producer). An organization is then a network of actors and transactions with every actor having a particular authority assigned on the basis of competence. Actors are assumed to exercise the authority with responsibility and act autonomously. The Ψ-theory consists in a set of four axioms and a theorem fully elicited in [8] and summarized next.

2.3   Operation and Transaction Axioms

In the Ψ-theory [10] – on which DEMO is based – the operation axiom [8] states that, in organizations, subjects perform two kinds of acts: production acts that have an effect in the production world or P-world and coordination acts that have an effect on the coordination world or C-world. Subjects are actors performing an actor role responsible for the execution of these acts. At any moment, these worlds are in a particular state specified by the C-facts and P-facts respectively occurred until that moment in time. When active, actors take the current state of the P-world and the C-world into account (Fig. 3). C-facts serve as agenda for actors, which they constantly try to deal with. In other words, actors interact by means of creating and dealing with C-facts. The production acts contribute towards the organization's objectives by bringing about or delivering products and/or services to the organization's environment and coordination acts are the way actors enter into and comply with commitments towards achieving a certain production fact [11].

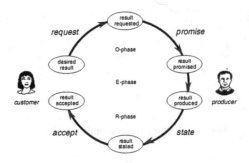

Fig. 2. Basic transaction pattern.

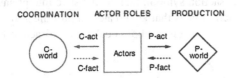

Fig. 3. Actors interaction with production and coordination worlds.

According to the Ψ-theory's transaction axiom the coordination acts follow a certain path along a generic universal pattern called transaction. The transaction pattern has three phases: (1) the order phase, were the initiating actor role of the transaction expresses his wishes in the shape of a request, and the executing actor role promises to produce the desired result; (2) the execution phase where the executing actor role produces in fact the desired result; and (3) the result phase, where the executing actor role states the produced result and the initiating actor role accepts that result, thus effectively concluding the transaction. This sequence is known as the basic transaction pattern and only considers the "happy case" where everything happens according to the expected outcomes. All these five mandatory steps must happen so that a new production fact is realized. In [11] we also find the universal transaction pattern that also considers many other coordination acts, including cancellations and rejections that may happen at every step of the "happy path" [11].

Even though all transactions go through the four – social commitment – coordination acts of request, promise, state and accept, these may be performed tacitly, i.e. without any kind of explicit communication happening. This may happen due to the traditional "no news is good news" rule or pure forgetfulness which can lead to severe business breakdown. Thus the importance of always considering the full transaction pattern when designing organizations. Transaction steps are the responsibility of two specific actor roles. The initiating actor role is responsible for the request and accept steps and the executing actor role is responsible for the promise, execution and state steps. These steps may not be performed by the responsible actor as the respective subjects, may delegate on another subject one or more of the transaction steps under their responsibility, although they remain ultimately responsible for such actions [11] (Fig. 2).

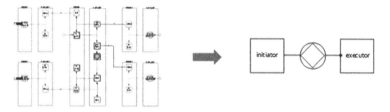

Fig. 4. The organization building block.

Using these two axioms we are able to form the organization building block depicted in Fig. 4. Every (elementary) actor role is the executor of exactly one transaction kind, and initiator of 0, 1 or more transaction kinds, and, next to the process interpretation of the transaction symbol, there is the state interpretation, the conceptual container of all coordination facts that are created in all transactions up to now. In the state interpretation, the transaction symbol is called a transaction bank.

2.4   Distinction Axiom

The distinction axiom from the Ψ-theory states that three human abilities play a significant role in an organization's operation: (1) the *forma* ability that concerns datalogical actions; (2) the *informa* that concerns infological actions; and (3) the performa that concerns ontological actions [8]. Regarding coordination acts, the performa ability may be considered the essential human ability for doing any kind of business as it concerns being able to engage into commitments either as a performer or as an addressee of a coordination act. When it comes to production, the performa ability concerns the business actors. Those are the actors who perform production acts like deciding or judging or producing new and original (non derivable) things, thus realizing the organization's production facts. The informa ability on the other hand concerns the intellectual actors, the ones who perform infological acts like deriving or computing already existing facts. And finally the forma ability concerns the datalogical actors, the ones who perform datalogical acts like gathering, distributing or storing documents and or data. The organization theorem states that actors in each of these abilities form three kinds of systems whereas the D-organization supports the I-organization with datalogical services and the I-organization supports the B-organization (from Business = Ontological) with informational services [11].

2.5   Problems of Traditional Approaches and Benefits of the EE Approaches

In Fig. 5 we illustrate the passing of a document from an activity A to an activity B, a pattern typically found in flowchart diagrams. Although there is no question that this illustrates a datalogical act, we can raise the question if it is really only a datalogical act? For example, if A delivers the document to B for B to archive it, it would be then a purely datalogical act. However if the purpose is to illustrate that A informs B about the content of the document upon the transfer, then it is also an infological act. If A is requesting B to take some decision based on the document it would be also an onto-logical act. For these reasons one cannot reach a conclusion without careful analysis of the case in question and we see how problematic can be the lack of a clear semantics in process diagramming. Current business process modeling approaches, like Flowchart, BPMN, EPC, and Petri Net reduce business processes to sequences of (observable) actions and results, thereby losing the essential deep structure (which is always a tree of transactions) and neglecting all tacitly performed coordination acts, and, therefore they are ambiguous (if not dangerous) for business process re-design and re-engineering. Even worse than these approaches are the function-oriented techniques (SADT, IDEF0) since by definition they reflect the personal interpretation of the modeler. Widespread approaches such as ArchiMate and BPMN suffer highly from the lacking of a solid formal theory behind them and from ambiguous semantics [12, 13]. The DEMO based approach however, is grounded in solid theory, and aims to allow the modeling of processes while capturing vital information of organizational responsibilities and information flows, normally neglected in other approaches.

Fig. 5. Document passing in a process.

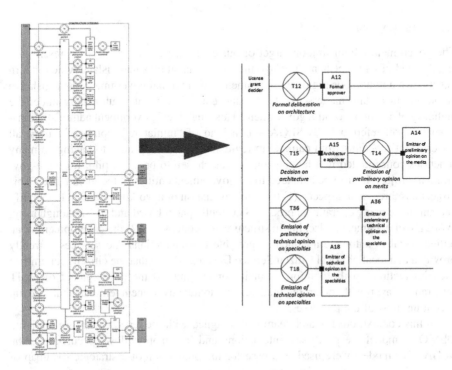

Fig. 6. City Hall licensing process - Tree of transactions.

With DEMO, a business process is modeled as a tree of transactions, as depicted in Fig. 6. This brings about a massive reduction of complexity when compared with the same content in a traditional flowchart approach. In the concrete complex example of a city hall licensing process reported in [14], we found, in the given documentation, 23 flowchart tasks spread over diagrams contained in 7 A4 pages, in 4 documents. Due to a lack of clear semantics of such a flowchart approach, direct interpretation of these flowcharts was either not easy or not possible. These flowcharts were accompanied by descriptions of the tasks contained in 21 A4 pages, also in 4 documents. From the total of 28 pages of content in diagrams and descriptions we could not have a succinct and crisp global view of these processes. By using DEMO, all this information, as well missing process information not written anywhere, was concisely summarized in a set of 38 transactions presented in 2 A4 pages, in the ATD (actors and transactions) view, or in 2.5 A4 pages, in the PSD (transactions process) view. Thanks to the clear semantics of DEMO and the natural devising of more precise and unambiguous names for the transactions and actor roles, the interpretation of DEMO's diagrams is much clearer and more precise than with the flowchart approach. If just looking at transaction and actor role names is not enough, one can look at the associated process description centered around the specified transactions that explains the meaning of all such transactions as well as process flow and inter-dependencies. This description occupies just 2.5 A4 pages [14].

3   Case Study

3.1   Introduction

The government administration target of our enterprise change project is present in a small island of a European archipelago that is dependent on a main island which in turn has its own autonomous regional government. The regional government has legislative autonomy in certain matters like health and education but it is ultimately under the authority of a national central government. This small local government administration – from now on referred to as SLGA – is a kind of "miniature" replica of almost all government functions from national to regional level and – thanks to having so many functions concentrated in a few persons – was chosen to be a test pilot for the e-government project, later to be extended to all government entities of the main island. This project has three main aspects: (1) the implementation of a workflow system to simplify and automate many operational processes currently paper based and/or – although using Word/Excel documents – lacking in structure and coherence; (2) the development of an online portal to automate as much as possible the interactions and services currently provided at a local physical Citizen Service Desk (CSD), so that the citizens can initiate such interactions in the comfort of their homes; and (3) the development of an IT integration layer with other regional and national government entities that end up executing most of the processes.

In this context, our research team was assigned with the responsibility of applying DEMO to model the processes, interactions and information flows occurring in the SLGA. Our models were used as a base for the production of a strategic roadmap of organizational changes that will have to occur for several alternative scenarios of e-government implementations, according to the possible levels of integration and change in current government entities and/or their IT systems. These models are also one of the central pieces of the tender document produced for the bidding stage of the project now approaching. The bidders will have to produce detailed technical designs and blueprints that will have to closely follow the structure of the models that we devised. This will facilitate a systematic insight and overview of the whole project as it is being bid and, later, implemented.

Our team comprised 4 DEMO experts, 2 working in the project full-time and 2 part-time – one 50 % and the other 25 % – totaling 55 man-days in a month of project execution. Many interviews were made to officials head of each of the SLGA's departments and also to most of the officials responsible for each unit of each department. Interviews were made both for information collection and model validation. A final global workshop with the presence of all interviewees – around 20 – was made for final validation where most models were deemed adequately correct and complete after some small corrections and additions. In the end we specified: 216 transactions – and their associated result types; and 232 fact types – these include classes/categories and fact types and exclude properties. We additionally specified 250 ontological transaction kinds that followed a certain repetitive pattern in certain departments and, because of that, were abstracted into a small subset of generic transactions of the above mentioned 216 transactions set. So, in fact, we specified almost 500 transaction kinds in this project. The huge complexity of the SLGA and the short time frame we had to completely model its organization lead to the need of having a very high throughput in model information collection, integration and validation. DEMO proved to indeed facilitate coherence, conciseness and comprehensiveness by making this enormous complexity intellectually manageable in a very short time frame. But we did face some problems in enabling these qualities of the method. The traditional official approaches to apply DEMO would not be feasible in the short time frame we had and current generally available DEMO supporting tools would not allow an effective and efficient propagation of model information and propagation of changes in individual model elements. We thus had to quickly create and apply new method steps and adequate tooling to support our needs. Wrapping up, the main contributions of our work are: (1) a practical validation of some of DEMO's qualities; (2) the specification of a generic and comprehensive DEMO based pattern that can be reused in other e-government transformation projects; and (3) the specification of more adequate method steps and tooling for the production of DEMO's Object Fact Diagram (OFD).

3.2   Center of Veterinary Care

The SLGA's CVC is the only veterinary care available on the island, and as such has to deal with many different matters, that in other places are handled by multiple entities such as pets and farm animals health care; activation, deactivation, reclassification, registry and other activities related to livestock farms as well as all relating legal aspects of the subject; livestock well being controll; identification and registration of livestock; livestock moving permits and control; financial support to the exercise of agricultural activity in the livestock sector decisions, sanitary certificates, vaccination, inspections, control of animal based or animal related goods; and citizen complaints relating pets or farm animals.

Almost all of these processes need to be reported to the Regional Veterinary Authority (RVA), as well as most of the incomes need also to be transfered to this entity. Although, in some cases, such as pets and farm animals health care, the income is local and therefore is transfered to the SLGA. The CVC also relates with the SLGA when it comes to transportation as it does not have own vehicles, therefore these need to be requested to the maintenance and fleet department of the SLGA. Overall the CVC executes over 60 processes and, in our project we applied the DEMO notion of transaction and modeled each of these processes as an individual transaction.

Although the transactions are executed in the CVC, the relating documentation must be scanned and forwarded, either by letter or by email, depending on the case, to the RVA and the originals must archived locally so that all records of all processes are kept. Besides this, there are also multiple interactions with both local, regional and national databases relating the animals that need to happen on most processes. Payments related to processes, that have a cost associated, take place in the CVC, are collected, grouped by entity and, monthly, are forwarded to the responsible entities either by a bank deposit or by a direct delivery depending if belonging to regional or local entity, having copies of all the receipts related to such payments the same

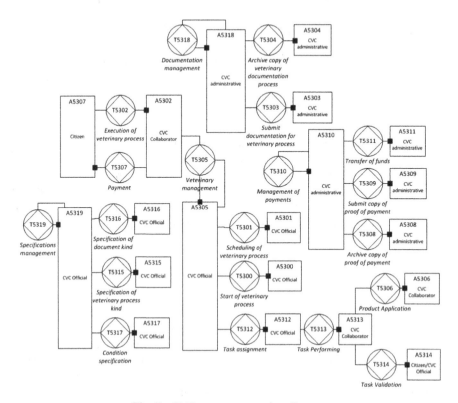

Fig. 7. CVC - actor transaction diagram.

destination. Considering the large amount of transactions that take place in the CVC our first step while specifying them was to try to find a pattern to abstract this complexity that would cover all ontological acts taking place in the CVC. In Fig. 7 we find Actor Transaction Diagram (ATD) of the CVC, where such pattern is specified.

In the identified transactions there are four that are not ontological. Archiving and submitting documents are datalogical but we explicitly specified them already at this stage as they will be a central aspect of the future workflow system to implement. The four transactions with the ID's 5315, 5316, 5317 and 5319 don't currently explicitly take place in the CVC in a structured way, but in a kind of ad hoc way. Namely, a few times per year procedural orders come from some government entity informing of a new process or a change in an existing process. Examples are changes in the required documents or new mandatory conditions for a given process. So the (around 60) kinds of processes currently handled at the CVC, respective rules to follow and associated costs are currently dispersed in many unstructured documents, such as law decree's, regional authorities recommendations and process forms. Our specification gives a needed structure to all this information and their specification. Future changes in the procedures and rules of the processes will be more easily integrated in the daily routine of the CVC workers thanks to the implementation of our specified pattern in the future workflow system.

We can notice four clear clusters in the ATD diagram. The first cluster is the already mentioned specifications management containing the needed transactions necessary to handle the information of the processes, that although non existing at the time, are foreseen as essential to determine each and every process that will take place in CVC in a future workflow implementation. In this cluster, besides the specifications management, we also have the specification of document kind to determine the relating veterinary process of each document, the specification of veterinary process kind to determine the existing different kind of processes and the condition specification to determine the conditions associated to each veterinary process.

The second cluster is the veterinary management cluster that contains the transactions needed to specify the day to day activities of the CVC collaborators. In this cluster we have the citizen requested executions of veterinary processes and their associated payments such as the treatment of an animal, as well as the CVC official induced veterinary processes such as a random control to a livestock transportation that takes place in the island. Most of these processes also require a proper scheduling of the CVC human resources and, in most cases, require the need for transportation that has to be requested to the maintenance and fleet department. Other transactions included in this cluster are the start of a veterinary process, that depending on the case can drag for a long time before its conclusion such as the licensing for livestock creation on a farm, the task assignments, the task performing, the product applications, such as vaccination campaigns, and, finally, the task validation, that can often be done by the CVC official, but in some special cases, is given by the citizen, usually a farmer, when the veterinary needs to perform a service at his or her farm.

The third cluster is the management of payments cluster, that deals with the financial part of the CVC. As part of the SLGA the CVC has no financial autonomy, therefore, the received income has to either be forwarded to the main RVA in cases such as an exploration licensing, or the SLGA in the case of local animal health care

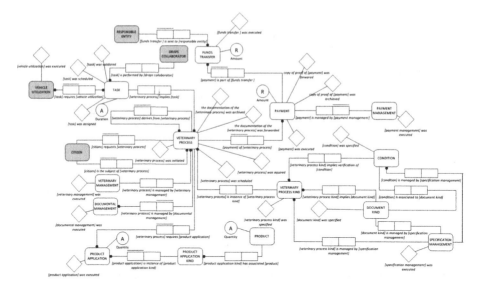

Fig. 8. CSD - object fact diagram.

services. As such, the other transactions present in this cluster are the transfer of funds, and the two datalogical transactions of copy of proof or payment submission and archive of the copy of proof or payment.

The final cluster is the documentation management cluster, that includes the two other datalogical transactions of document of veterinary process submission and archiving. These are again mandatory by the regional authority and therefore essential in the modeling.

In Fig. 8, we find the OFD which, thanks to previous explanations, should be mostly quite easy to understand. The central class of this OFD is the VETERINARY PROCESS. This VETERINARY PROCESS relates with the CITIZENS by either being requested by one or making him the subject of the process. Occasionally a VETERINARY PROCESS may lead to multiple other VETERINARY PROCESSES, and their execution will imply the realization of one or multiple TASKS that, on their hand will have associated a COLLABORATOR and may have a VEHICLE UTILI-ZATION. The VETERINARY PROCESS may also have associated one or multiple PRODUCT APPLICATIONS that, on their turn, are an instance of a PRODUCT APPLICATION KIND that has an associated PRODUCT. The VETERINARY PROCESS may also have a PAYMENT, that will be managed by the PAYMENT MANAGEMENT as it will be part of a FUND TRANSFER to a given ENTITY.

The VETERINARY PROCESS is also managed by both VETERINARY MAN-AGEMENT and DOCUMENTAL MANAGEMENT, and is an instance of a VETERI-NARY PROCESS KIND. This VETERINARY PROCESS KIND implies the verification of given CONDITIONS and the use of determined DOCUMENT KINDS. A CONDI-TION may also be associated to a DOCUMENT KIND, and all the VETERINARY PROCESS KIND, the CONDITION and the DOCUMENT KIND are managed by the SPECIFICATION MANAGEMENT thus concluding the OFD of the CVC.

4 Improving the DEMO WoW and Supporting Tools

In the organizational units information gathering process we stored all collected information in a Google spreadsheet shared between team elements. When we were in the process of producing the State Model, one of the issues we faced was frequent changes in class names while validating diagrams with the organizations collaborators, and subsequent need to change all related names of fact types and result types.

This would lead to a huge waste of time just in the renaming process. Taking advantage of the fact that we were working on a spreadsheet we developed a quick solution for our needs. We created a specific worksheet – named Facts – to store State Model elements. Excerpts of this worksheet – 3 main sets of columns – can be seen in Figs. 9, 10 and 11.

Columns Class_1 and Class_2 contain the class name that each particular fact type relates to. In the case of the specification of a class or a result type, only Class_1 is filled (in bold). Class_2 is filled whenever we have binary fact types (that were the ones with the most occurrences in our project). Examples are visible in Fig. 9. As one of our main goals was to facilitate name propagation, the only place that had the class name

	A Kind	B IDA	C IDB	D Class_1	E Class_2
1	Kind	IDA	IDB	Class_1	Class_2
2	CL	120	0	CITIZEN	
3	TF	121	0	CITIZEN	VETERINARY PROCESS
4	TF	122	0	CITIZEN	VETERINARY PROCESS
5	CL	123	0	VETERINARY PROCESS	

Fig. 9. Facts worksheet part 1 - classes and ID's.

F ID R	G Result type	H RT prefix	I RT class	J RT sufix
T5304	the documentation of the [veterinary process] was archived	the documentation of the	[veterinary process]	was archived
T5300	[veterinary process] was initiated		[veterinary process]	was initiated
T5303	the documentation of the [veterinary process] was forwarded	the documentation of the	[veterinary process]	was forwarded

Fig. 10. Facts worksheet part 2 - result type and it's construction.

K Fact_type	L FT_prefix	M FT_class_1	N FT_infix	O FT_class_2	P FT_sufix
[citizen] requests [veterinary process]		[citizen]	requests	[veterinary process]	
[citizen] is the subject of [veterinary process]		[citizen]	is the subject of	[veterinary process]	

Fig. 11. Facts worksheet part 3 - fact type and it's construction.

itself was the row containing the class definition. In all other cells we had that name passed by a reference to that cell, making all class name changes to propagate automatically to all result types and fact types referring to the class having the name changed. Namely, the specification of result types was done in the four columns visible on Fig. 10. The first column contained the result type formulation itself and it was automatically filled by a formula referring to the information in the other three columns containing, respectively, (1) the prefix (textual content before the referred class), (2) the class that the result type was related to and (3) the suffix. The column with the class name has a formula that fills the cells automatically when something was written in either the prefix or the suffix columns, and automatically changed the class name (referred in Class_1 column seen before) to lowercase adding the "["and"]" before and after the class name respectively.

For the binary fact types a similar solution was used, visible in Fig. 11, but, instead of four columns, we had the need for six: the fact type formulation itself, the prefix, the first class, the infix (textual content between classes), the second class and the suffix. We decided to include in this spreadsheet only the classes, result types and binary fact types. But this solution could be scaled to include other elements of the SM like properties and ternary fact types, just adding more columns to the table and applying similar formulas to the ones we specified.

As there was no need to propagate changes in the name of properties these were added manually in the diagrams and, when part of a fact type, inserted also manually in the formulation on the spreadsheet. For a more streamlined manipulation of the data in this spreadsheet we added three more features: one new column with the kind of fact type(CL for class, TR for result type and TF for fact type), made all the original class names bold to facilitate finding the correct place to proceed with a name change of a class and added a result ID to relate each fact type with the transaction whose result leads to the creation of instances of that fact type. In [15] one can find the live worksheet referred in this paper with the examples and formulas mentioned, used to automatically fill the result type and fact type formulations and class names.

This spreadsheet tool is one of the main contributions of our work. With it we had most of the content specified for our OFD diagrams with automatic propagation of name changes. But considering the huge amount of model elements involved, it became essential to have automatic synchronizing between the spreadsheet we were working with, and our diagrams in Microsoft Visio, an idea adapted from [16]. To realize such synchronization one needs first to export the spreadsheet to Excel format. Then, by selecting the "data" tab and clicking on the Link Data to Shapes, a new window will appear were one can select the Excel file.

The data importing process is rather straightforward, one simply has to select the file path, worksheet to use, the columns and rows and an identifier field. When this importing process is over, one will have an external data tab at the bottom of the screen with the source data from the selected worksheet as you can see in Fig. 13. To actually link the diagram shapes with the source data, one just places the shapes and fills them only with the unique row identifier. Then right clicking on the external data, using the "automatically link" option and matching the worksheet ID with the shape's label's ID. If done properly a chain symbol should appear in the left side of the linked rows as shown in Fig. 13. For a proper synchronization of the spreadsheet with the Visio

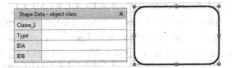

Fig. 12. Class shape.

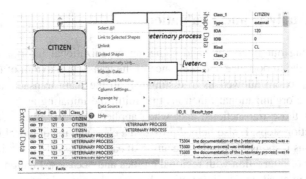

Fig. 13. Linking an excel spreedsheet with visio.

diagrams, we decided to use two ID's (IDA and IDB). This need existed due the direct matching between the column label in the spreadsheet and the corresponding Visio stencil shape label. We could not use the same row to synchronize classes and all their associated result types because we would need to have a huge stencil with many result type shapes that only differed in the shape's label name. We had the idea to give a new id for each result type, but, in this case we would lose the bond between the classes and related result types.

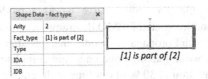

Fig. 14. Fact type shape.

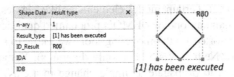

Fig. 15. Result type shape.

So the solution we could come up with was to add a second ID column (IDB) used just for result types. In rows specifying result types related to the same class the first ID (IDA) is always the same – the same as the class it is related to – and only IDB changes. The changes done to the original Visio DEMO OFD stencil shapes to accommodate the new ID's can be seen in Figs. 12, 14 and 15. This is an original and valuable contribution of our work as the Excel-Visio synchronizing solution presented in [16] was limited to ATD diagram data.

5   Results Analysis and Evaluation

While conducting our interviews and gathering organizational information we were able to witness the importance and relevance of the transaction axiom of the Ψ-theory. The transaction axiom allowed us to distinguish all the main transaction steps while clarifying all the organizational functions that fulfill the initiating and executing actor role of each of those transactions, and, when applied, identify all the delegations in specific transaction steps. It also helped provided a clear and unequivocal identification of the institutions responsible for the execution of each process step, specially useful in the many cases of interaction with regional entities and occasional national entities but also invaluable in clarifying the interactions between the multiple departments. Adding to the before mentioned benefits, the transaction axiom also aided in the process of specifying more precise transaction names, mostly by enhancing their comprehension.

The distinction axiom relevance was also noted in the process by providing the clear notions of what facts were really needed opposed to what documents or composite facts were normally asked for. To these facts we gave the name of *Informational Facts,* the information that is really needed, and while currently they will likely still be gathered asking for the same documentation, they may be, in the future, gathered directly from official online government databases through web-services thus facilitating and simplifying the processes.

Our improvised tooling solution to collect model data in a collaborative and integrated way was very useful to our DEMO practical project and it may also be to other colleagues allowing for effective collaborative work in model data gathering and synthesis and coherent and integrated DEMO diagramming. It saved us a huge amount of time in diagram editing and propagating model changes after validation rounds with officials from the SLGA while making coherence and completeness easier to achieve. We estimate that we spent around 30 % of our time in diagram creation and editing and that we would spend double of that time had we not used our method of collaboratively specifying transactions and facts and linking automatically all the data in the respective rows of a Google Spreadsheet to Visio diagrams. Such features were lacking in the modeling tools commonly used with DEMO. In our practical experience, a way to propagate changes done to certain model elements – like classes, our focus on this paper – is essential to maintain one of DEMO's proclaimed qualities: consistency. And this is currently not supported by the modeling tools we know to support DEMO, that we tested and analyzed. Neither Visio [17] (only diagrams), Xemod [18] and Model-World [19] have support to propagate a simple change in a class name to its related result and fact types. This could and should be easily implemented applying a similar method to the one we have presented here, and would be a relevant contribution for any project involving DEMO.

As a contribution to the practical method steps of DEMO, we find that while specifying fact types and result types any tool should provide a way to one easily just introduce the words of the formulations "surrounding" the referred classes and, for example, provide referral with auto-completion. We implemented such functionality in our collaborative spreadsheet with the automatically generated formulations based on string concatenation formulas provided by Google spreadsheet. We also contribute to the enterprise engineering community with method steps to integrate a spreadsheet with Visio diagrams allowing automatic synchronization between shapes and rows of the sheet. Just filling an ID in each shape instantiation in a diagram is sufficient so that, automatically, the DEMO id and the shape name or fact formulation of that shape automatically appear in the diagram in the next sync. Any slight change in names can be automatically reflected in several diagrams (e.g., a transaction name).

6   Conclusions

The amount of processes that take place in organizations may be a huge problem when it comes to modeling, but they may not differ much in their pattern, the generalization of the solution for the Center of Veterinary Care transaction patterns in the scope of e-government initiative allowed us to generalize over 60 processes into around ten transactions and it was possible to validate in interviews with officers that their operational processes fit in the generalized pattern. This sort of generalization can be easily applied to other contexts and have been inclusively applied in other departments of the SLGA with similar results/satisfaction.

The proposed method – directly supported by a prototype tool – to propagate changes in class names to derived result types and facts and then synchronize that data onto DEMO diagrams also proved to be an invaluable contribution in our work in both saving time in the production of diagrams and on keeping all the data coherent. The analysis we did on the CVC models shows how Ψ-theory and DEMO's "lenses" of transaction and distinction axioms contributed in this case-study to a concise and comprehensive view of the essential dynamic and static aspects of the interactions through the CVC.

In our project we identified nearly five hundred fact types in the SLGA. Keeping consistency with such numbers in not an easy task, especially considering all the information is not concentrated in one place, but frequently spread, redundantly in diagrams and tables – e.g. class names, our focus in this paper. Therefore, change propagation solutions like the one we provide are essential and quite useful to save huge amounts of time. For space reasons in this paper we focus on the OFD diagram, but this solution is easily applied for all DEMO diagrams and that is what we have done in our project.

As a final remark we found that using DEMO as the central method to gather specifications for our e-government project allowed a much more effective and successful outcome than approaches of traditional requirements engineering and quality management. SLGA had used these 2 approaches before we started our project and, in a much longer time frame than ours, the produced documents were ambiguous, incomplete and sometimes excessively detailed without need.

References

1. Dalal, S., Chhillar, R.S.: Case studies of most common and severe types of software system failure. Int. J. Adv. Res. Comput. Sci. Softw. Eng. **2**, 341–347 (2012)
2. Shull, F., Basili, V., Boehm, B., Brown, A.W., Costa, P., Lindvall, M., Port, D., Rus, I., Tesoriero, R., Zelkowitz, M.: What we have learned about fighting defects. In: Proceedings of 8th International Software Metrics Symposium, pp. 249–258 (2002)
3. Zeller, A., Hildebrandt, R.: Simplifying and isolating failure–inducing input. EEE Trans. Softw. Eng. **28**(2), 183–200 (2002)
4. Dalal, S., Chhillar, R.S.: Role of fault reporting in existing software industry. CiiT Int. J. Softw. Eng. Technol. **54**(12), 49–54 (2012)
5. Hoogervorst, J.A.P.: Enterprise Governance and Enterprise Engineering. Springer, Heidelberg (2009)
6. Dietz, J., Hoogervorst, J., Albani, A., Aveiro, D., Babkin, E., Barjis, J., Caetano, A., Huysmans, P., Iijima, J., van Kervel, S., Mulder, H., Land, M.O., Proper, H., Sanz, J., Terlouw, L., Tribolet, J., Verelst, J., Winter, R.: The discipline of enterprise engineering. Int. J. Organisational Des. Eng. (IJODE) **3**(1), 86–114 (2013)
7. Liles, D.H., Johnson, M.E., Meade, L.: Enterprise engineering: a discipline. In: 5th Industrial Engineering Research Conference (1996)
8. Dietz, J.L.G.: Enterprise Ontology: Theory and Methodology. Springer, Heidelberg (2006)
9. Dietz, J.L.G.: Architecture - Building strategy into design. Academic Service - Sdu Uitgevers bv (2008)
10. Dietz, J.L.G.: Is it PHI TAO PSI or bullshit? In: Presented at the Methodologies for Enterprise Engineering Symposium, Delft (2009)
11. Dietz, J.L.G.: On the nature of business rules. Adv. Enterp. Eng. I **10**, 1–15 (2008)
12. Dijkman, R.M., Dumas, M., Ouyang, C.: Semantics and analysis of business process models in BPMN. Inf. Softw. Technol. **50**, 1281–1294 (2008)
13. Ettema, R., Dietz, J.L.G.: ArchiMate and DEMO – Mates to Date? In: Albani, A., Barjis, J., Dietz, J.L.G. (eds.) Advances in Enterprise Engineering III. LNCS, vol. 34, pp. 172–186. Springer, Heidelberg (2009)
14. Aveiro, D., Pinto, D.: Devising DEMO guidelines and process patterns & validating comprehensiveness and conciseness. In: Presented at the Exploring Modelling Methods for Systems Analysis and Design (EMMSAD) 2014, Thessaloniki, Greece June (2014)
15. DEMO Project Method Suport Spreadsheet (2013). https://docs.google.com/spreadsheet/ccc?key=0AvE4a-Y0_lVMdHNUSmgwZmJzbGMzQV9oMmdBeUhYZFE#gid=1
16. Geskus, J., Dietz, J.L.G., Pluimert, N.: INQA, TU Delft: Electrical Engineering, Mathematics and Computer Science, TU Delft, Delft University of Technology: DEMO applied to Quality Management Systems (2008). http://resolver.tudelft.nl/uuid:7421072c-e61b-47b1-87a0-d2cf85050817
17. Microsoft: Visio Professional 2013 – business and diagram software - Office.com. http://office.microsoft.com/en-us/visio/visio-professional-2013-business-and-diagram-software-FX103472299.aspx
18. MPRISE: Xemod - Product overview. http://www.mprise.eu/xemod-product-overview.aspx
19. Bart-Jan Hommes: ModelWorld - Online Modeller and Repository. http://www.modelworld.nl/

Mastering Data-Intensive Collaboration and Decision Making: The Dicode Project

Nikos Karacapilidis[✉]

Computer Technology Institute and Press "Diophantus",
University of Patras, 26504 Rio Patras, Greece
nikos@mech.upatras.gr

Abstract. Many collaboration and decision making settings are nowadays associated with huge, ever-increasing amounts of multiple types of data, which often have a low signal-to-noise ratio for addressing the problem at hand. The Dicode project aimed at facilitating and augmenting collaboration and decision making in such data-intensive and cognitively-complex settings. To do so, whenever appropriate, it built on prominent high-performance computing paradigms and proper data processing technologies to meaningfully search, analyze and aggregate data existing in diverse, extremely large, and rapidly evolving sources. At the same time, particular emphasis was given to the deepening of our insights about the proper exploitation of big data, as well as to collaboration and sense making support issues. This chapter reports on the overall context of the Dicode project, its scientific and technical objectives, the exploitation of its results and its potential impact.

1 Introduction

Individuals, communities and organizations are currently confronted with the rapidly growing problem of information overload [1]. An enormous amount of content already exists in the digital universe (i.e. information that is created, captured, or replicated in digital form), which is characterized by high rates of new information that is being distributed and demands attention. This enables us to have instant access to more information (that is of interest) than we can ever possibly consume. As pointed out in a recent IDC's White Paper [2], the amount of information created, captured, or replicated exceeded available storage for the first time in 2007, while the digital universe is expanding by a factor of 10 every five years.

People have to cope with such a diverse and exploding digital universe when working together; they need to efficiently and effectively collaborate and make decisions by appropriately assembling and analyzing enormous volumes of complex multifaceted data residing in different sources [3–5]. For instance, imagine:

- A community of clinical researchers and bio-scientists, supported in their scientific collaboration by a system that allows them to easily examine and reuse heterogeneous clinico-genomic data and information sources for the production of new insightful conclusions or the formation of reliable biomedical knowledge, without having to worry about the method of locating and assembling these huge quantities of data (clinical and genomic data, molecular pathways, DNA sequence data, etc.).

© Springer-Verlag Berlin Heidelberg 2015
A. Fred et al. (Eds.): IC3K 2013, CCIS 454, pp. 21–36, 2015.
DOI: 10.1007/978-3-662-46549-3_2

- Or a community of clinicians, radiologists, radiographers, patients and pharma-researchers being able to contribute more effectively to clinical decisions and drug testing by combining heterogeneous, collaboratively annotated datasets from patient results (e.g. blood tests, physical examinations, free text journals from patients on their experience from treatment) and different scan modalities (e.g. X-Ray, Static and Dynamic MRI), without having to be anxious about tracking the data and their provenance through the complex decision making process, and the handling of the associated multimedia material.
- Or even, a marketing and consultancy company being able to effortlessly forage the Web (blogs, forums, wikis, etc.) for high-level knowledge, such as public opinions about its products and services; it is thus able to capture tractable, commercially vital information that can be used to quickly monitor public response to a new marketing launch; having the means to meaningfully filter, collate and analyse the associated findings; and use the information to inform new strategy.

The goal of the Dicode project [6] was to turn this vision into reality. The project was funded by the European Commission under the FP7 Work Programme (contract number: FP7-ICT-257184 - http://dicode-project.eu). It started on September 1st, 2010 and its duration was 36 months. The partners of the Dicode consortium were: Computer Technology Institute & Press "Diophantus" (project coordinator, Greece), University of Leeds (United Kingdom), Fraunhofer-Gesellschaft zur Förderung der angewandten Forschung e.V. (Germany), Universidad Politécnica de Madrid (Spain), Neofonie Gmbh (Germany), Image Analysis Ltd (United Kingdom), Biomedical Research Foundation - Academy of Athens (Greece), and Publicis Frankfurt Zweigniederlassung der PWW GmbH (Germany).

This chapter describes the overall context of the Dicode project (Sect. 2), its scientific and technical objectives (Sect. 3), as well as the exploitation of its results and its potential impact (Sect. 4).[1]

2   Overall Project Context

Many collaboration and decision making settings are nowadays associated with huge, ever-increasing amounts of multiple types of data, obtained from diverse sources, which often have a low signal-to-noise ratio for addressing the problem at hand. These data may also vary in terms of subjectivity and importance, ranging from individual opinions and estimations to broadly accepted practices and trustable measurements and scientific results. Their types can be of diverse level as far as human understanding and machine interpretation are concerned. At the same time, the associated data are in most cases interconnected, in a vague or explicit manner.

Additional problems start when we want to consider and exploit accumulated volumes of data, which may have been collected over a few weeks or months, and meaningfully analyze them towards making a decision. Admittedly, when things get complex, we need to identify, understand and exploit data patterns; we need to

---

[1] A shorter version of this chapter appears in [6].

aggregate appropriate volumes of data from multiple sources, and then mine them for insights that would never emerge from manual inspection or analysis of any single data source. In other words, the pathologies of big data are primarily those of analysis. The way that data will be structured for query and analysis, as well as the way that tools will be designed to handle them efficiently are of great importance and certainly set a big research challenge.

In the settings under consideration, "big data" analytics technology currently receives much criticism, in that it does not provide proper insight into what the data means. To make sense of big data and come with discoveries that help improve decision making in practical contexts, human intelligence should be also exploited. We need to provide the appropriate ways to nurture and capture this human intelligence in order to extract the necessary insights and improve the way machines deal with complex situations.

Fig. 1. The Dicode services exploit the cloud computing paradigm and build on the synergy of machine and human reasoning.

Taking the above issues into account, the Dicode project aimed at facilitating and augmenting collaboration and decision making in data-intensive and cognitively-complex settings. To do so, whenever appropriate, it built on prominent high-performance computing paradigms and proper data processing technologies to meaningfully search, analyze and aggregate data existing in diverse, extremely large, and rapidly evolving sources. At the same time, particular emphasis was given to the deepening of our insights about the proper exploitation of big data, as well as to collaboration and sense making support issues. Building on current advancements, the solution proposed by the Dicode project brings together the reasoning capabilities of both the machine and the humans (Fig. 1).

It can be viewed as an innovative "workbench" incorporating and orchestrating a set of interoperable services that reduce the data-intensiveness and complexity overload at critical decision points to a manageable level, thus permitting stakeholders to be more productive and effective in their work practices. Services that were developed and integrated in the context of the Dicode project are released under an open source license.

The achievements of the Dicode project were validated through three use cases:

- **Clinico-Genomic Research Assimilator.** The need to collaboratively explore, evaluate, disseminate and diffuse relative scientific findings and results is more than profound today. Towards this objective, Dicode elaborated an integrated clinico-genomic (tacit) knowledge discovery and decision making use case that targets the identification and validation of predictive clinico-genomic models and biomarkers [7].
- **Trial of Clinical Treatment Effects.** The goal of this case (which has been expanded in the second year of the project to cover broader clinical trials, not just for Rheumatoid Arthritis) was to facilitate the process of making clinical decisions in drug trials by combining datasets from patient results (blood tests, physical examinations) and the different scan modalities (X-Ray, Static and Dynamic MRI scan images) to reveal the effectiveness of a drug within a trial.
- **Opinion Mining from unstructured Web 2.0 data.** Through this case, we validated the Dicode services for the automatic analyses of the voluminous amount of unstructured information existing on the Web, especially in the highly dynamic social media space. Data for this case were primarily obtained from spidering the most popular social Web sites making use of APIs from various Web 2.0 platforms [8].

3 Scientific and Technical Objectives

The project's objectives have been fully accomplished through an evolutionary approach characterised by:

- the active engagement of all stakeholders (technical partners and use case representatives) in the specification, design and evaluation of the foreseen technological solutions throughout the project;
- the adoption of an incremental development approach, which ensured that end users can experiment with the Dicode services from the early stages of the project (operational prototype versions of the Dicode services were available at the end of the first year of the project, enhanced versions were delivered in month 24, final versions were ready in month 33);
- the continuous refinement of user requirements through testing (involving users from all three use cases), and
- the early availability of an operational integrated suite of services, which facilitated trials and proof-of-concept purposes, enabled proper exploitation and dissemination activities, and ensured project sustainability.

The association between the project's objectives and the project's milestones is illustrated in Fig. 2. As shown, "Laying Foundations", "Integration, Validation and

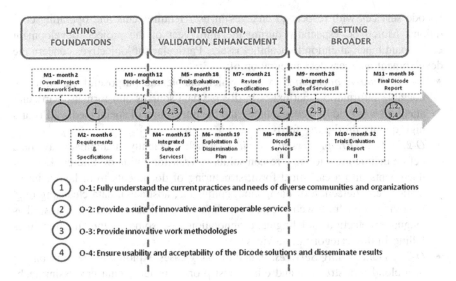

Fig. 2. S&T objectives, project's milestones and goals set for each year of the Dicode project.

Enhancement" and "Getting Broader" was the overall goal for each year of the project, respectively. As justified in the following, the Dicode project successfully reached these goals.

In particular, the project's scientific and technical objectives were:

- *O-1: To fully understand the current practices and needs of diverse communities and organizations as far as data-intensive and cognitively-complex collaboration and decision making is concerned.* Three representative use cases were continuously elaborated throughout the project. Related settings were also considered, aiming to reveal practices and needs associated with both large data sets and real-time data [9]. The accomplishment of this objective was critical for the applicability of the Dicode approach in a wide variety of settings.

This objective was of high importance throughout the project. Thoroughly considering the feedback from the two evaluation rounds of Dicode services across the project's use cases, an analysis of the lessons learned was documented and services' specifications were revised to inform the final iteration of development. A much deeper understanding of the use cases' differences and similarities, as well as of their potential to explore the full range of Dicode services, was achieved through close collaboration between technical partners and end users.

- *O-2: To provide a suite of innovative, adaptive and interoperable services (both at a conceptual and a technical level) that satisfies the full range of the associated requirements.* The development of Dicode services facilitated and augmented collaboration, sense-making and decision-making in data-intensive and cognitively-complex settings, while also serving the underlying requirements of capturing, delivering and analyzing pertinent information (Fig. 3). Dicode services are running on the Web. Throughout the project, much attention was given to the adaptability of

Dicode services with respect to changes in user requirements and operating conditions. Moreover, especially during the third year of the project, development efforts paid much attention to usability issues. Particular sub-objectives concern the development and seamless integration of:

- *O-2.1: Data Acquisition Services,* which enable the purposeful capturing of tractable information that exists in diverse data sources and formats. Particular attention was paid to web resources and the integration of social media APIs and high quality third-party feeds.

- *O-2.2: Data Pre-processing Services,* which efficiently manipulate raw data before their storage to the foreseen solution. Transformation of different kinds of documents into a canonical form, structuring of documents from layout information (e.g. detection of navigation, comments, abstracts), data cleansing (e.g. removing noise from web pages, discarding useless database records), as well as language detection and linguistic annotations are some of the functionalities falling in this category of services.

- *O-2.3: Data Mining Services,* which in many cases exploit and are built on top of a cloud infrastructure and other most prominent large data processing technologies to offer functionalities such as high performance full text search, data indexing, classification and clustering, directed data filtering and fusion, and meaningful data aggregation. Advanced text mining techniques, such as named entity recognition, relation extraction and opinion mining, help to extract valuable semantic information from unstructured texts. Intelligent data mining techniques elaborated include local pattern mining and similarity learning.

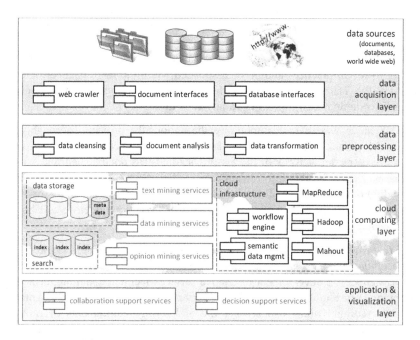

Fig. 3. The Dicode architecture and suite of services.

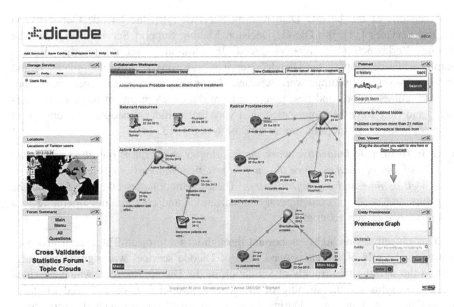

Fig. 4. Integrating Dicode services: An instance of the Dicode Workbench.

- *O-2.4: Collaboration Support Services,* which facilitate the synchronous and asynchronous collaboration of stakeholders through adaptive workspaces, efficiently handle the representation and visualization of the outcomes of the data mining services (through alternative and dedicated data visualization schemas), and accommodate a series of actions for the appropriate handling of data in each use case.
- *O-2.5: Decision Making Support Services,* which augment (both individual and group) sense-making and decision-making by supporting stakeholders in locating, retrieving and arguing about relevant information and knowledge, as well as by providing them with appropriate notifications and recommendations.

This objective was of paramount importance for the success of the project. Taking into account feedback from the two evaluation rounds of the project, as well as recommendations of the Project officer and Project Reviewers, the final operational versions of the Dicode Workbench and integrated Dicode services were developed and tested across the project's use cases (Fig. 4). Much attention was given to the openness of the Dicode solution, in order to augment exploitation purposes. An appropriate infrastructure of in-house computer clusters for running large scale data mining experiments and testing prototype implementations, as well as data collections for benchmarking based on textual and structured data, were set and maintained. Standards and guidelines for the development of Dicode services - aiming at ensuring interoperability between the services to be developed and reusability of them through diverse scenarios of use – were defined and revised upon the evolution of the project. Issues around both the conceptual and technical integration of the full range of Dicode services were thoroughly elaborated to upgrade user experience. According to the workplan, the final versions of the Dicode Data Mining Services [10], the Dicode Collaboration Support Services [11], the Dicode Decision Making Support Services [11], and the Dicode Workbench [12] were produced. In addition, a set of practical lessons learned while developing the Dicode's services and using them in data-intensive and cognitively-complex settings were reported. These lessons concern experiences, concrete recommendations and best practices from the development of the project's services, and they have been presented in a way that could aid people who engage in various phases of developing similar kind of systems [13, 14].

- *O-3: To provide innovative work methodologies that exploit the abovementioned suite of services and advance the current practices in terms of efficiency, creativity, as well as time and cost effectiveness.* These methodologies take into account the nature and needs of contemporary organisations and communities operating in a knowledge-driven economy.

This objective was highly important throughout the project. The established consensus on the role of the envisioned suite of Dicode services was significantly augmented through the two rounds of validation of the Dicode services, which provided valuable insights for the shaping of novel methodologies to be followed in stakeholders' daily work practices. During these evaluation rounds, a long and diverse set of end users tested the Dicode solution (services and Workbench) and provided valuable feedback by pointing out both strengths and weaknesses. These were considered through various real-world scenarios, which actually constituted the base for the definition of Dicode's innovative work methodologies [7, 8]. The proposed methodologies reflect our experiences gained from the overall validation of the project's results and provide useful suggestions and insights to relevant communities and organizations.

- *O-4: To ensure usability and acceptability of the above services and work methodologies through their validation in real use cases, and disseminate the project's results by dedicated actions.*

This objective was also of paramount importance for the success of the project. Two rounds of evaluation of the Dicode Workbench and integrated services through the project's use cases were performed. Properly formulated metrics and questionnaires were employed to analyse the feedback received. Appropriate video-casts – based on everyday user stories from user communities, developers and early adopters – were prepared for each use case. The parameters assessed for each service concerned their acceptability, ease of use, usability, and overall quality [7, 8].

In addition, a comprehensive exploitation and dissemination plan has been produced, ensuring the impact and sustainability of the Dicode outcomes. Initial dissemination and exploitation activities included the development of a corporate identity of the project, the set-up of a web portal, and initial public relations efforts. A significant number of publications have resulted out of joint work among consortium members. These publications appear in international scientific journals and proceedings of international peer-reviewed scientific conferences and workshops (a detailed list of Dicode's dissemination activities appears at http://dicode-project.eu/index.php?q=news). Presentations of project-related work were also given in some of the top technology and marketing conferences. Moreover, Dicode organized four scientific workshops, one in the context of the world leading conference on collaboration support (CSCW 2012), another in the context of the best European conference on machine learning and knowledge discovery (ECML-PKDD 2012), a third one at the leading international conference on knowledge engineering and knowledge management (EKAW 2012), and a fourth one at world leading conference on hypertext and social media (Hypertext 2013). A series of exploitation activities has been also carried out, especially during the last two years of the project. Each Dicode partner put much effort in developing a concrete and realistic exploitation strategy (see next section). Several success stories concerning exploitation of Dicode results, development of strategic partnerships with industry and co-operation with other EU projects have been already reported.

4 Exploitation of Results and Potential Impact

The combination of academic and industrial partners within the Dicode consortium was perfectly suited for working with existing customers and collaborators in a variety of industry and academic segments to develop the Dicode platform for market use. Suitable targets were defined in the early stages of the project. As each target has specific needs which can be met through the technology developed in the Dicode project, partners in the project consortium were involved in cultivating and extending ties to their existing customer base to keep these key assets informed of project developments. The project partners also organized dedicated demonstrations of running prototypes and scenarios of use for key persons in the target organisations.

Figure 5 gives an overview of Dicode's target groups for the exploitation of the project's foreground. In the public sector, the focus lies on public services, public health and e-Science. In the private sector, advertising and communication, media and medicine are the main target areas. In the IT industry, Dicode caters to service integrators, service developers and consultancy companies.

To ensure the sustainability of the Dicode project, each consortium partner formulated a detailed exploitation plan and carried out a set of associated activities, based on modern marketing and communication best practices. Market entry strategies followed in the context of the Dicode project included:

- The definition of appropriate targets, both public and private entities, and partners in the network of the consortium partners who have an interest in the outcomes of the Dicode project, and are also suitable for obtaining first experiences and willing to be used as success stories;
- Building out additional use-cases to fit the needs of the defined targets;
- Strategic partnerships with established players in the market.

The key success indicators of the Dicode project, together with their high effects and actions taken towards ensuring their accomplishment are summarized in Table 1. The final results of the Dicode project advance the state-of-the-art in approaches on (i) the proper exploitation of big data (dealing with the "big data fallacy" issue) and the integrated consideration of data mining and sense-making issues, (ii) recommender systems, with respect to recommendations in heterogeneous, multi-faceted data and the identification of hidden links in complex data types, (iii) understanding text to

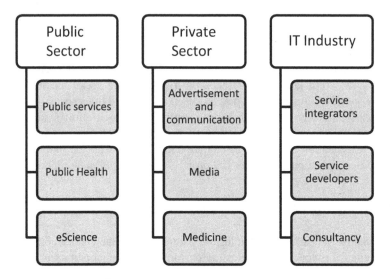

Fig. 5. Dicode targets a wide audience.

drastically reduce the annotation effort for extracting relations, (iv) opinion mining by considering opinion statements as *n*-ary relations and apply the highly scalable methodology implemented for their recognition, (v) Web 2.0 collaboration support tools in terms of interoperability with third party tools and integration of appropriate reasoning and data mining services, and (vi) decision making support applications, by integrating knowledge management and decision making features as well as by building on the synergy of human and machine argumentation-based reasoning.

Such advancements have shaped innovative work methodologies for dealing with the problems of information overload and cognitive complexity in diverse collaboration and decision making contexts. Adopting the proposed solution, both individual and collaborative sense making are augmented through the meaningful exploitation of prominent data processing and data analysis technologies. The Dicode solution is user-friendly and built on the synergy of human and machine intelligence. Adopting open standards, and in accordance with EU's recent initiatives on Open Systems and Data, the Dicode project has the potential of forming a rich ecology of domain specific and non-specific extensions. The Dicode platform allows for external data service providers to supply information, as well as for external developers to supply additional modules and applications, which are tailored to evolving market conditions. Finally, it enables diverse public and private entities to aggregate, structure, semantically enrich and analyse vast amounts of information. This turns the problem of information overload into a benefit of structured data, which can be used as the basis for decisions of better quality. Simply put, the Dicode solution is able to turn information growth into economic growth.

In particular, the potential impact of the Dicode project (including the socio-economic impact and the wider societal implications so far) concerns:

Table 1. Success indicators and actions to ensure them.

Key success indicators	Actions
Deployment of the Dicode framework is not too costly → high rate of Dicode technology adoption	Adoption of standards, exploitation of existing prototypes and background technology, open-source policy
High level of the Dicode framework's acceptance by users involved in Dicode's use cases → increased users' productivity and creativity	Early and continuous involvement of end users in the development and evaluation of the Dicode platform
Acceptance of the Dicode framework by users outside the Dicode consortium → recognition of the Dicode platform's value from relevant groups and communities	Exploitation of Dicode partners' liaisons with scientific and business stakeholders; Dicode workshops and diverse dissemination activities; relevant market watch
Adaptability and proven portability of the Dicode framework in a wide range of application domains → acceptance of the Dicode framework by industry and academia	Generic and flexible development approach; show cases at relevant scientific and business events and stakeholders; Dicode scientific workshops; scientific dissemination activities; market analysis
Foundations for high-performance scalable data mining in the cloud computing initiative and related open source community → proven added value of the Dicode framework	Adoption of open source principles and concepts; advancement of cloud computing paradigm; dissemination activities related to dedicated workgroups and communities
Hit the optimal market → strengthened EU leadership in the domain of intelligent information management	Development of a detailed and coherent exploitation strategy and thorough consideration of associated perspectives

Better leveraging of human skills, improved quality and quantity of output and reduced time and cost allowing users to concentrate on more creative and innovative activities.

- The Dicode integrated suite of services and corresponding work methodologies facilitate and enhance the integration and aggregation of different stakeholders' perspectives across different collaboration and decision making activities, by explicitly addressing their knowledge and social dynamics. The Dicode platform is able to augment the creativity of stakeholders (stakeholders save time by skipping unnecessary tasks, accomplishing trivial tasks faster, while the platform provides a remedy to the information and cognitive overload). Stakeholders may easily customize the Dicode platform and concentrate on more creative and innovative activities.
- The Dicode platform enables new working practices for stakeholders involved in data-intensive and/or cognitively-complex settings. It has followed a component-based approach, based on open standards. This allows for further development by using and adapting existing modules, or developing new ones to cover the needs of related contexts.

Increased ability to identify and respond appropriately to evolving conditions (e.g. in finance, epidemiology, environmental crises …) faster and more effectively. Reinforced ability to collaboratively evolve large-scale, multi-dimensional models from the integration of independently developed datasets.

- In Dicode, machine-tractable knowledge concerning the full lifecycle of collaboration and decision making is accumulated and maintained. Consequently, the Dicode platform augments the productivity of stakeholders, e.g. by enabling them to easily locate and meaningfully reuse existing content. This affects both individuals and the workgroups they belong to.
- The Dicode platform improves the quality and quantity of the collaboration process. Since needs and user types evolve over time, the platform can be easily customized and adapted to address diverse needs and user types.
- The Dicode platform enhances collaboration between individual stakeholders through the meaningful integration and aggregation of independently developed applications (and associated datasets), which allows for a quicker consensus in the decision making process.
- The Dicode platform allows for external data service providers to supply information, as well as for external developers to supply additional modules and applications, which are tailored to evolving market conditions.

Higher levels of information portability and reuse by creating an ecology of systems and services that are dynamic, interoperable, trustworthy and accountable by design.

- The Dicode platform advances the state-of-the-art in information portability and reuse by considering interoperability issues, while also fostering standards-based integration and exploitation of information resources across organisational boundaries.
- The Dicode platform has been developed using existing standards and exploiting existing open source software.
- The Dicode project has developed a large-scale data processing platform. This platform allows for diverse data processing modules to be integrated through appropriate interfaces.
- The Dicode platform exploits, whenever appropriate, a cloud computing environment, which allows for improved information portability and reuse.
- The Dicode platform is web-based. This allows system independence for the end user.
- During the development of the Dicode platform, strong cooperation with committees and organizations which set standards in the fields of cloud computing was held. A series of contributions to free software projects concerning large-scale data processing in a cloud environment have been performed.
- Being designed with "openness" in mind, the Dicode platform is able to create a rich ecology of domain specific and non-specific extensions.

Increased EU competitiveness in the global knowledge economy by fostering standards-based integration and exploitation of information resources and services across domains and organisational boundaries.

- The global knowledge economy demands no barriers to entry. Accordingly, the Dicode platform has been developed by adopting open standards. Additionally, the platform allows for easy sharing of data and information. This enables the creation of marketplaces for information and information suppliers. Data rich applications can be implemented more quickly due to easy access of data through a shared environment.
- The web-based development of the Dicode platform, together with the exploitation of the cloud in some of its modules, allows for global access to innovative data processing services. Moreover, the platform can be easily adapted for international use (i.e. no cultural barriers to entry). The above may reduce fixed costs for companies using the Dicode platform, allowing them to invest more resources and money into their core line of business activities, thus providing them a competitive advantage in the international marketplace (i.e. no financial and technological barriers).
- The Dicode platform allows public and private entities to aggregate, structure and analyse vast amounts of information. This turns the problem of information overload into a benefit of structured data which can be used as the basis for better and quicker decisions. The Dicode platform helps stakeholders enrich current information, and turns the problem of information overload into knowledge discovery.

Strengthened EU leadership at every step of the computer-aided information and knowledge management lifecycle, creating the conditions for the rapid deployment of innovative products and applications based on high quality content.

- The European IT landscape is generally comprised of small and medium sized enterprises (SMEs). For many SMEs, it is difficult to develop new, data rich applications from scratch, basically due to the associated high investment costs. Open Source solutions, such as the Dicode platform, will reduce the barriers for SMEs in the development and hosting of data rich applications.
- In Europe, there are many different languages. For developing an application which can handle and process text sources from different EU countries, it is necessary to use different language dependent modules. Data and text processing standards, as supported by the Dicode platform, allow for the simple replacement of compatible modules which switch from one language to another (plug and play integration).
- Based on the existing Dicode infrastructure and services, new applications can be developed in less time. This yields to quicker "time to market" and faster return-on-investment due to decreased development costs.
- The Dicode platform is able to assist European companies in making better decisions quicker, based on the largest data set possible. As much of the data on the Web is text, Dicode solutions for issues such as sentiment analysis, opinion mining, data mining, trend mining etc. will continue to grow in importance for decision makers.

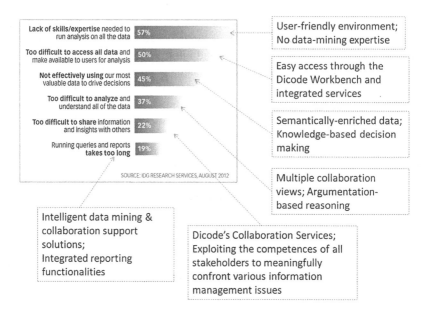

Fig. 6. Big Data challenges and Dicode responses.

5   Discussion and Conclusions

The Dicode platform enables a meaningful aggregation and analysis of Big Data in complex settings. The proposed solution (infrastructure and services) builds on a holistic approach where decision support technologies have been evolved and inter-related in order to efficiently and effectively address the requirements of the knowl-edge-intensive organization [15]. The Dicode outcomes enable new working practices that turn the problem of information overload and cognitive complexity into the benefit of knowledge discovery. This is achieved through properly structured data that can be used as the basis for more informed decisions. Simply put, the Dicode approach is able to turn information growth into knowledge growth; it improves the quality of collab-oration within a Web community, while enabling its users to be more productive and focus on creative activities.

The Dicode approach provides responses to all six Big Data challenges identified in a recent White Paper [16]. Specifically, as shown in Fig. 6, through the user-friendly environment and the integrated approach provided in Dicode, users do not need to possess any particular skills and data mining expertise to run analysis of data; the Dicode Workbench and its integrated services enable easy access to related data, thus making them available for further analysis; the Dicode approach exploits semantically-enriched data and associated knowledge to effectively use the most valuable data and drive decisions; through alternative collaboration views and associated reasoning mechanisms [11], the Dicode Collaboration and Decision Making Support Services facilitate data analysis and understanding; in addition, the abovementioned services exploit the competences of all stakeholders to meaningfully confront diverse information management issues; finally, a set of intelligent functionalities and solutions offered in Dicode expedite the running of queries and production of reports.

As a last note, we point out that the overall Dicode approach is fully in line with a set of imperatives concerning challenges and opportunities with Big Data, which are reported in another White Paper authored by 21 prominent researchers [17]. Specifically, the Dicode platform enables stakeholders "to run heterogeneous workloads on a single infrastructure that is sufficiently flexible to handle all these workloads"; it is "designed explicitly to have a human in the loop", thus enabling "humans to easily detect patterns that computers algorithms have a hard time finding"; it provides "supplementary information that explains how each result was derived, and based upon precisely what inputs"; it offers "a rich palette of visualizations", which are "important in conveying to the users the results of the queries in a way that is best understood in the particular domain".

References

1. Eppler, M., Mengis, J.: The concept of information overload: a review of literature from organization science, accounting, marketing, MIS, and related disciplines. Inf. Soc. 20(5), 325–344 (2004)
2. IDC. The Diverse and Exploding Digital Universe. White Paper, March 2008. www.idc.com
3. Economist. A Special Report on managing information: Data, data everywhere. Economist (2010)
4. Hara, N., Solomon, P., Kim, S.L., Sonnenwald, D.H.: An emerging view of scientific collaboration: Scientists' perspectives on collaboration and factors that impact collaboration. J. Am. Soc. Inform. Sci. Technol. 54, 952–965 (2003)
5. Shim, J.P., Warkentin, M., Courtney, J.F., Power, D.J., Sharda, R., Carlsson, C.: Past, present and future of decision support technology. Decis. Support Syst. 33, 111–126 (2002)
6. Karacapilidis, N. (ed.): Mastering Data-Intensive Collaboration and Decision Making: Cutting-edge research and practical applications in the Dicode project, Studies in Big Data Series, vol. 5, Springer (2014)
7. Tsiliki, G., Kossida, G.: Clinico-genomic research assimilator: a dicode use case. In: [6], pp. 165–180 (2014)
8. Löffler, R.: Opinion mining from unstructured web 2.0 data: a dicode use case. In: [6], pp. 181–200 (2014)
9. Lau, L., Yang-Turner, F., Karacapilidis, N.: Requirements for big data analytics supporting decision making: a sensemaking perspective. In: [6], pp. 49–70 (2014)
10. Friesen, N., Jakob, M., Kindermann, J., Maassen, D., Poigné, A., Rüping, S., Trabold, D.: The dicode data mining services. In: [6], pp. 89–118 (2014)
11. Tzagarakis, M., Karacapilidis, N., Christodoulou, S., Yang-Turner, F., Lau, L.: The dicode collaboration and decision making support services. In: [6], pp. 119–139 (2014)
12. de la Calle, G., Alonso-Martínez, E., Rojas-Vera, M., García-Remesal, M.: Integrating dicode services: the dicode workbench. In: [6], pp. 141–164 (2014)
13. Friesen, N., Kindermann, J., Maassen, D., Rüping, S.: Data mining in data-intensive and cognitively-complex settings: lessons learned from the dicode project. In: [6], pp. 201–212 (2014)

14. Christodoulou, S., Tzagarakis, M., Karacapilidis, N., Yang-Turner, F., Lau, L., Dimitrova, V.: Collaboration and decision making in data-intensive and cognitively-complex settings: lessons learned from the dicode project. In: [6], pp. 213–226 (2014)
15. Karacapilidis, N.: An Overview of Future Challenges of Decision Support Technologies. In: Gupta, J., Forgionne, G., Mora, M. (eds.) Intelligent Decision-Making Support Systems: Foundations, pp. 385–399. Applications and Challenges, Springer-Verlag, London, UK (2006)
16. SAS. Data Visualization: Making Big Data Approachable and Valuable. White Paper (2013). http://www.sas.com/content/dam/SAS/en_us/doc/whitepaper2/sas-data-visualization-market pulse-106176.pdf
17. Computing Community Consortium - Computing Research Association. Challenges and Opportunities with Big Data: A community white paper developed by leading researchers across the United States. White Paper, February 2012. http://www.cra.org/ccc/files/docs/init/ bigdatawhitepaper.pdf

Knowledge Discovery and Information Retrieval

Integration of Text Mining Taxonomies

Katja Pfeifer[✉] and Eric Peukert

SAP AG, Chemnitzer Str. 48, 01187 Dresden, Germany
{katja.pfeifer01,eric.peukert}@sap.com

Abstract. Text mining services can be used to extract and categorize entities from textual information on the web. Merging results from multiple services could improve extraction quality. This requires to have an integrated extraction taxonomy and corresponding mappings between individual taxonomies that are used for categorizing extracted information. However, current ontology matching approaches cannot be applied since the available meta data within most taxonomies is weak.

In this article we propose a novel taxonomy alignment process that allows us to automatically identify equal, hierarchical and associative mappings and integrate those mappings in a global taxonomy. We broadly evaluate our matching approach on real world service taxonomies and compare to state-of-the-art approaches.

Keywords: Instance-based matching · Text mining · Taxonomy alignment

1 Introduction

Analysts estimate that up to 80 % of all business relevant information within companies and on the web is stored as unstructured textual documents [1]. Being able to exploit such information for example for market analysis, trending or web monitoring is a competitive advantage for companies. To support the extraction of information from unstructured text, a multitude of text mining techniques were proposed in literature (see [2]) and some were publicly made available as Web Services (e.g., [3,4]). These services are able to classify text documents, recognize entities and relationships or identify sentiments. Individual services often have specific strengths and weaknesses. By combining them the overall extraction quality and amount of supported features can be increased [5].

Unfortunately, merging the results from multiple extraction services is problematic since individual services rely on different taxonomies or sets of categories to classify or annotate the extracted information (e.g., entities, relations, text categories). To illustrate the problem we show the results of extracting entities from a news text in Fig. 1.

Entities have been annotated by several text mining services (OpenCalais [3], Evri [6], AlchemyAPI [4], FISE [7]) that rely on different taxonomies to annotate found entities. For instance the text sequence *Airbus* is annotated with three

© Springer-Verlag Berlin Heidelberg 2015
A. Fred et al. (Eds.): IC3K 2013, CCIS 454, pp. 39–55, 2015.
DOI: 10.1007/978-3-662-46549-3_3

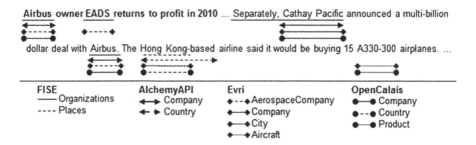

Fig. 1. Analysis of a business news by several named entity recognition services (retrieved on March 9, 2011).

different entity types: *Organization* (by FISE), *Company* (by AlchemyAPI and OpenCalais) and *AerospaceCompany* (by Evri).

To be able to combine and merge extraction results from multiple services a mapping between different taxonomy types and a merged taxonomy is required. Finding such mappings manually is not feasible as the taxonomies can be very large and evolve over time (e.g., AlchemyAPI uses a taxonomy with more than 400 entity types). Unfortunately applying existing (semi-)automatic ontology and schema matching techniques [8,9] does not provide the requested quality since the available meta data within existing service taxonomies is weak (i.e., no descriptions are available, the taxonomies have a flat structure). Moreover, existing matching approaches are not able to identify relations between the taxonomy types (i.e., if two types are equal or if one type is a subtype of the other).

To overcome those limitations, we introduce a novel taxonomy alignment process that enables the merging of taxonomies for text mining services. The following contributions are made within this article:

- We introduce a novel taxonomy alignment approach that is based on generated instances of input taxonomies.
- In particular, a novel metric for instance-based matchers is proposed that is able to identify equal, hierarchical and associative mappings.
- Based on the automatically computed mappings a cluster-based taxonomy merging process is described.
- The taxonomy alignment process is compared to state-of-the-art instance-based alignment methods. For evaluation, reference mappings between a number of real-world text mining services and their taxonomies were created through an online survey with numerous participants.

The remainder of the article is structured as follows: In Sect. 2 we formally describe the problem and introduce the notation being used within this article. Section 3 introduces our taxonomy alignment process and presents the metric for instance-based matching as well as the combined matching strategy used within our process. The experimental setup and the results of our evaluation can be found in Sects. 4 and 5. We introduce a taxonomy merging approach that makes use of the introduced taxonomy alignment process in Sect. 6 before we review related work in Sect. 7. Section 8 closes with conclusions and an outlook to future work.

2   Problem Description

Combining the results of multiple text mining services is promising as it can increase the quality and functionality of text mining. This requires us to have a mapping between the underlying taxonomies of the individual extraction services. However, finding such a mapping is challenging. A review of existing text mining services and their taxonomies revealed that the taxonomies differ strongly in granularity, naming and their modeling style. Many taxonomies are only weakly structured and most taxonomy types are lacking any textual description. Therefore manually defining a mapping between text-mining taxonomies is a complex, challenging and time consuming task.

Within this article we want to apply ontology- and schema matching techniques [8,9] to automatically compute mappings between text mining taxonomies. Matching systems take a source and a target ontology as input and compute mappings (alignments) as output. They employ a set of so called matchers to compute similarities between elements of the source and target and assign a similarity value between 0 and 1 to each identified correspondence. Some matchers primarily rely on schema-level information whereas others also include instance information to compute element similarities. Typically, the results from multiple of such matchers are combined by an aggregation operation to increase matching quality. In a final step a selection operation filters the most probable correspondence to form the final alignment result.

Unfortunately existing matching approaches solve the challenges of matching text mining taxonomies only partly. Schema-based matchers can only be applied to identify mappings between equal concepts (e.g., by using a name-matcher) as the scarcity of broader meta data disables the use of more enhanced matchers (e.g., retrieving hierarchical mappings through the comparison of the taxonomy structure). Instance-based approaches are mainly limited to equal mappings. The few instance-based approaches that support hierarchical mappings still suffer from limited accuracy as we show in our evaluation (see Sect. 7 for a broader review of related work).

To overcome the aforementioned limitations, we proposed an instance enrichment algorithm in [10] that populates the taxonomy types with meaningful instances. This allows us to apply instance-based matchers and similarity metrics like Jaccard and Dice [11,12] to identify mapping candidates. Since those metrics can only be used to identify equality mappings we introduce a novel metric that allows to identify hierarchical and associative mappings like broader-than, narrower-than or is-related to. We integrate the instance enrichment and instance matching together with some optimizations in a novel taxonomy alignment process that we describe below.

To sharpen the description of our contributions, we formalize the problem. The overall goal of the taxonomy alignment process is to integrate the taxonomies $T_1, T_2, ..., T_n$ of the text mining services $S_1, S_2, ..., S_n$ into one global taxonomy $\mathcal{G}$. We make the assumption that each service S_i uses its own taxonomy T_i to classify the text mining results. In order to align two taxonomies T_s and T_t mappings between the types of the taxonomies need to be identified.

A mapping M is a triple (T_{sj}, T_{tk}, R) in which $R \in \{\equiv, <, >, \sim\}$ indicates a relation between a type $T_{sj} \in \mathcal{T}_s$ and a type $T_{tk} \in \mathcal{T}_t$. $(T_{sj}, T_{tk}, \equiv)$ means that the taxonomy types T_{sj} and T_{tk} are equivalent, $(T_{sj}, T_{tk}, <)$ indicates that T_{sj} is a subtype of T_{tk} (i.e., T_{sj} is narrower than T_{tk}), $(T_{sj}, T_{tk}, >)$ is the inverse subsumption relation (i.e., T_{sj} is broader than T_{tk}). $(T_{sj}, T_{tk}, \sim)$ represents an associative relation (e.g., *car* and *truck* are associated). The set of instances annotated by a type T_{ij} is specified by $I(T_{ij})$, its cardinality by $|I(T_{ij})|$. When matching two dissimilar taxonomies we speak of inter-matching whereas matching the types of a taxonomy with itself $(\mathcal{T}_s = \mathcal{T}_t)$ is called intra-matching. Since equal mappings are not relevant in the intra-matching case the set of relevant relations is $R \in \{<, >, \sim\}$.

3 Taxonomy Alignment Process

Initially, the overall taxonomy alignment process is described. Section 3.2 introduces our new metric that is applied within the alignment process. The matching process applying instance- and schema-based matching is detailed in Sect. 3.3.

3.1 Overall Alignment Process

The general taxonomy alignment process is depicted in Fig. 2. The overall idea is to retrieve mappings for the taxonomy types by a matching process and subsequently integrate those mappings to form a global taxonomy $\mathcal{G}$ (see Sect. 6 for the integration). This taxonomy $\mathcal{G}$ reflects all types of the individual taxonomies $\mathcal{T}_i$ and the relations between the particular types (expressed in the mappings). Additionally, the mappings can optionally be cleaned (e.g., by detecting cycles within the graph) and complemented by new mappings (e.g., by exploiting the given hierarchical structure) in mapping rewrite steps as done by existing ontology matching tools like ASMOV [13]. In order to integrate n taxonomies $\binom{n}{2}$ inter-matching processes and n intra-matching processes are applied within our taxonomy alignment process. Each of these inter-matching processes takes two taxonomies as input and identifies equivalence, hierarchical and associative mappings between the types of these taxonomies. The intra-matching processes discover hierarchical and associative mappings within one taxonomy in order to validate and correct/enhance the existing taxonomy structures.

The inter-matching process is implemented by a combined matcher consisting of a schema-based and an instance-based matcher. The schema-based matcher exploits the names of the taxonomy types (e.g., $T_1.a$ and $T_2.i$ in Fig. 2) and is able to identify candidates for equivalence mappings. If sufficient meta data is available for the taxonomies, the schema-based matcher can be extended with matchers that additionally take into account the descriptions or the structures of the input taxonomies. The instance-based matcher exploits the instances of the taxonomy types to identify mapping candidates. The instances of the taxonomy types are retrieved by a new iterative instance enrichment algorithm that was presented in [10]. Furthermore the instance-based matcher applies a novel

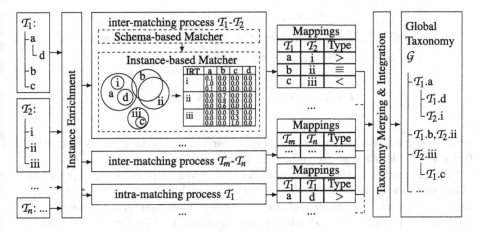

Fig. 2. Taxonomy alignment process.

similarity metric – the intersection ratio triple (IRT) – that allows to identify equivalence, hierarchical as wells as associative relations between the taxonomy types. We will present the metric in Sect. 3.2 and give details on the inter- and intra-matching process in Sect. 3.3.

The intra-matching process uses a slightly adjusted version of the instance-based matcher. A combination with a schema-based matcher is not necessary as equivalence mappings are irrelevant here. The results of the intra-matching process can be used to bring structure into flat taxonomies and check and correct given taxonomy structures.

3.2 IRT Metric

In this section, we present our novel similarity metric for instance-based matchers that is able to indicate equivalence, hierarchical and associative relations between the elements of two taxonomies T_s and T_t. Additionally it allows to identify hierarchical and associative relations within one taxonomy, when used with slightly changed parameters.

It is a common technique within instances-based matchers to rate the similarity of two taxonomy elements $T_{sj} \in T_s$ and $T_{tk} \in T_t$ by analyzing instance overlaps and to represent them by a similarity metric. We propose a novel metric that consists of three single values to represent equivalence, hierarchical and associative relations. The metric adopts the corrected Jaccard coefficient presented by [11]:

$$JCcorr(T_{sj}, T_{tk}) = \frac{\sqrt{|I(T_{sj}) \cap I(T_{tk})| \times (|I(T_{sj}) \cap I(T_{tk})| - c)}}{|I(T_{sj}) \cup I(T_{tk})|}$$

In contrast to the original Jaccard coefficient, that is the ratio of the instance intersection size and the size of the union of the instances, the corrected Jaccard

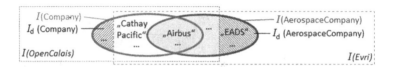

Fig. 3. Example for quality restrictions.

coefficient considers the frequency of co-occurring instances with its correction factor c. For details how to configure c please refer to [11].

We rely on this basic metric as it allows us to deal with possible data sparseness of the instances determined with our instance enrichment process. Additionally, the instances retrieved from text mining services have some quality restrictions that need to be handled. Text mining faces the problem of potentially being inaccurate. Thus, the instances can include false positives (i.e., instances having been extracted wrongly) and for some services miss false negatives (e.g., instances that should be extracted, but having eventually only been extracted by some services).

In order to handle these quality restrictions, we propose an extension of the corrected Jaccard metric as follows: We introduce a weakening factor w that reduces a negative effect of instances only found by one of the services. The factor is trying to correct the influence of the false positives and negatives of the extraction process. Therefore the set of distinct instances $I_d(T_{sj})$ and $I_d(T_{tk})$ that were only extracted by one of the services (independent from the entity type assigned to them) are integrated in the corrected Jaccard factor weakened by w:

$$JCcorr^+(T_{sj}, T_{tk}) = \frac{\sqrt{|I(T_{sj}) \cap I(T_{tk})| \times (|I(T_{sj}) \cap I(T_{tk})| - c)}}{|I(T_{sj}) \cup I(T_{tk})| - w\,|I_d(T_{sj})| - w\,|I_d(T_{tk})|}$$

with $I_d(T_{sj}) \subseteq I(T_{sj}) \backslash \bigcup_{A \in T_t} I(A)$, $I_d(T_{tk}) \subseteq I(T_{tk}) \backslash \bigcup_{B \in T_s} I(B)$ and $0 \leq w \leq 1$

Figure 3 exemplarily depicts the interrelationships between the quality restrictions (e.g., "EADS" as false negative annotation for OpenCalais) and the distinct instances (data was retrieved from Fig. 1).

The similarity value retrieved by the $JCcorr^+$ coefficient enables decisions on the equality of two taxonomy types. If the value is close to 1 it is likely that the type T_{sj} is equal to T_{tk}, if the value is 0, the two taxonomy types seem to be unequal. However, the similarity value does not provide an insight into the relatedness of the two types, when the value is neither close to 1 nor 0. Let us consider the type *Company* and the type *AerospaceCompany*. The extended corrected Jaccard value would be very small – only those company instances of the *Company* type that are aerospace companies might be in the intersection, whereas the union set is mainly determined by the instance size of the type *Company*. In order to detect subtype and associative relations we introduce two more measures $JCcorr^+_{T_{sj}}$ and $JCcorr^+_{T_{tk}}$ rating the intersection size per type:

$$JCcorr^+_{T_{sj}}(T_{sj}, T_{tk}) = \frac{\sqrt{|I(T_{sj}) \cap I(T_{tk})| \times (|I(T_{sj}) \cap I(T_{tk})| - c)}}{|I(T_{sj})| - w\,|I_d(T_{sj})|}$$

$$JCcorr^+_{T_{tk}}(T_{sj}, T_{tk}) = \frac{\sqrt{|I(T_{sj}) \cap I(T_{tk})| \times (|I(T_{sj}) \cap I(T_{tk})| - c)}}{|I(T_{tk})| - w\,|I_d(T_{tk})|}$$

These coefficients are the ratio of the intersection size of the instance sets of the two elements T_{sj} and T_{tk} and the size of one of the instance sets (the instance set $I(T_{sj})$ and $I(T_{tk})$ respectively). All three intersection values together ($JCcorr^+$, $JCcorr^+_{T_{sj}}$, $JCcorr^+_{T_{tk}}$) form the intersection ratio triple (IRT). We can monitor the following states for the values of the IRT metric:

- If all three values are very high, it is very likely that the elements for which the measures were calculated are equal, i.e., the mapping $(T_{sj}, T_{tk}, \equiv)$ can be derived.
- If $JCcorr^+_{T_{sj}}$ is high and the difference $\text{diff}_{T_{tk}}$ of $JCcorr^+$ and $JCcorr^+_{T_{tk}}$ is close to zero, it is an indication that the element T_{sj} is a subtype of T_{tk}, i.e., the mapping $(T_{sj}, T_{tk}, <)$ can be derived.
- If $JCcorr^+_{T_{tk}}$ is high and the difference $\text{diff}_{T_{sj}}$ of $JCcorr^+$ and $JCcorr^+_{T_{sj}}$ is close to zero, it is an indication that the element T_{tk} is a subtype of T_{sj}, i.e., the mapping $(T_{sj}, T_{tk}, >)$ can be derived.
- If none of the three states above yields, but at least one of the IRT-values is clearly above zero the elements T_{sj} and T_{tk} are associated, i.e., the mapping $(T_{sj}, T_{tk}, \sim)$ can be derived.

The IRT metric can also be applied for intra-matching processes. However, the weighting factor is set to 0, i.e., the corrected Jaccard coefficient (and the modified corrected Jaccard coefficients for the second and the third value of the IRT) is used in fact. In the following we show how our novel metric is used within our combined matcher.

3.3   The Matching Process

As already described we use a complex matching strategy that combines both schema-based and instance-based matcher in a single matching process. The combination strategy is visualized in Fig. 4.

The strategy consists of a number of operators that are commonly used in schema matching such as selection (Sel), aggregation (Agg) and matching (mat). Moreover two additional operators ($Trans$ and $Diff$) are included that are needed for processing the IRT matcher results. The process starts by executing the schema- and our instance-based matcher (mat_{schema} and mat_{inst}). They take as input the two taxonomies T_s and T_t and calculate a similarity matrix consisting of $|T_s| \times |T_t|$ entries (Sim and Sim_{IRT}). Each entry of the Sim-matrix is a value between 0 and 1 with 0 representing low and 1 representing high similarity between two pairs of elements from the input taxonomies. The similarity values of this matrix are calculated by a simple name-matcher as proposed

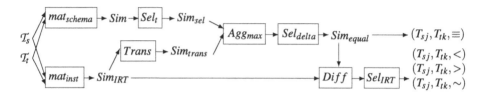

Fig. 4. Combined matching strategy.

in COMA++ [14]. In contrast to that, the entries of the Sim_{IRT}-matrix are composed of the three values computed by our IRT metric (see an exemplary IRT-matrix in Fig. 2).

For equal mappings, we trust in the most likely matching candidates identified by the schema-based matcher. As discussed, the naming of taxonomy types is typically clear and precise and therefore name-matchers tend to have a very high precision. With a selection operation Sel_t the most probable matching candidates are extracted. This operation sets all matrix entries below a given threshold to 0 and all others to 1. We pick a high selection threshold (0.8) to minimize the chance to select wrong mappings.

To simplify the combination of the Sim_{IRT} matrix and the Sim_{sel} matrix, the Sim_{IRT} matrix is transformed by a transformation operation $Trans$. It maps the three IRT values to one value that expresses the probability that the two taxonomy elements are equal. Different transformation operations are possible. A trivial transformation operation $trans_{triv}$ just takes the first IRT value (the extended corrected Jaccard coefficient $JCcorr^+$) or the average of all three values. However, such a trivial transformation may lead to false positive equal mappings since some identified candidates may rather be subtype mappings. As already mentioned in Sect. 3.2 a very low difference value $\mathrm{diff}_{T_{sj}}$ and $\mathrm{diff}_{T_{tk}}$ respectively, may indicate a hierarchical relation. We therefore propose a transformation that lowers the similarity values for such cases:

$$trans = trans_{triv} - corr_{sub}$$

$$corr_{sub} = \begin{cases} 0 & \text{max diff of IRT values} < 0.2 \\ z \cdot e^{-\lambda \cdot \mathrm{diff}_{T_{sj}}} & JCcorr^+_{T_{sj}} < JCcorr^+_{T_{tk}} \quad \text{with } \lambda > 0 \text{ and } 0 \le z \le 1 \\ z \cdot e^{-\lambda \cdot \mathrm{diff}_{T_{tk}}} & JCcorr^+_{T_{sj}} > JCcorr^+_{T_{tk}} \end{cases}$$

The transformation relies on an exponential function to weight the influence of the difference values ($\mathrm{diff}_{T_{sj}}$ or $\mathrm{diff}_{T_{tk}}$) on the transformation result. In particular when the three IRT values are not very close to each other (i.e., having a maximal difference greater than 0.2) the exponential function is applied. The subtype correction $corr_{sub}$ has the biggest value if the difference is zero and then exponentially decreases to zero. The λ value defines how strong the value decreases. Example: With $\lambda = 20$ and a difference value of 0.05 the value $trans_{triv}$ is decreased by 0.368. For $\lambda = 100$ the decrease is only 0.007. The correction value can be further adapted by a weight z that can be based on the value of $JCcorr^+_{T_{sj}}$ and $JCcorr^+_{T_{tk}}$ respectively.

The selected similarity matrix Sim_{sel} is combined with the transformed similarity matrix Sim_{trans} of the instance-based matcher with a MAX-Aggregation operation Agg_{max}. For each pair of entity pairs the maximum of the two matrix entries (one entry from the Sim_{sel} and one from Sim_{trans} matrix) is taken. The result of the mapping aggregation still contains up to $|T_s| \times |T_t|$ correspondences. From these correspondences the most probable ones need to be selected. A number of selection techniques have been proposed in literature (see [14]). We apply the MaxDelta selection from [14] in Sel_{delta} since it has shown to be an effective selection strategy. MaxDelta takes the maximal correspondence within a row (or column) of a similarity matrix. Additionally, it includes correspondences from the row (or column) that are within a delta-environment of the maximal correspondence. The size of the delta environment depends on the value of the maximal element for each row (or column). Both sets of maximal correspondences for each row and correspondences for each column are intersected to get the final selection result Sim_{equal}. Finally, equality mappings are created from the selected matrix Sim_{equal} for each matrix entry above a given threshold.

Subtype and associative mappings are directly derived from the Sim_{IRT} matrix. However, all equality mapping candidates are eliminated from the matrix ($Diff$) before a fine granular selection operation Sel_{IRT} is applied. Sel_{IRT} derives subtype mappings if $JCcorr^+_{T_{sj}}$ (or $JCcorr^+_{T_{tk}}$) is above a given threshold and if $\text{diff}_{T_{tk}}$ (or $\text{diff}_{T_{sj}}$) is smaller than a distance threshold. All remaining matrix entries that are not selected as subtype mappings but indicate a certain overlap of the instances are categorized as associative mappings if one of the three IRT values is significantly above zero.

The presented strategy can be adaptively fine-tuned by analyzing the results of the schema-based matcher. Differing strength and performance of the extraction services for which taxonomies are matched can be identified. For instance, if the text mining service S_s is consistently stronger than the service S_t, we can observe the following: The instance set $I(T_{tk})$ is included in the instance set $I(T_{sj})$ even if the two taxonomy types T_{sj} and T_{tk} are identical (i.e., the schema-based matcher indicates an equivalence relation). For those cases a transformation which corrects subtypes is not recommended. Additionally the selection thresholds can be adapted by observing the instance-matching values for which equivalence relations hold.

4   Experimental Setup

Before we present the results of our experiments in matching entity taxonomies of text mining services in Sect. 5, we give an overview of the experimental setup. The goal of the experiments was to evaluate if our automatic matching approach is applicable for matching taxonomies of text mining services and if our novel metric performs better than traditional approaches. All datasets and manually created gold standards are available upon request.

4.1   Dataset

We evaluated our approach on three entity taxonomies of public and well known text mining services, that are OpenCalais [3], AlchemyAPI [4] and Evri [6]. We only considered the taxonomies that are provided for English text. The entity taxonomy of OpenCalais is documented on the service website and consists of 39 main entity types that are partially further specified with predefined attributes (e.g., the entity Person has the attributes PersonType, CommonName, Nationality). In total it contains 58 entity types. AlchemyAPI documented its entity types classified in a two-level hierarchy on the service website. We observed that not all types AlchemyAPI extracts are listed on the service website. That is why we extended the taxonomy with types having been extracted during the instance enrichment process. All together the taxonomy then consists of 436 types. Evri does not provide an overview of the entity types the service can extract. However, it was possible to extract information via service calls. The Evri taxonomy constructed from the service calls is made up of 583 types.

4.2   Gold Standard

So far no mappings between the taxonomies of text mining services existed. In order to evaluate the quality of the mappings retrieved with our approach, we manually produced a gold standard in [10] through an online evaluation.

We use three values to rate the quality of the retrieved mappings compared to the gold standard: precision, recall and F-measure. Precision is the ratio of accurately identified mappings (i.e., the ratio of the retrieved mappings being in the gold standard and the retrieved mappings). Recall marks the ratio of mappings within the gold standard that were identified by the matcher. The F-measure is the harmonic mean of precision and recall and is a common metric to rate the performance of matching techniques. We consider a matcher to be as good as the F-measure is.

4.3   Matcher Configurations

We experimented with different configurations of our instance-based matcher and determined the best setting - a Jaccard correction factor $c = 0.6$ and a weight w to 0.95 (i.e., integrated the instances only retrieved by one of the services to five percent into the calculations). We achieved good results with a transformation operation using the average of the three IRT values slightly corrected by the exponential function as given in Sect. 3.3. We scaled this correction down or rather ignored it, when observing strongly differing service strength (that was the case, when matching the taxonomy of the OpenCalais service with the taxonomies of the weaker services AlchemyAPI and Evri). The selection threshold for retrieving equality mappings was set to 0.2 when used stand alone and to 0.5 when used in the combined matcher. For the subtype selection operation we used a threshold of 0.65 and a distance threshold of 0.05 within inter-matching processes and a threshold of 0.9 and 0.001 within intra-matching processes.

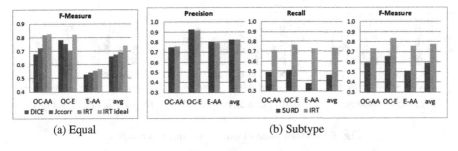

Fig. 5. Comparison of similarity metrics.

We compared our instance-based matching approach and the IRT metric to common metrics of instance-based matching systems: for equality mappings we compared against the Dice and the corrected Jaccard metric, for hierarchical mappings against the SURD metric. The selection thresholds of Dice and corrected Jaccard were set to those values for which the highest average F-measure could be retrieved (Dice: 0.1, corrected Jaccard with correction factor 0.8: 0.05). For SURD we used the threshold proposed in [15] – ratios below 0.5 are low values, ratios above 0.5 are high values. Independent from the used metric the instance intersections were determined by comparing the strings of the instances and only accepting exact matches for the intersection. Moreover, the Sel_{delta} selection techniques described in Sect. 3.3 was applied in all cases.

5 Experimental Results

In the following we present our experimental results proving that our approach is applicable for matching taxonomies of text mining services. We start comparing the IRT metric to state-of-the-art metrics for instance-based matching in Sect. 5.1. Afterwards we rate the performance of the overall intra- and inter-matching processes in Sect. 5.2.

5.1 Comparison of Similarity Metrics

We compared the IRT metric to Dice, corrected Jaccard and SURD and analyzed the performance regarding the identification of equal and subtype mappings (see Sect. 4.3 for the matcher configurations). The results of the comparison are depicted in Fig. 5, in which *OC-AA* indicates the matching process between the OpenCalais and the AlchemyAPI taxonomy, *OC-E* between OpenCalais and Evri, *E-AA* between Evri and AlchemyAPI and *avg* the average between the three values.

Figure 5(a) shows the F-Measure for retrieving equality mappings. We were able to slightly increase the average F-measure compared to the classical metrics Dice and corrected Jaccard. When individually setting the threshold (e.g., by using the schema-based matcher as indicator) the F-measure as well as precision

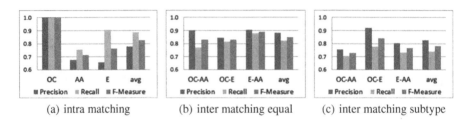

(a) intra matching   (b) inter matching equal   (c) inter matching subtype

Fig. 6. Performance of our matching approach.

and recall can be again increased (IRT ideal). Independent from the specific metric used the performance for the matching process between Evri and AlchemyAPI is worse than the other two matching processes. Reasons for this are on the one hand relatively few instances used for the matching and on the other hand the big performance difference of the two services. We detected that in average equal types only have 30 % in common and it is therefore very hard to detect all mappings correctly.

Figure 5(b) presents the results for the identification of subtype mappings. One can see, that the IRT metric can significantly raise the recall (nearly 30 %) by keeping the same good precision like the SURD metric. Thereby the F-measure can be increased by nearly 20 % which proves that our IRT metric is suited much better for the matching of text mining taxonomies.

5.2   Overall Matching Process

We applied the instance enrichment algorithm, the IRT metric and the combined matching strategy for the intra- and the inter-matching processes for the three services and their taxonomies. The performance results are given in Fig. 6. We compared the mapping results of the intra-matcher to the relations given within the taxonomy structure (Fig. 6(a)). Our approach covered exactly the relations given within the OpenCalais taxonomy. On the contrary, the mappings retrieved by our matching approach and the relations of the AlchemyAPI and Evri taxonomy differed. However, this discrepancy is not a result of the inability of our approach, but rather an indication that the taxonomies are not structured accurately. *AircraftDesigner* is for example listed as a *Person* subtype in the taxonomy used by AlchemyAPI. In practice aircraft designing companies instead of persons are annotated with this type. On the other hand, the flat structure of the taxonomies ignores relations within the subtypes of an entity. *USPresident* and *Politician* are both subtypes of *Person* (which is given in the taxonomy) and the former is in addition a subtype of the latter (this information was retrieved by our approach, but is not represented in the taxonomy). The results show that overreliance on the given taxonomy structures is not reasonable. Instead our approach should be used to validate and correct the taxonomy structure.

The results for the inter-matching processes clearly show that a combination of schema- and instance-based matcher improves quality. The F-measure has

Fig. 7. Taxonomy examples and mapping.

been raised by more than 15 % compared to the instance-based matcher only approach (see Fig. 5). An average F-measure of 85 % for equal (Fig. 6(b)) and 77 % for subtype (Fig. 6(c)) shows that an automatic matching of text mining taxonomies is possible. We observed that in average 63 % of the wrong subtype mappings and 16 % of the missed subtype mappings can be traced back to instance scarcity (i.e., have five or less instances in the intersection). One quarter of the missed equal mappings result from instance scarcity too. Increasing the amount of instances (e.g., by allowing more iterations in the instance enrichment process) and adapting the parameters for each matching process separately (e.g., by using the name-matcher as an indication for the thresholds) quality can be increased.

6   Cluster-Based Taxonomy Merging

To be able to make use of the identified taxonomy mappings the individual taxonomies need to be merged to build a global taxonomy. That global taxonomy could serve as an interface to the combined extraction service. Given a number of taxonomies $T_1, T_2, \ldots, T_n$, a mapping between each possible pair of input taxonomy needs to be computed. The example in Fig. 7 consists of three simple taxonomies that need to be merged and the computed set of mappings. Each mapping entry in the table consists of source, target, relation and confidence. The matching process from above is able to retrieve the mappings and assign confidences to each computed mapping entry. Due to the automatic instance enrichment and matching not all possible mappings are in fact identified and some might be incorrect or questionable like $A.Company > C.Car$.

In the example three equal mappings have been identified, that is between the $A.Person$ category from Taxonomy A and $B.Person$ category from Taxonomy B, between $A.Politician$ and $B.Person\_Political$ and between $A.Product$ and $B.Product$. Moreover, a number of subtype mappings have been found. In particular with subtype mappings it often happens that a category is a subtype of multiple other categories which is natural. The US-President is both a Person and a Politician. The equal and subtype mappings are complemented by related mappings. Only one related mapping is listed in the example. Finally also intra

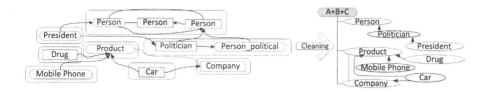

Fig. 8. Clustering and Merging.

mappings were computed that later help to structure the final merged taxonomy and correct the given taxonomy structures. In the following a process is described that builds an integrated global taxonomy that can be exposed for the merged service and that can be used for annotating results.

1. From the found mappings a graph is build. For each type of mapping a different edge is used in the example Fig. 8 (a undirected edge for equal, a directed edge for subtype and a dotted undirected one for associated). Obviously there is still some ambiguity that needs to be resolved and the different categories need to be merged.
2. We identify strongly connected components in the graph of equal matches. The output are clusters of strongly connected elements (illustrated by the grey ovals in the left side of Fig. 8). These clusters build the new categories. Cluster names are assigned from the category that collected most instances in the instance enrichment example.
3. If a cluster also consists of subtype matches then such matches are changed to equal matches.
4. The edges from categories within a cluster pointing to categories outside of a cluster are now changed to edges between the clusters instead of the individual categories. Multiple similar edges between clusters are replaced by one. However the single edge remembers how many and which representatives formally existed.
5. In the next step, the transitivity relation is exploited to remove unnecessary subtype matches between clusters. That means, if a cluster A is subtype B and B is subtype C then all subtype edges between A and C can be removed.
6. The remaining edges between the clusters are further cleaned. If there is a subtype and an equal edge between clusters - then the subtype edge is taken and the equal edge is removed. If there are equal edges between clusters the edge is replaced by related edges. If subtype edges point in both directions then the edge with smaller edge count is removed.

There can be further rules to correct the graph such as removing possible cycles of subtype edges. The final merged taxonomy A+B+C is shown in Fig. 8. Note, that this taxonomy is not a tree, it is rather a graph of categories. The mapping result from above can also be filtered with higher thresholds to remove some mappings with low confidence. The final merged taxonomy structure then has higher credibility. However, the number of categories could increase and the final taxonomy might be less structured. For each service that should be merged the merged taxonomy needs to be recomputed. This ensures that the order of addition has no influence on the result. Note that for each cluster in the taxonomy, the original cluster is retained as a mapping between the individual taxonomies and the global taxonomy. First evaluations of the cluster-based merging process are promising and a first use case to illustrate the value of computing taxonomy alignments between extraction services was implemented with a web news analysis application [10].

7 Related Work

A number of matching systems have been developed that are able to semi-automatically match meta data structures like taxonomies, ontologies or XSD schemata (see [9, 16]). Most of these systems rely on schema-based matching techniques, that consider names, structure or descriptions of elements for matching. For some test-cases they are able to identify equal mappings as we show in our evaluation. However, schema-based techniques are not suited to generate subtype or associative mappings when dealing with flat taxonomies.

A number of existing matching systems like QuickMig [17], COMA++ [14], RiMOM [18] or Falcon [19] rely on instance-based matching techniques to find further correspondences when schema-based matchers are not sufficient. Some of them look for equality of single instances [17–19], others employ metrics that rely on the overlap of instance sets [14]. The latter rely on similarity metrics like Jaccard, corrected Jaccard, Pointwise Mutual Information, Log-Likelihood ratio and Information Gain (see [11]). Massmann and Rahm [12] apply the dice metric to match web directories from Amazon and Ebay. All of these similarity metrics can only be applied to retrieve equal mappings. Moreover, they only perform well when instance sets are quite similar and strongly intersect. They do not consider inaccurate and incomplete instances, like we do with our IRT metric.

The PARIS system [20] employs a probabilistic approach to find alignments between instances, relations and classes of ontologies. The system is mainly able to identify equivalence relations but the authors also introduce an approach to find subclass relations. However, they neither presented how to apply this approach in order to decide for equivalence or subtype relations of classes nor have they evaluated the identification of subclasses. Reference [15] recently proposed a metric of two coefficients to resolve the question how to identify hierarchical relationships between ontologies. This metric is similar to our IRT metric, but does not consider failures within the instances. Moreover, due to relying on only two values and basic heuristics this metric is more inaccurate than the IRT metric presented in this article. By relying on three coefficients we can further refine relationships and besides identifying equivalence and hierarchical relations also identify associative relations between the types of two taxonomies which can not be done with metrics proposed so far. Moreover, we are the first to apply ontology matching techniques for matching text mining taxonomies.

For merging taxonomies most approaches apply merging based on an algorithm where one structure is integrated into another one like PORSCHE [21] or ATOM [22]. The order of merging with these approaches is crucial and can change the resulting global structure. In contrast to that, cluster-based merging approaches do not rely on the order of merges and always reflect an optimal global schema as was also done in ARTEMIS [23] or by Dragut et al. [24] when integrating web data source schemata. Our approach goes beyond existing work since it includes subtype edges and internal structure in the merge process which creates a global schema of higher quality similar to MOMIS [25].

8   Conclusions and Future Work

In this article we presented a number of contributions that help to automatically match and integrate taxonomies of text mining services and therewith enable the combination of several text mining services. In particular we proposed a general taxonomy alignment process that applies a new instance-based matcher using a novel metric called IRT. This metric allows us to derive equality, hierarchical and associative mappings. Our evaluation results are promising, showing that the instance enrichment and matching approach returns good quality mappings and outperforms traditional metrics. Furthermore, the matching process again indicated that the results of different text mining services are very different, i.e., the instances of semantically identical taxonomy types are only partly overlapping (partly only 5 % of the instances overlap). This emphasizes the results from [5] that the quality and quantity of text mining can be increased through the aggregation of text mining results from different services. The presented taxonomy alignment process will allow us in future to automate the matching of text mining taxonomies and subsequently the automatic merging of text mining results from different services (see [26]).

References

1. Grimes, S.: Unstructured data and the 80 percent rule. Clarabridge Bridgepoints (2008). http://breakthroughanalysis.com/2008/08/01/unstructured-data-and-the-80-percent-rule/
2. Hotho, A., Nürnberger, A., Paaß, G.: A brief survey of text mining. LDV Forum **20**(1), 19–62 (2005)
3. OpenCalais: Calais Homepage. March 2013. http://www.opencalais.com/
4. AlchemyAPI: AlchemyAPI Homepage. March 2013. http://www.alchemyapi.com/
5. Seidler, K., Schill, A.: Service-oriented information extraction. In: Proceedings of the Joint EDBT/ICDT Ph.D. Workshop 2011, pp. 25–31 (2011)
6. Evri: Evri Developer Homepage. June 2012. http://www.evri.com/developer/
7. FISE: Furtwangen IKS Semantic Engine project page. March 2013. http://wiki.iks-project.eu/index.php/FISE
8. Euzenat, J., Shvaiko, P.: Ontology Matching. Springer, New York (2007)
9. Rahm, E., Bernstein, P.A.: A survey of approaches to automatic schema matching. VLDB J. **10**, 334–350 (2001)

10. Pfeifer, K., Peukert, E.: Mapping text mining taxonomies. In: KDIR 2013 Proceedings, Scitepress, Portugal (2013)
11. Isaac, A., van der Meij, L., Schlobach, S., Wang, S.: An empirical study of instance-based ontology matching. In: Aberer, K., Choi, K.-S., Noy, N., Allemang, D., Lee, K.-I., Nixon, L.J.B., Golbeck, J., Mika, P., Maynard, D., Mizoguchi, R., Schreiber, G., Cudré-Mauroux, P. (eds.) ASWC 2007 and ISWC 2007. LNCS, vol. 4825, pp. 253–266. Springer, Heidelberg (2007)
12. Massmann, S., Rahm, E.: Evaluating instance-based matching of web directories. In: WebDB 2008 Proceedings (2008)
13. Jean-Mary, Y.R., Shironoshita, E.P., Kabuka, M.R.: Ontology matching with semantic verification. Web Semant. 7(3), 235–251 (2009)
14. Do, H.H., Rahm, E.: COMA - a system for flexible combination of schema matching approach. In: VLDB Proceedings (2002)
15. Chua, W.W.K., Kim, J.J.: Discovering cross-ontology subsumption relationships by using ontological annotations on biomedical literature. In: ICBO. CEUR Workshop Proceedings, vol. 897 (2012)
16. Shvaiko, P., Euzenat, J.: A survey of schema-based matching approaches. In: Spaccapietra, S. (ed.) Journal on Data Semantics IV. LNCS, vol. 3730, pp. 146–171. Springer, Heidelberg (2005)
17. Drumm, C., Schmitt, M., Do, H.H., Rahm, E.: QuickMig: automatic schema matching for data migration projects. In: CIKM'07 Proceedings (2007)
18. Li, J., Tang, J., Li, Y., Luo, Q.: RiMOM: a dynamic multistrategy ontology alignment framework. TKDE 21(8), 1218–1232 (2009)
19. Hu, W., Qu, Y.: Falcon-AO: a practical ontology matching system. Web Semant. 6(3), 237–239 (2008)
20. Suchanek, F.M., Abiteboul, S., Senellart, P.: Paris: probabilistic alignment of relations, instances, and schema. In: Proceedings of the VLDB Endowment, vol. 5(3), pp. 157–168 (2011)
21. Saleem, K., Bellahsene, Z., Hunt, E.: PORSCHE: performance oriented schema mediation. Inf. Syst. 33, 637–657 (2008)
22. Raunich, S., Rahm, E.: ATOM: automatic target-driven ontology merging. In: ICDE Proceedings, pp. 1276–1279 (2011)
23. Castano, S., Antonellis, V.D., Vimercati, S.D.C.D., Melchiori, M.: An xml-based integration scheme for web datasources. Ingénierie des Systèmes d'Information 6(1), 99–122 (2001)
24. Dragut, E.C., Wu, W., Sistla, A.P., Yu, C.T., Meng, W.: Merging Source Query Interfaces on Web Databases. In: ICDE Proceedings, vol. 46 (2006)
25. Beneventano, D., Bergamaschi, S., Guerra, F., Vincini, M.: The MOMIS approach to information integration. In: ICEIS 2001 Proceedings, Setubal, Portugal, vol. 1, pp. 194–198 July 2001
26. Pfeifer, K., Meinecke, J.: Identifying the truth - aggregation of named entity extraction results. In: iiWAS'13 Proceedings, ACM, Austria (2013)

A Multi-dimensional Model
for Computer-Aided Measuring Planning
(CAMP) in Digital Manufacturing

Xiaoqing Tang and Zhehan Chen[✉]

School of Mechanical Engineering and Automation,
Beihang University, Beijing, China
tangxq@buaa.edu.cn, chenzh_buaa@163.com

Abstract. Measuring process based on laser tracker, iGPS and other digital measurement instruments becomes more and more important in digital manufacturing. Traditionally, a measuring process is planned based on human knowledge in planning strategies, measuring regulations, instruments, measuring operations and historical data. Relatively, computer aided measuring planning (CAMP) makes the process formatted, and enables manufacturers to improve measurement accuracy and efficiency and reduce the cost. To implement CAMP for digital measuring process, the concept of general measurement space (GMS) is proposed, and a multi-dimensional model (MDM) is built to support the structural storing and handling of the key information for measuring process. The model has multi attributes in three dimensions, which describe and classify multi-source and heterogeneous data related to a measuring process. Finally, the GMS's characteristics matrix is constructed, providing a feasible way to evaluate measurement plans based on measuring process knowledge.

Keywords: Multi-dimension model · Computer-aided measuring planning · General measurement space · Digital measurement instrument · Digital manufacturing

1 Introduction

Digital measurement technologies have been increasingly and widely employed, which provide more efficient and highly precise approaches for inspection and quality assurance in digital manufacturing process, and thus have drawn significant attention from manufacturers [1–5]. Digital metrologies are based on laser tracker, photogrammetry, iGPS and other digital measurement instruments; while improving the forms of measuring and inspecting, their applications also bring a new principle: measurement is not just an operation for geometrical dimensions inspection, but becomes the eyes of entire production process for digital data transferring, collecting and quality assurance.

In comparison to the traditional measurement approaches, digital metrologies have the attributes in both progressiveness and complexity [6]; in order to meet the requirements of measurement accuracy, measuring time and total cost for a given task, it is necessary to plan the measuring process based on the human knowledge in planning strategies, measurement instruments, historical data and measuring operations, followed

© Springer-Verlag Berlin Heidelberg 2015
A. Fred et al. (Eds.): IC3K 2013, CCIS 454, pp. 56–65, 2015.
DOI: 10.1007/978-3-662-46549-3_4

by making out a measurement plan for guiding the measuring process. The most important factor during measurement planning is accuracy, which is consisting of trueness and precision. The trueness reflects the systematic errors and the precision reflects the random errors [7]. For applications of digital metrologies, measurement plan is significantly important for ensuring the accuracy, validity and creditability of measuring results that is output by the measuring process.

Measurement planning based on the historical knowledge is aimed at ensuring the measurability of measuring process, which is determined by a number of manifold factors involved in the process. The relations between measurability and process factors are the foundation of measurement plan decision, and they are embedded in historical data of measuring process. Therefore, data collection from measuring process and structural storage and analysis are critical for measurement planning in digital manufacturing. However, there is a lack of research on how to integrate the complex historical information from measuring process, and to build up the relationship between measurability and process factors; and thus measurement planning without the support of historical information is still non-formatted and uncertain.

In order to integrate, express and take full reuse of historical information for measurement planning, a multi-dimensional measuring process model based on general measurement space (GMS) is proposed; a methodology, which is based on the GMS model, for formally integrating and expressing different measuring process information is presented and discussed; finally, the GMS's characteristics matrix is constructed, providing a feasible way of evaluating and optimizing measurement plans in digital manufacturing.

2 Related Works

Digital measurement has become one of the critical parts in product manufacturing. As a typical application form, instruments such as laser trackers, iGPS are largely used to measure the position and orientation of complex components during fixtures calibration and components alignment by Boeing, Airbus, Rolls-Royce and other manufacturers [8–10]. Widely employments accumulate a large amount of process information for knowledge integration.

In order to optimize the measurement plan of measuring process, the majority of the research focuses on precision analysis of digital instruments. Jamshidi analyzed the relation between precision and physical structure of iGPS, and compared it with laser tracker [11–13]; Muelaner proposed a mathematical model for evaluating the measurement capability of different instruments, which is able to support instruments selection for measurement planning [14]; on the basis of principle analysis, Du conducted research on how the precision of iGPS measurement field is affected by its deployment types [15]; Wang investigated the error sources of large scale measurement system and proposed a method for uncertainty evaluation [16]. Their research usually established the link between precision and parameters of special instrument, but was only a part of the measuring process and not enough for measurement planning.

For knowledge management of measuring process, Maropoulos firstly proposed the concept of metrology process model [17, 18]; then, Chen used a measurement field model with input and output to express digital measuring process, analyzed its attributes and discussed the evaluation method [19]; subsequently, a measurement data model based on key measurement characteristics was given for unifying, storing and managing the measuring process data in product assembly [20].

In summary, the research and applications of digital metrologies are still in their infancy; the lack of investigations on measuring process model and information integration causes that, a large amount of historical information and knowledge of measuring process haven't been fully discovered, managed and reused. Data collecting and information modeling and analyzing of measuring process have significances for measurement planning and the development of digital metrologies and in manufacturing.

3   Multi-dimensional Model of Measuring Process

Measuring process consists of four stages: demands analysis, process planning, data collection and result output; in order to describe the useful information in measuring process, firstly, on the foundation of object-oriented modeling methodologies, measurement space is considered as an object with functions of receiving demands, executing commands and outputting result; therefore, measuring process can be described by behaviors of measurement space, and information in the process can be described by attributes of measurement space.

Traditional measurement space is considered to be a three-dimensional geometrical space; in order to embed the manifold elements, general measurement space (GMS) model with three dimensions and ten attributes is proposed as the model of measuring process, as shown in Fig. 1; it integrates five key elements in measuring process: people, machine, material, method and environment.

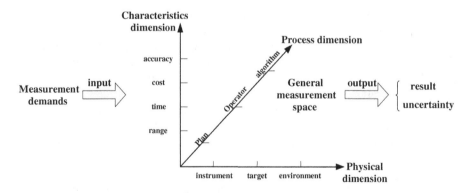

Fig. 1. Definition of general measurement space.

Attributes of GMS are classified into three dimensions: (1) Physical dimension includes basic elements of measuring process such as instrument, target and environment; (2) Process dimension includes additional elements for implementing the behaviours and functions of measuring process, such as plan, operator and algorithm; (3) Characteristics dimension includes key characteristics of measuring process, such as accuracy, cost, time and range, these characteristics are determined by attributes of physical and process dimensions. Based on the GMS model, measuring process can be described as: demands inputting, GMS constructing, characteristics evaluating and result outputting.

3.1   Attributes in Physical Dimension

In physical dimension, instrument attribute is used to describe information and knowledge of digital measurements, such as laser tracker, photogrammetry, iGPS, and laser radar and so on. Measurement targets of GMS are usually comprised of optical target points (OTPs) on the surfaces of different features and structures. Environment attribute is used to describe temperature, humidity, air pressure and other factors that influence on the measuring process. Contents of attributes in physical dimension are shown in Fig. 2.

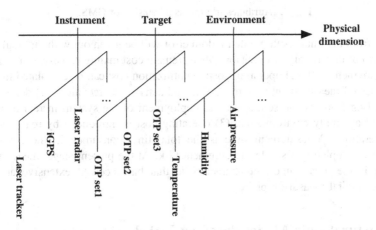

Fig. 2.   Attributes in physical dimension of GMS.

3.2   Attributes in Process Dimension

Attributes of plan, operator and algorithm in process dimension determine the operation mechanism of GMS. Process plan file is used to describe processes, steps of measuring process; a measuring step is mainly comprised of geometrical features and OTPs. Additionally, referred standards and specifications and product model are also included in the plan file. Operator attribute distinguishes different workers by their work number, skills, technical levels and other features, technical level of operator will affect the measurement results to a certain extent. Algorithms, such as auto measuring algorithm, abnormal point judgment algorithm, data fusion algorithm and so on, are called from the algorithm database; different algorithms which implement the same function will lead to different result. Contents of attributes in process dimension are shown in Fig. 3.

3.3   Attributes in Characteristics Dimension

In characteristics dimension, accuracy describes the measured systematic error and random error of any point in GMS, systematic error of measuring process can be

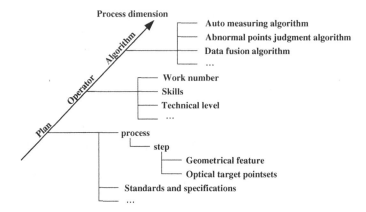

Fig. 3. Attributes in process dimension of GMS.

removed from the final result, while random error will be given out with the final result in the form of uncertainty or precision. Measurement cost mainly consists of utilization cost, deployment cost and operating cost: (1) utilization cost can be calculated in terms of the selected measurement system's value and activity depreciation; (2) deployment cost is arisen from by the setting-up and deployment of the system in real manufacturing and assembly environments; (3) operating cost is introduced by real measurement operations. Measurement time is the total time consumed in the measuring process for completing a single measurement task. At the present stage, measurement accuracy is the most important characteristic that has received extensive attention during the digital measuring process.

4   Structural Data Model Based on MDM

The GMS model has classified and described the multi-source information in measuring process. On the foundation of this, it is necessary to build structural data models to integrate and express the information with a formatted form, which is easy to be used by program for knowledge-based reasoning, and then realizing the CAMP of measuring process for measurement planning.

4.1   Data Structure of Physical Dimension

Information in physical dimension includes instrument knowledge, target knowledge and environment knowledge, which can be transformed to structured information. Through analyzing their properties and relations among those properties, an Entity-Relation diagram is built to give the information model of knowledge in physical dimension, as shown in Fig. 4.

Entity of instrument has properties of manufacturer, model, type, number, name, precision, calibration cycle, calibration date, cost, purchase date and others; it is contained in both instrument database and knowledge in physical dimension. Entity of target has

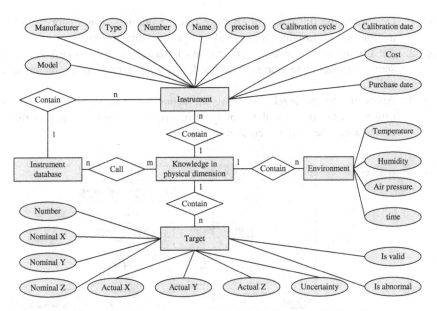

Fig. 4. Entity-Relation diagram of knowledge in physical dimension.

properties of number, nominal value of its coordinate, actual value of its coordinate, uncertainty, is abnormal, is valid and others, it is also contained in knowledge in physical dimension. Entity of environment has properties of temperature, humidity, air pressure, time and others, it is contained in knowledge in physical dimension; the time properties is used to record the measuring time of other properties of environment.

4.2   Information Model of Process Dimension

Knowledge about process plan, operator and algorithm are the main contents in process dimension. Manufacturers are usually skilled to build valid databases for storing, managing and calling resources of their employee and algorithm, because of that those information is easy to be transformed to structured data. By contrast, a process plan file is usually complex and mixed document, which contains lots of structured and non-structured information. In order to unify those information in process dimension for knowledge discovering and reusing, a unified information model based on standards of XML, I++ and DMIS is proposed, the model uses a tree structure and embeds the information of operator and algorithm into itself, as shown in Fig. 5.

4.3   Data Structuring for Characteristics Dimension

Accuracy, cost, time and range are four main characteristics of GMS, and their value determine that if the GMS meet the requirements of measurement task. Data of these attributes is usually continuous, and has some uncertainty and error. Therefore, a method is discussed as follows, for transforming those continuous data to discrete and structured data.

Step 1: Determine the full range of characteristics' value based on historical information and knowledge in physical and process dimensions.

Step 2: Set the threshold value of different levels based on experience and expert scoring method, and then give out the levels with its range of characteristics' value.

Step 3: According to the levels, mapping the actual value of characteristics from original data to level data, and form structured data in characteristics dimension. The flow is depicted in Fig. 6.

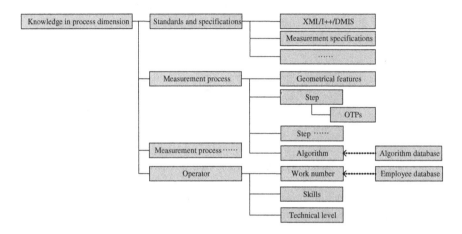

Fig. 5. Unified information model of knowledge in process dimension.

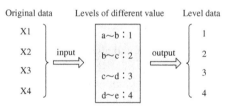

Fig. 6. Method of transforming characteristics knowledge to structured data.

5   Characteristics Matrix for Measurement Planning

To make out a reasonable measurement plan based on multi-dimensional model with historical information includes three steps:

(1) Construct process model and integrate process information for knowledge discovering and storing;

(2) Reveal and present the mapping relationship between characteristics and the basic attributes in physical and process dimensions;

(3) Determine the values of transfer factors from basic attributes to characteristics based on historical knowledge;

(4) Calculate and evaluate the measurement capability of the GMS according to the given deployments, and provide guides and optimization decisions for measurement planning.

The third and fourth sections have given out a measuring process model and structured forms of process knowledge. Thus, in this section, a characteristics relation matrix (CRM) of GMS will be discussed for expressing the relations between characteristics and basic attributes.

The CRM includes two matrixes: characteristics value matrix and characteristics weights matrix, as shown in Fig. 7. In the value matrix, element aij reflects the influence of the jth basic attribute on the ith characteristic, while in the weights matrix, element vij is the weights of aij in all six ai, and

$$\sum_{j=1}^{6} v_{ij} = 1 \tag{1}$$

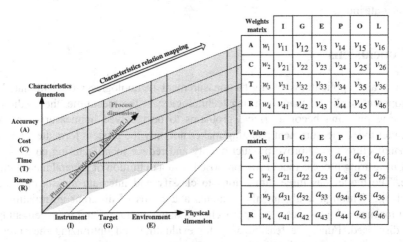

Fig. 7. Characteristics relation matrix of GMS.

The final capability of the GMS is the sum of four characteristics, and is defined as Measurement Capability Index (MCI), which is derived as:

$$MCI = \sum_{i=1}^{4} w_i \cdot C_i \tag{2}$$

where, the factor C_i is the value of the ith characteristic, and the factor w_i is the weights of the ith characteristic in MCI. For different measurement tasks, the importance rank of four characteristics may be not the same, as a result, the value of w_i will be decided in the actual process based on the specific demands.

On the basis of the CRM, the relationship between MCI and all basic attributes can be presented as followed:

$$MCI = \sum_{i=1}^{4} w_i \cdot \sum_{j=1}^{6} v_{ij} \cdot a_{ij} \tag{3}$$

Equation (3) provides a way for calculating the MCI based on the historical knowledge and actual measurement plan prior to executing the measuring process; for calculating one of those characteristics, it is only required to focus on the referred line of the CRM; taking the accuracy prediction as an typical example, the relationship between accuracy and basic attributes can be expressed as:

$$Accuracy = C_1 = \sum_{j=1}^{6} v_{1j} \cdot a_{1j} \tag{4}$$

The relationship expressed by Eq. (4) is the foundation of accuracy prediction. Then, it is necessary to determine the weights v_{1j} and value a_{1j} for each pair of accuracy-attribute.

6 Conclusions

Research and applications of digital measurement technologies have stimulated the development of digital manufacturing technologies; in the meantime, the methods of measuring planning become critical problems to be resolved in measuring process. Computer-aided measurement planning makes not only the process formatted, and also for manufacturers to improve product quality and reduce manufacturing cost.

A measuring process model was proposed with a definition of general measurement space. The model has three dimensions to classify and integrate measuring process knowledge. Through analyzing the contents and forms of different information collected for measuring process, the approaches of data structuring and management have been discussed. Finally, a feasible way for evaluating and optimizing measurement plan based on characteristics matrix was explored.

Future work will focus on historical data collecting, database constructing and evaluation of characteristics relation matrix, in order to improve the methodology of computer-aided digital measurement planning.

References

1. Jody, M., Amir, K., et al.: Measurement assisted assembly and the roadmap to part-to-part assembly. In: Proceedings of DET 2011 7th International Conference on Digital Enterprise Technology (2011)
2. Muelaner, J.E., Maropoulos, P.G.: Design for measurement assisted determinate assembly (MADA) of large composite structures. J. CMSC 5(2), 18–25 (2010)

3. Du, F.Z., Chen, Z.H.: Research on the implementation technologies of measurement driven aircraft sub-assembly digital joining system. Aeronaut. Manufact. Technol. **17**, 52–55 (2011)
4. Wang, Z., Liang, M., et al.: High accuracy mobile robot positioning using external large volume metrology instruments. Int. J. Comput. Integr. Manuf. **24**(5), 484–492 (2011)
5. Liu, S.L., Luo, Z.G., et al.: 3D measurement and quality evaluation for complex aircraft assemblies. Acta Aeronaut. et Astronaut. Sin. **34**(2), 409–418 (2013)
6. Peggs, G.N., Maropoulos, P.G., et al.: Recent developments in large-scale dimensional metrology. Proc. Inst. Mech. Eng. Part B: J. Eng. Manuf. **223**, 571–595 (2009)
7. DE-DIN. Accuracy (trueness and precision) of measurement methods and results - Part 4: Basic methods for the determination of the trueness of a standard measurement method (ISO 5725-4:1994) (2003)
8. Yu, Y., Tao, J., et al.: Assembly technology and process of Boeing 787 jet. Aeronaut. Manufact. Technol. **14**, 44–47 (2009)
9. Jamshidi, J., Kayani, A., et al.: Manufacturing and assembly automation by integrated metrology systems for aircraft wing fabrication. Proc. Inst. Mech. Eng. Part B: J. Eng. Manuf. **224**(25), 25–36 (2010)
10. Jayaweera, N., Webb, P., et al.: Measurement assisted robotic assembly of fabricated aero-engine components. Assembly Autom. **30**(1), 56–65 (2010)
11. Muelaner, J.E., Wang, Z., et al.: Study of the uncertainty of angle measurement for a rotary-laser automatic theodolite (R-LAT). Proc. Inst. Mech. Eng. Part B: J. Eng. Manuf. **223**, 217–229 (2009)
12. Maisano, D.A., Jamshidi, J., et al.: Indoor GPS: system functionality and initial performance evaluation. Int. J. Manuf. Res. **3**(3), 335–349 (2008)
13. Maisano, D.A., Jamshidi, J., et al.: A comparison of two distributed large-volume measurement systems: the mobile spatial coordinate measuring system and the indoor global positioning system. Proc. Inst. Mech. Eng. Part B: J. Eng. Manuf. **223**(B3), 511–521 (2009)
14. Muelaner, J.E., Cai, B., et al.: Large volume metrology instrument selection and measurability analysis. Proc. Inst. Mech. Eng. Part B J. Eng. Manuf. **224**(6), 853–868 (2010)
15. Du, F.Z., Chen, Z.H., et al.: Precision analysis of iGPS measurement field and its application. Acta Aeronaut. et Astronaut. Sin. **33**(9), 1737–1745 (2012)
16. Wang, Q., Zissler, N., et al.: Evaluate error sources and uncertainty in large scale measurement systems. Robot. Comput. Integr. Manuf. **29**(1), 1–11 (2013)
17. Maropoulos, P.G., Zhang, D., et al.: Key digital enterprise technology methods for large volume metrology and assembly integration. Int. J. Prod. Res. **45**(7), 1539–1559 (2007)
18. Maropoulos, P.G., Guo, Y., et al.: Large volume metrology process models: a framework for integrating measurement with assembly planning. CIRP Ann. – Manufact. Technol. **57**, 477–480 (2008)
19. Chen, Z.H., Du, F.Z.: Research on key measurement field building technologies for aircraft digital assembly. Aeronaut. Manufact. Technol. **22**, 44–47 (2012)
20. Chen, Z.H., Du, F.Z., et al.: Key measurement characteristics based inspection data modeling for aircraft assembly. Acta Aeronaut. et Astronaut. Sin. **33**(11), 2143–2152 (2012)

The Application of Support Vector Machine and Behavior Knowledge Space in the Disulfide Connectivity Prediction Problem

Hong-Yu Chen[1], Kuo-Tsung Tseng[2], Chang-Biau Yang[1(✉)], and Chiou-Yi Hor[1]

[1] Department of Computer Science and Engineering, National Sun Yat-sen
University, Kaohsiung 80424, Taiwan
cbyang@cse.nsysu.edu.tw
[2] Department of Shipping and Transportation Management, National Kaohsiung
Marine University, Kaohsiung 81157, Taiwan
tsengkt@nkmu.edu.tw

Abstract. In this paper, we apply support vector machine (SVM) and
behavior knowledge space (BKS) to the disulfide connectivity prediction problem. The problem aims to establish the disulfide connectivity
pattern of the target protein. It is an important problem since a disulfide bond, formed by two oxidized cysteines, plays an important role
in the protein folding and structure stability. The disulfide connectivity
prediction problem is difficult because the number of possible patterns
grows rapidly with respect to the number of cysteines. We discover some
rules to discriminate the patterns with high accuracy in various methods.
Then, the pattern-wise and pair-wise BKS methods to fuse multiple classifiers constructed by the SVM methods are proposed. Finally, the CSP
(cysteine separation profile) method is also applied to form our hybrid
method. We perform some simulation experiments with the 4-fold cross-
validation on SP39 dataset. The prediction accuracy of our method is
increased to 69.1 %, which is better than the best previous result 65.9 %.

Keywords: Disulfide bond · Cysteine · Connectivity pattern · Support
vector machine · Behavior knowledge space

1 Introduction

A *disulfide bond*, also called *SS-bond* or *SS-bridge*, is a single covalent bond which
is usually formed from the oxidation of two thiol groups (-SH). The transformation is described as

$$2RSH \rightarrow RS\text{-}SR + 2H^+ + 2e^- \tag{1}$$

where R represents the carbon-containing group of atoms.

In proteins, only the thiol groups of cysteine residues can form the disulfide bonds by oxidation. The goal of the *disulfide connectivity prediction* (DCP)
problem is to figure out which cysteine pair would be cross-link from all possible

© Springer-Verlag Berlin Heidelberg 2015
A. Fred et al. (Eds.): IC3K 2013, CCIS 454, pp. 66–79, 2015.
DOI: 10.1007/978-3-662-46549-3_5

candidates. It may be conducive to the solution of protein structure prediction problem if precise disulfide connectivity information is available.

There are two main ways for connectivity pattern prediction in previous works, pair-wise and pattern-wise. The pair-wise method focuses on the bonding potential of each cysteine pair, and encodes the target based on cysteine pairs. The pattern-wise method makes a comprehensive survey of whole connectivity pattern and usually ranks the connectivity patterns by their possibilities, so the prediction ability may be limited to the diversity of patterns in a training set.

The prediction task is difficult because the number of possible connectivity patterns grows rapidly with respect to the number of cysteines. Most previous studies are limited by the number of disulfide bonds from two to five. It is well known that the number of possible patterns is given as follows:

$$N = \frac{C_2^{2B} \times C_2^{2B-2} \times \ldots \times C_2^2}{B!} = (2B - 1)!! \tag{2}$$

where B denotes the number of disulfide bonds in the protein. For instance, if we have known which cysteines are oxidized in advance, $N = 945$ when $B = 5$, and N is up to 10395 when $B = 6$.

Some statistical analyses [1–4] have been applied to the disulfide connectivity prediction problem. Many researchers tried to solve the problem with machine learning methods, such as neural network (NN) [5–10] and support vector machine (SVM) [2,11–17].

Before 2005, many studies [5,8] were devoted to the connectivity prediction, but most of their accuracies are below 50 %. In 2005, Zhao et al. [18] utilized the global information of a protein, called *cysteine separation profile* (CSP), which represents the separations among all oxidized cysteines in a protein sequence. In 2007, Lu et al. [2] proposed a novel concept of the CP_2 representation, which uses every two cysteine pairs (four cysteines) as one sample, and applied the genetic algorithm (GA) to the optimization of feature selection.

In 2012, Wang et al. [19] built a hybrid model based on SVM and the weighted graph matching, with accuracy 65.9 %. They extracted different feature sets depending on whether the number of disulfide bonds in a protein is odd or even. The main difference of the feature sets for the two submodels is the secondary structure information around the oxidized cysteines.

The rest of this paper is organized as follows. We introduce some preliminary knowledge, including support vector machine and behavior knowledge space in Sect. 2. In Sect. 3, we present our hybrid method for solving the DCP problem. The experimental results are given in Sect. 4, and we also describe the performance comparison between our method and some previous works. Finally, our conclusion is given in Sect. 5.

2 Preliminary

In this section, we introduce some background knowledge used in this paper, including support vector machine and behavior knowledge space.

2.1   Support Vector Machine

Support Vector Machine (SVM) is a machine learning method for classification and regression. It was first introduced by Vapnik [20] in 1999. SVM seeks to create a hyperplane to discriminate different labels of the vectors in the training set and utilizes the model to predict the labels of other data. To discover the discriminative features is the key point for applying SVM. Figure 1 shows an example of the SVM solution with maximum *margin* which means the distance between the hyperplane and the given objects.

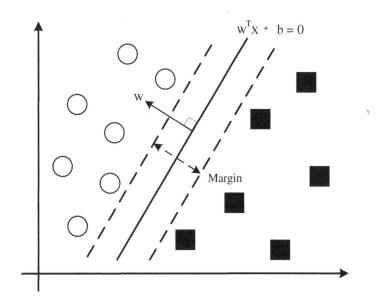

Fig. 1. An example of the SVM solution with maximum margin.

For SVM implementation, we use the LIBSVM package [21] which is an easy-to-use tool for *support vector classification* (SVC) and *support vector regression* (SVR). The SVC function classifies the data with their probabilities, and the SVR function generates the regression value of each target data element.

2.2   Behavior Knowledge Space

Behavior knowledge space (BKS) [22] is a kind of method for fusing multiple classifiers. It is a table look-up approach for estimating the probability of every vote combination. Assume there are m classifiers composing an ensemble for a classification task of n labels. The BKS table contains n^m entries, the number of all possible combinations of m classifiers' outputs. And each entry records the distribution of n true labels in the training set.

Table 1 illustrates an example of the BKS table for the 3-label classification problem with two classifiers. The 'C1' and 'C2' represent the predicted outputs from the two classifiers, and the entries below them are all possible prediction combinations. Cells below 'Real label', 'L1', 'L2', and 'L3', are the distribution of the true labels associated with the predicted label vectors. For example, when 'C1'='L1' and 'C2'='L3', the fused answer should be 'L2' since it is the most possible label. And, if we have 'C1'='L3', 'C2'='L2', the fused answer should go to 'L3'.

Table 1. An example of the BKS table.

Predicted label		Real label		
C1	C2	L1	L2	L3
L1	L1	**23**	8	2
L1	L2	**5**	0	4
L1	L3	2	**7**	1
⋮	⋮	⋮	⋮	⋮
L3	L2	1	1	**5**
L3	L3	1	3	**12**

3   Algorithms for Connectivity Prediction

We observe that the prediction accuracies of Chung *et al.* [23] and Wang *et al.* [19] are 63.5 % and 65.9 %, respectively. It may be hard to find more features with good discrimination capability for a single SVM method in the connectivity prediction. However, we may get better accuracies if we fuse the advantages of the multiple models.

Our method utilizes BKS to fuse the results obtained from SVM models. The features and cysteine-pair representation we adopted are inspired by Wang *et al.* [19] and Lu *et al.* [2]. In addition, we also combine the CSP method [18] to our hybrid method.

3.1   Feature Extraction

In the past, the bonding states of each cysteine pair are usually used to describe the disulfide pattern and used as the samples of SVM. Lu *et al.* [2] call it as the CP_1 representation. Lu *et al.* further proposed a novel concept of the CP_2 representation which use every two cysteine pairs (four cysteines) as the samples. In our method, we adopt the features used by Wang *et al.* [19]. In addition, we encode the CP_2 representation as the *permutation order*, which is also included in our feature set. The definition of the permutation order is given as follows.

Permutation Order: This feature implies the order of feature extraction in each cysteine window. For every cysteine-pair combination in the CP_2 representation, we encode the samples in three permutations illustrated in Table 2. For example, C_1-C_3-C_2-C_4 means that the first and third cysteines form a disulfide bond in these four cysteines, and the second and fourth form the other bond. This bond pattern is represented by the feature vector $(0.25, 0.75, 0.5, 1)$.

Table 2. The feature vector of the permutation order.

Permutations	Feature vector
C_1-C_2-C_3-C_4	(0.25, 0.5, 0.75, 1)
C_1-C_3-C_2-C_4	(0.25, 0.75, 0.5, 1)
C_1-C_4-C_2-C_3	(0.25, 1, 0.5, 0.75)

3.2 SVM Method

We implement three SVM models with different feature sets, CP_1F_{521}, CP_1F_{623} and CP_2Label_2, as shown in Table 3. These features are encoded by the segments of every cysteine pair. The cysteine segment is a window centering at a target cysteine. Many previous works [2,4,8,9,12,14,16,17,24–26] also adopted the similar idea of the window approach. Here we set the window size to 13. In other words, $2k + 1 = 13$. So there are 521 features in CP_1F_{521} and 623 features in CP_1F_{623}.

Table 3. The feature sets used in our three models.

Feature	size	M^a	M^b	M^c
Distance of cysteines	1	Y	Y	Y
Cysteine order	2		Y	
Protein weight	1		Y	
Protein length	1		Y	
Amino acid composition	20		Y	
PSSM around cysteine	$(2k + 1) \times 20 \times 2$	Y	Y	Y
Secondary structure around cysteine	$(2k + 1) \times 3 \times 2$		Y	
Permutation order	4			Y

[a] CP_1F_{521} model.
[b] CP_1F_{623} model.
[c] CP_2Label_2 model.

Table 4. The details of the probability intervals for the BKS method, where B denotes the number of bonds in a target protein.

B	Type of BKS	Probability intervals
2	Pattern-wise	(0, 0.15, 0.2, 0.25, 0.3, 0.35, 0.5, 1)
3	Pattern-wise	(0, 0.25, 0.5, 1)
4	Pair-wise	(0, 0.1, 0.2, 0.3, 0.4, 0.5, 1)
5	Pair-wise	(0, 0.1, 0.2, 0.3, 0.4, 0.5, 1)

3.3 BKS Method

We adopt the concept of the behavior knowledge space (BKS) to fuse the above SVM classifiers. We design two BKS models, pattern-wise BKS and pair-wise BKS, combined with the probability intervals, where the probabilities are obtained from the prediction of SVM classifiers. The details of the probability intervals for the proteins with various number of disulfide bonds are shown in Table 4.

Pattern-Wise BKS Method. After the two classifiers CP_1F_{521} and CP_1F_{623} finish the pattern prediction, the probability of each bonding pattern is obtained. Then, the pattern-wise BKS is constructed according to the prediction probabilities. We adopt the pattern-wise BKS method for the prediction of proteins with two or three bonds. Table 5 illustrates an example of the partial pattern-wise BKS table for 2-bond proteins. For example, in the second row, the probabilities of the predicted pattern 1-1-2-2 for the two classifiers locate in $(0.15, 0.2)$. In this case, 5, 3 and 1 proteins have the true patterns 1-1-2-2, 1-2-1-2 and 1-2-2-1, respectively. Thus, the fused answer is decided to be 1-1-2-2.

We set the threshold of the patterns supported in the pattern-wise BKS table to 2, and reject to give an answer in the case below the threshold. Table 6 shows some examples for 3-bond proteins whose prediction can be corrected by the pattern-wise BKS method.

Table 5. An example of the partial pattern-wise BKS table for 2-bond proteins.

CP_1F_{521}	Interval	CP_1F_{623}	Interval	1-1-2-2	1-2-1-2	1-2-2-1
1-1-2-2	$(0.15, 0.2)$	1-1-2-2	$(0, 0.15)$	0	1	0
1-1-2-2	$(0.15, 0.2)$	1-1-2-2	$(0.15, 0.2)$	5	3	1
1-1-2-2	$(0.15, 0.2)$	1-1-2-2	$(0.2, 0.25)$	4	0	0
1-1-2-2	$(0.15, 0.2)$	1-1-2-2	$(0.25, 0.3)$	0	0	0
1-1-2-2	$(0.15, 0.2)$	1-1-2-2	$(0.3, 0.35)$	0	0	0
1-1-2-2	$(0.15, 0.2)$	1-1-2-2	$(0.35, 0.5)$	1	0	0
1-1-2-2	$(0.15, 0.2)$	1-1-2-2	$(0.5, 1)$	0	0	0

Table 6. Examples for 3-bond proteins corrected by the pattern-wise BKS method.

Proteins	Real patterns	CP_1F_{521}	CP_1F_{623}	Predicted by BKS
CXOA_CONMA	1-2-3-1-2-3	1-2-1-3-2-3	1-2-1-3-2-3	1-2-3-1-2-3
HST1_ECOLI	1-2-3-1-2-3	1-2-1-3-2-3	1-2-1-3-2-3	1-2-3-1-2-3
HCYA_PANIN	1-1-2-2-3-3	1-1-2-2-3-3	1-1-2-3-3-2	1-1-2-2-3-3
CXOB_CONST	1-2-3-1-2-3	1-2-1-3-2-3	1-2-1-3-2-3	1-2-3-1-2-3

Pair-Wise BKS Method. The pattern-wise BKS method is not suitable for the prediction of all proteins. The number of all possible combinations of patterns grows rapidly with respect to the number of bonds, so the number of the training samples is relatively not enough. We propose the pair-wise BKS method for the prediction of proteins with four or five bonds. The pair-wise BKS table records the ratio of the pairs truly bonded or not in various probability intervals from the two classifiers, CP_1F_{521} and CP_2Label_2. Table 7 shows an example of the partial pair-wise BKS table for 5-bond proteins. For every cysteine pair, we advisably adjust the original probability from CP_1F_{521} method according to the ratio of the truly bonded pairs in the pair-wise BKS table. Table 8 illustrates the adjustment rules. Eventually, the predicted pattern is derived from the top N maximum weighted graph matching by the adjusted weighted matrix until the matching pattern belongs to a real pattern in PDB dataset according to our statistics, where the probabilities obtained by SVM classifiers are input as the edge weights in graph matching.

Table 7. An example of the partial pair-wise BKS table for 5-bond proteins.

Pairs from CP_1F_{521}	Pairs from CP_2Label_2	Truly bonded	Not bonded
(0.3, 0.4)	(0, 0.1)	0	0
(0.3, 0.4)	(0.1, 0.2)	0	1
(0.3, 0.4)	(0.2, 0.3)	6	5
(0.3, 0.4)	(0.3, 0.4)	4	6
(0.3, 0.4)	(0.4, 0.5)	6	6
(0.3, 0.4)	(0.5, 1)	1	12

Table 8. The adjustment rules for the pair-wise BKS method.

Truly bonded	Not bonded	Adjustment ratio
0	0	1
[1, 10)	0	4
≥ 10	0	8
0	[1, 10)	0.25
0	≥ 10	0.125
x	x	1
[x, 2x)	x	2
$\geq 2x$	x	4
x	[x, 2x)	0.5
x	$\geq 2x$	0.25

3.4   Hybrid Method

Instead of a large amount of features used by the SVM method, Zhao *et al.* [18] adopted only one feature, CSP (cysteine separations profile), to achieve nearly 50 % accuracy in the dataset with insufficient information. The CSP of protein x with $2n$ oxidized cysteines (n disulfide bonds) is defined as

$$CSP_x = (\delta 1, \delta_2, \dots, \delta_{2n-1}) = (\rho_2 - \rho_1, \rho_3 - \rho_2, \dots, \rho_{2n} - \rho_{2n-1}) \qquad (3)$$

where ρ_i denotes the sequence position of the ith oxidized cysteine in the protein and δ_i denotes the separation distance between oxidized cysteines i and $i+1$.

The divergence (D) of two CSPs for two proteins x and y is defined [18] as follows:

$$D = \sum_{i=1}^{i=2n-1} |\delta_{x,i} - \delta_{y,i}|. \qquad (4)$$

It has been shown that the CSP is an important global feature for the disulfide connectivity prediction, so we also combine the CSP method into our hybrid method. Figure 2 exhibits the flow chart of our work. Our hybrid method for predicting the disulfide connectivity pattern is described as follows.

Algorithm. The hybrid method.
Input: A protein sequence and the bonding states of its all cysteines.
Output: The predicted disulfide connectivity pattern.
Case 1: For a 2-bond or 3-bond protein.

- Step 1.1: Apply the maximum weighted graph matching algorithm to derive the pattern from the CP_1F_{521} method for 2-bond proteins (the CP_1F_{623} method for 3-bond proteins). If the normalized weight of one pattern is greater than or equal to 0.5, report this pattern as the predicted pattern.
- Step 1.2: If the condition meets a predefined threshold in the pattern-wise BKS method, report this pattern as the predicted pattern.
- Step 1.3: If the minimum divergence obtained by the CSP search is less than or equal to a predefined threshold, report this pattern as the predicted pattern.
- Step 1.4: For the remaining, take the original maximum weighted pattern from the CP_1F_{521} method for 2-bond proteins (the CP_1F_{623} method for 3-bond proteins) as the predicted result.

Case 2: For a 4-bond or 5-bond protein.

- Step 2.1: Apply the maximum weighted graph matching algorithm to derive the pattern from the CP_1F_{521} method. If the normalized weight of one pattern is greater than or equal to 0.5, report this pattern as the predicted pattern.
- Step 2.2: If the minimum divergence obtained by the CSP search is less than or equal to a predefined threshold, report this pattern as the predicted pattern.
- Step 2.3: Adjust the original weight (probability) of each pair from the CP_1F_{521} method according to the pair-wise BKS table and report the pattern derived from weighted graph matching algorithm as the predicted result.

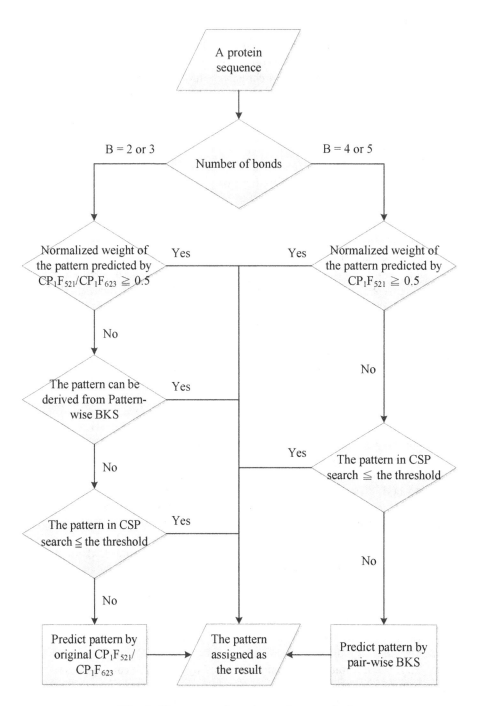

Fig. 2. The system flow chart of our method.

4 Experimental Results

In this section, we introduce our testing dataset and performance evaluation criteria of the disulfide connectivity prediction. In addition, we show the experimental results of various methods.

4.1 Dataset

For the fair comparison of the prediction accuracy with previous works, we use SP39 dataset, which is the same dataset adopted in some previous works, for our training and testing. Table 9 illustrates the summary of SP39 dataset. This dataset contains 446 proteins with two to five disulfide bonds, derived from the SWISS-PROT release no. 39. It was first used by Vullo and Frasconi [10]. We also use the same way as Wang et $al.$'s [19] to divide SP39 dataset into four subsets for the 4-fold cross-validation. The sequence identity of proteins between any two subsets is less than 30 %.

Table 9. The summary of SP39 dataset.

	Number of proteins by the number of bonds					Number of cysteines	
	$B = 2$	$B = 3$	$B = 4$	$B = 5$	$B = 2 \cdots 5$	Oxidized	Total
SP39[a]	156	146	99	45	446	2742	4401

[a] Defined by Vullo and Frasconi [10].

4.2 Performance Evaluation

The definition of k-fold cross-validation is given as follows. A dataset D is divided into k subsets $D_1, D_2, \ldots, D_k$, which are disjoint to each other. Each time, we take a subset D_i, $1 \leq i \leq k$, as the testing set and use the other $k - 1$ subsets for training. Repeat this procedure k times until each subset is tested once. Here, we adopt the 4-fold cross-validation. For the measurement of the performance in connectivity pattern prediction, the accuracy is calculated as follows:

$$Q_p = \frac{C_p}{T_p}, \tag{5}$$

where C_p denotes the number of proteins whose connectivity patterns are correctly predicted, and T_p is the total number of proteins for testing.

4.3 Results

In the CP_1F_{521} method, combined by the SVM method with the maximum weighted graph matching [27], we discover that the prediction accuracy is very high when the normalized weight of one predicted pattern is greater than or equal to 0.5 (half). Table 10 shows the ratio and accuracies of these patterns. In other words, the confidence of such prediction is very high. Thus, in Step 1.1 or Step 2.1 of our method, the answer is settled down for these predictions.

Table 10. The accuracy and ratio of the predicted patterns, whose normalized weights are greater than or equal to 0.5 (half) by CP_1F_{521} method in SP39 dataset.

	$B = 2$	$B = 3$	$B = 4$	$B = 5$	$B = 2 \cdots 5$
Accuracy (Q_p)	100	93.0	93.9	92.9	96.0
Ratio in the dataset	37.8	29.5	33.3	31.1	33.4

Table 11. The accuracy and ratio of the pattern-wise BKS method in SP39 dataset when the threshold is set to 2.

	$B = 2$	$B = 3$
Accuracy (Q_p)	94.0	95.7
Ratio in the dataset	53.2	31.5

Table 12. The Q_p and ratio of the divergence of the CSP in SP39 dataset.

	$B = 2$		$B = 3$		$B = 4$		$B = 5$	
CSP	Q_p	Ratio	Q_p	Ratio	Q_p	Ratio	Q_p	Ratio
0	100	13	100	5	100	2	N/A	0
≤ 5	90	38	96	34	100	24	100	11
≤ 10	78	54	77	57	84	31	100	13
≤ 15	71	67	72	66	73	37	100	13
≤ 20	73	71	72	66	52	55	62	29
≤ 25	71	76	71	66	48	59	53	33

We use the BKS as a supporting role in our method. For the two kinds of the BKS, the performance of the pattern-wise BKS is more effective than the pair-wise BKS. Table 11 illustrates the details of the pattern-wise BKS method when we set the threshold to 2 for 2-bond and 3-bond proteins.

Our hybrid method is combined by the SVM method and the BKS method. In addition, we also take the CSP method into consideration. Table 12 shows the Q_p and ratio of the divergence of the CSP. Take 3-bond proteins as an example. There are 34 % proteins with CSP search less than or equal to 5, and the Q_p of these proteins reaches up to 96 %. According to the observation, we set the applicable thresholds of CSP to pick out the patterns as results. Here, we set the threshold of CSP to 0, 5, 10, and 15 for proteins with two to five bonds, respectively.

Table 13 shows the Q_p of our methods and some previous works in SP39 dataset. The accuracies of the three SVM models are derived from the patterns with the maximum weighted graph matching. However, we find that it is hard to improve the accuracy by one single SVM model. Although the performance of CP_2Label_2 is not better than CP_1F_{521} or CP_1F_{623}, CP_2Label_2 provides the effect for pair-wise BKS since CP_2Label_2 represents another concept of pair extraction. Eventually, the prediction accuracy of our hybrid method with SVM and BKS reaches 65.9 %, and up to 69.1 % combined with CSP method.

Table 13. The Q_p of our methods and previous works in SP39 dataset.

Method	$B = 2$	$B = 3$	$B = 4$	$B = 5$	$B = 2 \cdots 5$
CSP[a]	72.4	54.1	33.3	17.8	52.2
Wang's method[b]	84.0	60.3	55.6	44.4	65.9
CP_1F_{521}	84.0	53.4	55.6	46.7	63.9
CP_1F_{623}	78.2	60.3	53.5	44.4	63.5
CP_2Label_2	75.0	49.3	52.5	40.0	58.1
CP_1F_{521} + BKS	84.0	56.8	55.6	55.6	65.9
CP_1F_{521} + BKS + CSP	84.0	64.4	57.6	57.8	69.1

[a]Proposed by Zhao et al. [18].
[b]Proposed by Wang et al. [19].

5 Conclusion

According to the study of Wang et al. [19], which focuses SVM models on varied features, and the concept of different cysteine-pair representations proposed by Lu et al. [2], we do many integrated experiments. However, the improvement of the pure SVM methods is not so significant although the SVM method is still relatively better, compared with other machine learning methods. Some studies [28,29] combine the SVM method with CSP or sequence alignment to raise the accuracy. The key step of the CSP method and the sequence alignment method is to search for a good template set. However, the accuracy of these two methods depends on the pattern varieties in the template set.

We think that the design of hybrid methods is the trend in the disulfide connectivity prediction problem. In this paper, we gather some statistics about the disulfide bonds, and have successfully found some rules to discriminate the patterns with high accuracy in several methods. Furthermore, we adopt the pattern-wise and pair-wise BKS methods to fuse multiple SVM models, and use the predicted patterns from the original SVM method for the rest of proteins.

In the future, we may examine our hybrid method to other datasets, and explore more methods for fusing multiple classifiers, such as the weighted majority vote. We may try the CSP method with the inter-bond template dataset to explore more possibilities of the development with the concept of subpatterns.

Acknowledgements. This research work was partially supported by the National Science Council of Taiwan under contract NSC 100-2221-E-242-003.

References

1. Harrison, P.M., Sternberg, M.J.E.: Analysis and classification of disulphide connectivity in proteins: the entropic effect of cross-linkage. J. Mol. Biol. **244**(4), 448–463 (1994)

2. Lu, C.-H., Chen, Y.-C., Yu, C.-S., Hwang, J.-K.: Predicting disulfide connectivity patterns. Proteins Struct. Funct. Genet. **67**, 262–270 (2007)
3. Mirny, L.A., Shakhnovich, E.I.: How to derive a protein folding potential? a new approach to an old problem. J. Mol. Biol. **264**(5), 1164–1179 (1996)
4. Rubinstein, R., Fiser, A.: Predicting disulfide bond connectivity in proteins by correlated mutations analysis. Bioinformatics **24**(4), 498–504 (2008)
5. Baldi, P., Cheng, J., Vullo, A.: Large-scale prediction of disulphide bond connectivity. In: Saul, L., Weiss, Y., Bottou, L. (eds.) Advances in Neural Information Processing Systems, vol. 17, pp. 97–104. MIT Press, Cambridge (2005)
6. Cheng, J., Saigo, H., Baldi, P.: Large-scale prediction of disulphide bridges using kernel methods, two-dimensional recursive neural networks, and weighted graph matching. Proteins Struct. Funct. Genet. **62**, 617–629 (2006)
7. Fariselli, P., Riccobelli, P., Casadio, R.: Role of evolutionary information in predicting the disulfide-bonding state of cysteine in proteins. Proteins Struct. Funct. Genet. **36**, 340–346 (1999)
8. Ferre, F., Clote, P.: Disulfide connectivity prediction using secondary structure information and diresidue frequencies. Bioinformatics **21**(10), 2336–2346 (2005)
9. Martelli, P.L., Fariselli, P., Malaguti, L., Casadio, R.: Prediction of the disulfide-bonding state of cysteines in proteins at 88 % accuracy. Protein Sci. **11**, 2735–2739 (2002)
10. Vullo, A., Frasconi, P.: Disulfide connectivity prediction using recursive neural networks and evolutionary information. Bioinformatics **20**(5), 653–659 (2004)
11. Chen, Y.-C., Lin, Y.-S., Lin, C.-J., Hwang, J.-K.: Prediction of the bonding states of cysteines using the support vector machines based on multiple feature vectors and cysteine state sequences. Proteins Struct. Funct. Genet. **55**, 1036–1042 (2004)
12. Chen, Y.-C., Hwang, J.-K.: Prediction of disulfide connectivity from protein sequences. Proteins Struct. Funct. Genet. **61**, 507–512 (2005)
13. Frasconi, P., Passerini, A., Vullo, A.: A two-stage svm architecture for predicting the disulfide bonding state of cysteines. In: Proceedings of the IEEE Workshop on Neural Networks for Signal Processing, pp. 25–34 (2002)
14. Jayavardhana Rama, G.L., Shilton, A.P., Parker, M.M., Palaniswami, M.: Prediction of cystine connectivity using svm. Bioinformation **1**(2), 69–74 (2005)
15. Liu, H.-L., Chen, S.-C.: Prediction of disulfide connectivity in proteins with support vector machine. J. Chin. Inst. Chem. Eng. **38**(1), 63–70 (2007)
16. Tsai, C.-H., Chen, B.-J., Chan, C.-H., Liu, H.-L., Kao, C.-Y.: Improving disulfide connectivity prediction with sequential distance between oxidized cysteines. Bioinformatics **21**(24), 4416–4419 (2005)
17. Vincent, M., Passerini, A., Labbe, M., Frasconi, P.: A simplified approach to disulfide connectivity prediction from protein sequences. BMC Bioinform. **9**(1), 20 (2008)
18. Zhao, E., Liu, H.-L., Tsai, C.-H., Tsai, H.-K., Chan, C.-H., Kao, C.-Y.: Cysteine separations profiles on protein sequences infer disulfide connectivity. Bioinformatics **21**(8), 1415–1420 (2005)
19. Wang, C.-J., Yang, C.-B., Hor, C.-Y., Tseng, K.-T.: Disulfide bond prediction with hybrid models. In: Proceedings of the 2012 International Conference on Computing and Security (ICCS 2012), Ulaanbaatar, Mongolia, July 2012
20. Vapnik, V.N.: The Nature of Statistical Learning Theory. Springer, Heidelberg (1999)
21. Chang, C.-C., Lin, C.-J.: LIBSVM: A library for support vector machines (2001). http://www.csie.ntu.edu.tw/cjlin/libsvm

22. Raudys, S., Roli, F.: The behavior knowledge space fusion method: analysis of generalization error and strategies for performance improvement. In: Windeatt, T., Roli, F. (eds.) MCS 2003. LNCS, vol. 2709, pp. 55–64. Springer, Heidelberg (2003)
23. Chung, W.-C., Yang, C.-B., Hor, C.-Y.: An effective tuning method for cysteine state classification. In: Proceedings of National Computer Symposium, Workshop on Algorithms and Bioinformatics, Taipei, Taiwan, 27–28 November 2009
24. Chen, G., Deng, H., Gui, Y., Pan, Y., Wang, X.: Cysteine separations profiles on protein secondary structure infer disulfide connectivity. In: 2006 IEEE International Conference on Granular Computing, pp. 663–665, May 2006
25. Chuang, C.-C., Chen, C.-Y., Yang, J.-M., Lyu, P.-C., Hwang, J.-K.: Relationship between protein structures and disulfide-bonding patterns. Proteins Struct. Funct. Genet. **53**, 1–5 (2003)
26. Jones, D.T.: Protein secondary structure prediction based on position-specific scoring matrices. J. Mol. Biol. **292**(2), 195–202 (1999)
27. Fariselli, P., Casadio, R.: Prediction of disulfide connectivity in proteins. Bioinformatics **17**(10), 957–964 (2001)
28. Chen, B.-J., Tsai, C.-H., Chan, C.-H., Kao, C.-Y.: Disulfide connectivity prediction with 70 % accuracy using two-level models. Proteins Struct. Funct. Genet. **64**, 246–252 (2006)
29. Chen, Y.-C.: Prediction of Disulfide Connectivity from Protein Sequences. Ph.D. dissertation, National Chiao Tung University, Hsinchu, Taiwan (2007)

Two-Tier Machine Learning Using Conditional Random Fields with Constraints

Sebastian Lindner[(⊠)]

illucIT Software GmbH, Würzburg, Germany
lindner@illucit.com
http://www.illucit.com

Abstract. This paper shows a novel approach of two-tier machine learning to locate bibliographic references in HTML and separate them into fields. First it is demonstrated, how Conditional Random Fields (CRFs) with constraints can be used to split bibliographic references into fields e.g. authors and title. Therefore a unique feature set, constraints and a method for automatic keyword extraction are introduced. The output of this CRF for tagging bibliographic references, Part Of Speech (POS) analysis and Named Entity Recognition (NER) build the first tier and their output is used to locate the bibliographic reference section in the first place. For this the documents are split into blocks, which are then used for classification. For this task a Support Vector Machines (SVM) approach is compared with another one using a CRF. We demonstrate this two-tier approach archives very good results, while the reference tagging approach is able to compete with other state-of-the-art approaches.

Keywords: Classification · Conditional random fields · Constraint-based learning · Information extraction · Information retrieval · Machine learning · References parsing · Semi-supervised learning · Support vector machines

1 Introduction

Due to the increasing number of scientific publications, there is a growing demand to search for similar works and compare new results with previous ones.

There already are certain online publishing systems and special search engines like Google Scholar[1] or CiteSeerX[2] that allow this kind of research. To support searches e.g. for specific authors, first of all the reference section has to be located within all indexed documents. After that, each reference has to be divided into a set of fields e.g. author or journal title. In the remaining part of this paper, we will therefore use the terms labeling, tagging and splitting into fields as synonyms. Because of the diversity of the content and the corresponding reference sections this process is not easy to automate.

---

[1] http://scholar.google.com.
[2] http://citeseerx.ist.psu.edu/index.

© Springer-Verlag Berlin Heidelberg 2015
A. Fred et al. (Eds.): IC3K 2013, CCIS 454, pp. 80–95, 2015.
DOI: 10.1007/978-3-662-46549-3_6

Initial approaches to cope with this kind of data mining task e.g. the previous version of CiteSeer used rule based algorithms. CiteSeer therefore applied heuristic rules to big sets of data and became a well-known search engine for references [1]. But the sets of rules were difficult to maintain and to adjust to other domains. In contrast to that, machine learning techniques are much more adaptable to other reference locating and labeling domains [2]. Generally speaking, supervised machine learning algorithms use labeled training material to build a statistical model, which is then used to tag or label further data. First machine learning approaches used Hidden Markov Models (HMMs) for this task [3]. Nowadays CRFs are most popular, because they allow the use of dependend features and joint inference over the whole reference and so achieve better results [4].

In cooperation with Springer Science+Business Media we develop the web platforms SpringerMaterials[3] and SpringerReference[4] for online document publishing. Those platforms have a great amount of bibliographic data with a variety of citation styles. Because of these different citation styles, using some random part of the available references for training to label some other part of the references would not lead to a good labeling performance. However, since the content is organized in books and subject areas, training of separate models can be done with regard to this segmentation. Because the generation of labeled training data for each of this content partitions is a time consuming job, the focus of this paper lies on cases where only a few training instances are available.

Due to the limited amount of training instances, a Conditional Random Field and additional prior knowledge in form of constraints about the bibliography domain in addition to a few labeled instances are used for training. This CRF can then easily be adapted to other domains (citation styles) by generating a few new labeled instances for training and changing the constraints. That way, a new CRF model can be trained for each book or subject area. This results in a significant improvement in overall labeling accuracy.

Afterwards, we explain how the results of this reference labeling process can then be used to locate the reference section in the first place. Normally, the locating of the reference section would be the first step in extracting bibliographic references from documents. But we demonstrate how results of the reference tagging can even be of value for the reference locating process. Therefore, we reuse the features and the output of the reference labeling process in addition to the results of a Named Entity Recognition (NER) analysis, a Part Of Speech (POS) tagging step and the generation of some additional features to locate the reference section in a document with a machine learning algorithm. We compare the use of a Conditional Random Field (CRF) and a Support Vector Machine (SVM) for this task.

The results in this paper proof that the mentioned reference locating process has a very good accuracy, while the reference tagging can compete with other state-of-the-art approaches and even outperform them. Since the features and the output of the reference parsing step are used to locate the reference section in an article this step is introduced first.

---

[3] http://www.springermaterials.com.
[4] http://www.springerreference.com.

So in the first part of this paper, we demonstrate how references can be parsed. Therefore, first of all an extended constraint model for CRFs is introduced. Next, all used features and constraints needed for this task are described. In addition to that, it is shown how automatically extracted constraints can be used to construct new features for learning.

In the second part of the paper, we propose a new method for the classification of HTML blocks into those containing a reference and those which do not contain a reference. Therefore, a novel feature set for two machine learning algorithms to classify these blocks is introduced. Next, the results for the reference parsing and location step are shown and compared to other previous approaches. Last but not least, we draw conclusions and propose several topics for further research.

2   Reference Parsing

First of all, we formally introduce the task of reference parsing. It is to assign the correct labels $\mathbf{y} = \{y_1, y_2, \ldots y_n\}$ to tokens of an input sequence $\mathbf{x} = \{x_1, x_2, \ldots x_n\}$. Tokens are thereby generated by splitting a reference string on whitespace. For example, the token *Meier* in an input sequence should receive the label *AUTHOR*. For examples of tagged references see Table 1.

This also is a common task in other research areas like Part Of Speech tagging and semantic role labeling [5]. So all techniques shown in this paper can similarly be used in other fields of research as well.

Table 1. Examples of tagged references.

(a)   \<**author**\>N. Benvenuto and F. Piazza.\</**author**\>
\<**title**\>On the Complex Backpropagation Algorithm.\</**title**\>
\<**journal**\>IEEE Transactions on Signal Processing,\</**journal**\>
\<**volume**\>40(4)\</**volume**\>\<**pages**\>967-969,\</**pages**\>
\<**date**\>1992.\</**date**\>

(b)   \<**author**\>C. Jay, M. Cole, M. Sekanina, and P. Steckler.\</**author**\>
\<**title**\>A monadic calculus for parallel costing of a
functional language of arrays.\</**title**\>
In \<**editor**\>C. Lengauer, M. Griebl, and S. Gorlatch, editors,\</**editor**\>
\<**booktitle**\>Euro-Par'97 Parallel Processing,\</**booktitle**\>
\<**volume**\>volume 1300\</**volume**\> of LNCS,
\<**pages**\>pages 650-661.\</**pages**\>
\<**publisher**\>Springer-Verlag,\</**publisher**\>\<**date**\>1997.\</**date**\>

2.1   CRFs with Extended Generalized Expectation Criteria

Linear-chain Conditional Random Fields (CRFs) are a probabilistic framework that uses a discriminative probabilistic model over an input sequence $\mathbf{x}$ and an output label sequence $\mathbf{y}$ as shown in Eq. 1.

$$p_\lambda(\mathbf{y}|\mathbf{x}) = \frac{1}{Z(\mathbf{x})} exp\left(\sum_i \lambda_i F_i(\mathbf{x}, \mathbf{y})\right) \tag{1}$$

In case of a linear chain CRF $F_i(\mathbf{x}, \mathbf{y})$ are a number of feature functions and $Z(\mathbf{x})$ is a normalization factor as described in [6].

Since the goal is to train a model with as few labeled training instances as possible, constraints are used to improve labeling results. An existing concept of Conditional Random Fields [7] is therefore extended to allow more complex constraints.

Given a set of training data $T = \left\{\left(\mathbf{x}^{(1)}, \mathbf{y}^{(1)}\right), \ldots, \left(\mathbf{x}^{(m)}, \mathbf{y}^{(m)}\right)\right\}$ and an additional set of unlabeled references U, the goal of training a Conditional Random Field with extended constraints is to learn the parameters λ_i by maximizing Eq. 2. While training, these additional unlabeled references are used to calculate a distance between the current labeling results on these unlabeled references and the provided constraints.

$$\Theta(\lambda, T, U) = \sum_i log p_\lambda\left(\mathbf{y}^{(i)}|\mathbf{x}^{(i)}\right) - \frac{\sum_i \lambda_i^2}{2\sigma^2} - \delta D(q\|\hat{p_\lambda}), \tag{2}$$

where q is a given target distribution and

$$\hat{p_\lambda} = p_\lambda(y_k|f_1(\mathbf{x}, k) = 1, \ldots, f_m(\mathbf{x}, k) = 1) \tag{3}$$

with all $f_i(\mathbf{x}, k)$ being feature functions that only depend on the input sequence. The first term in Eq. 2 is the log-likelihood used by most CRF implementations for training and the second term is a Gaussian prior for regularization. $D(q\|\hat{p_\lambda})$ is a function to calculate a distance between the provided target probability distribution and the distribution calculated with the help of the unlabeled training data. To penalize differences in these distributions, δ is thereby used as a weighting factor [8]. This is the part of the equation that takes the constraints into account during training.

In this paper we compare three different distance metrics against each other. The first one is the Kullback-Leibler (KL), the second the L_2 distance and the third one is a range-based version of the L_2-distance. In the last case, target probability ranges can be specified instead of one specific value. If the calculated probability then falls within this range, the calculated distance is 0.

As implementation we used MALLET [9], which implements CRFs with Generalized Expectations (GE) as described in [7]. Instead of restricting the target probability distribution to the use of only one feature (see Eq. 4), we extended this functionality to support multiple features as well (see Eq. 3). This way more complex constraints can be defined, which in return improved our labeling results.

$$\hat{p_\lambda} = p_\lambda(y_k|f_i(\mathbf{x}, k) = 1) \tag{4}$$

An example of such a constraint could be that if a name appears at the beginning of a document, it should be labeled *AUTHOR*.

2.2 CRF Features for Reference Parsing

The following enumeration briefly lists the different categories of features used by the reference tagging process. We therefore use a set of binary feature functions similar to the ones used by ParsCit [10].

1. **Word based Features:** Indicate the presence of some significant predefined words like '*No.*', '*et al*', '*etal*', '*ed.*' and '*eds.*'.
2. **Dictionary based Features:** Indicate whether a dictionary contains a certain word in the reference string like for author first- and lastnames, months, locations, stop words, conjunctions and publishers.
3. **Regular Expression based Features:** Indicate whether a word in the reference string matches a regular expression like ordinals (e.g. 1st, 2nd...), years, paginations (e.g. 200–215), initials (e.g. J.F.) and patterns that indicate whether a word contains a hyphen, ends with punctuation, contains only digits, digits or letters, has leading/trailing quotes or brackets or if the first char is upper case.
4. **Keyword Extraction based Features:** Indicate whether a word is in a list of previously extracted keywords for a certain label.

For the keyword extraction process the GSS measure [11] was used. GSS is thereby defined as

$$GSS\left(t_k, c_i\right) = p\left(t_k, c_i\right) \cdot p\left(\overline{t_k}, \overline{c_i}\right) - p\left(t_k, \overline{c_i}\right) \cdot p\left(\overline{t_k}, c_i\right), \qquad (5)$$

where $p\left(\overline{t_k}, c_i\right)$ for example is the probability that given a label c_i, the word t_k does not occur under this label.

Table 2 shows a brief excerpt of the extracted keywords. As one can see, many useful keywords could be automatically extracted for the labels.

In addition to the already mentioned features, a window feature with size 1 and a feature that indicates the position of each word in the reference string was used for training the CRF. Window feature in this case means that features of one token (word) are transferred to its neighbors.

Table 2. Automatically extracted keywords with their corresponding label.

Keyword	Label
Proceedings	BOOKTITLE
Conference	BOOKTITLE
pp	PAGES
Press	PUBLISHER
ACM	JOURNAL
Journal	JOURNAL
University	INSTITUTION
Proc	BOOKTITLE
Vol	VOLUME

2.3 CRF Constraints for Reference Parsing

In order to use Conditional Random Fields with extended Generalized Expectations as shown in Eq. 2, constraints must be defined for the reference labeling task. The following enumeration shows the types of constraints used and names a few examples of them:

1. **Constraints that Depend on a Single Feature Function:** e.g. extracted keywords for a label should be tagged with that exact label or words that match a year pattern should definitely be labeled *YEAR*
2. **Constraints that Depend on Multiple Feature Functions:** e.g. words at the beginning of the reference string and are contained in the dictionary of names should be labeled *AUTHOR*. At the end of the reference string they should be labeled *EDITOR*. Also a number right to word 'No.' should receive the label *VOLUME*.

An extensive list of all used constraints, their assigned probability distributions and other extracted keywords can be found in our previous paper [12].

3 Reference Locating

Existing algorithms for extracting information from HTML usually depend on the Document Object Model (DOM) structure of the document. A related field of work that has recently been studied by several researches is the automatic extraction of records from web pages. Such records could for example be the items in an online shopping cart. Most of them are based on identifying similar DOM tree structures in the web page. Using visual information, tree matching and tree alignment, Zhai and Liu were able to successfully extract such structured records [13]. Fontan et al. even extended these approaches to extract sub-records [14].

On the other hand, data mining algorithms on scanned documents mostly analyze geometric features and layout. Either a top-down approach to recursively divide the whole document into smaller sections e.g. using X-Y cut [15], or a bottom-up approach to cluster more and more small components together [16] is used. Both of these processes terminate when some criteria is met.

In this part of the paper, two machine learning techniques for the identification of a reference section in an HTML document are introduced. Therefore, the performances of a Conditional Random Field (CRF) and a Support Vector Machine (SVM) are compared against each other. First of all the HTML is partitioned into units belonging together (blocks) based on the HTML DOM (Document Object Model) and visual layout information. After that, features for each of these blocks are extracted and used for classification into reference and non-reference blocks.

Zou et al. followed a similar approach by first extracting zones from HTML and then training a SVM with features from these zones [2]. Since we used our own dataset from SpringerReference, there are some different preconditions in comparison to the approach of Zou et al. They used the MEDLINE 2006 database for reference section locating experiments. In their approach the output of the SVM and the corresponding confidence values for the classification results are used in an equation to determine the best candidates for the first and last reference entry (see Eq. 6).

$$[t_F^*, t_L^*] = \arg\max_{t_F, t_L} \prod_{t_F \leq i \leq t_L} P(c_i = R)$$
$$\prod_{0 \leq j < t_F, t_L < j \leq N} (1 - P(c_j = R)), \qquad (6)$$

where t_F and t_L are the locations of the first and last reference, respectively. N ist the total number of child zones and $P(c_k = R)$ is the probability of the kth child being a reference zone [2].

Because not all of our documents even have a reference section or have multiple reference sections in one single document, their approach is not directly applicable. Their 'repair step' can also not be usefully applied if a document only contains one reference entry. In addition to that, their approach uses features like the left and top position of extracted zones for training. Since some of our articles have multiple reference sections these features can not be transferred to our domain. If we had a further reading section after the first paragraph in a long document, the location information of this block would not be different in comparison to other normal blocks. So the position of a block with a reference in the document is not significant for the learning process.

3.1   Extraction of Blocks

Table 3 shows an example of two references (a) and their corresponding source code (b). As one can see in this case an unordered list is used to arrange the two references. But there are many other ways to structure the same information, for example by using a table <table>, a new paragraph <p> for each entry or just line breaks <br> between text.

In order to get blocks of content belonging together, one can not just use the DOM and extract all elements that have no child elements. For example in Table 3(b) the title of the two citations is italic. Extracting elements without children as blocks would lead to very small blocks, which have too few relevant information for a correct classification.

To cope with that, a bottom-up approach is used to successively merge inline DOM nodes. These are nodes that do not introduce line breaks. For this we use the HTML parser jsoup[5]. This parser has a predefined lists of tags that introduce line-breaks like <div> or <p> and those, which are inline elements like <i>, <span> or simple text. In the example of Table 3(b) the italic titles would be merged with the surrounding rest of the reference string, because the text and the italic HTML node are inline elements. The <li> element however, introduces a line-break. So each of these two references would end up in an own block. Figure 1 shows examples of extracted blocks. If the whole document

---

[5] http://jsoup.org/.

Table 3. (a) Example of a reference section, (b) HTML source code of the reference section, (c) Named Entity Recognition results for a reference entry, (d) Part Of Speech tagging results for a reference entry (e) Part Of Speech tagging results for a random paragraph, (f) CRF tagging results for a random paragraph.

(a) 1. McDuffie, H. H., Dosman, J. A., Semchuk, K. M., Olenchock, S. A., & Sentihilselvan, A. (1995). *Agricultural health and safety: Workplace, environment and sustainability.* Boca Raton, FL: Lewis. 2. Messing, K. (1998). *One-eyed science: Occupational health and women workers* Philadelphia: Temple University Press.

(b)
```
<ul>
    <li>1. McDuffie, H. H., Dosman, J. A., Semchuk, K. M.,
    Olenchock, S. A., &amp; Sentihilselvan, A. (1995).
    <i>Agricultural health and safety: Workplace, environment
    and sustainability</i>. Boca Raton, FL: Lewis.</li>
    <li>2. Messing, K. (1998). <i>One-eyed science: Occupa-
    tional health and women workers</i>. Philadelphia:
    Temple University Press.</li>
</ul>
```

(c) McDuffie, H. H., Dosman, **J. A.**, **Semchuk, PERSON** K. M., **Olenchock PERSON**, S. A., & Sentihilselvan, A. **(1995) DATE**. *Agricultural health and safety: Workplace, environment and sustainability* **Boca Raton LOCATION, FL LOCATION: Lewis. PERSON**

(d) McDuffie NNP, H. NNP H. NNP, Dosman NNP, J. NNP A. NNP, Semchuk NNP, K. NNP M. NNP, Olenchock NNP, S. NNP A. NNP, & CC Sentihilselvan NNP, A. NN (1995) CD. *Agricultural NNP health NN and CC safety NN*: *Workplace NNP, environment NN and CC sustainability NN*. Boca NNP Raton NNP, FL NN: Lewis NNP.

(e) The DT appearance NN of IN any DT coast NN, the DT Arctic NNP included VBD, depends VBZ on IN the DT alterations NNS that WDT have VBP occurred VBN to TO the DT geologic JJ base NN it PRP inherited VBD.

(f) The TITLE appearance TITLE of TITLE any TITLE coast TITLE, the TITLE Arctic TITLE included TITLE, depends NOTE on NOTE the NOTE alterations NOTE that NOTE have NOTE occurred NOTE to NOTE the NOTE geologic NOTE base NOTE it NOTE inherited NOTE.

basically only contains text and is structured by <br> and inline elements that use Cascading Style Sheet (CSS) information, the extraction of the blocks can be quite difficult.

3.2 Reference Block Classification

First of all, the features for the classification process are introduced. Afterwards it is shown, how two different machine learning techniques can be used to classify blocks into those, which contain a reference and those which do not contain a reference. Therefore, a SVM approach is presented and then compared to a CRF for classification.

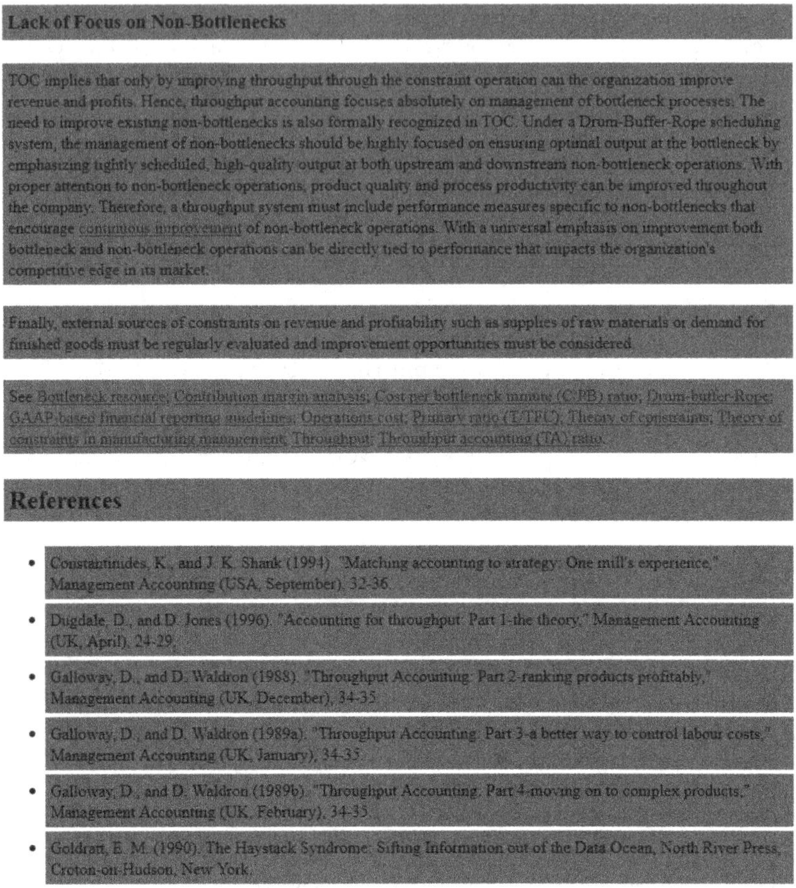

Lack of Focus on Non-Bottlenecks

TOC implies that only by improving throughput through the constraint operation can the organization improve revenue and profits. Hence, throughput accounting focuses absolutely on management of bottleneck processes. The need to improve existing non-bottlenecks is also formally recognized in TOC. Under a Drum-Buffer-Rope scheduling system, the management of non-bottlenecks should be highly focused on ensuring optimal output at the bottleneck by emphasizing tightly scheduled, high-quality output at both upstream and downstream non-bottleneck operations. With proper attention to non-bottleneck operations, product quality and process productivity can be improved throughout the company. Therefore, a throughput system must include performance measures specific to non-bottlenecks that encourage continuous improvement of non-bottleneck operations. With a universal emphasis on improvement both bottleneck and non-bottleneck operations can be directly tied to performance that impacts the organization's competitive edge in its market.

Finally, external sources of constraints on revenue and profitability such as supplies of raw materials or demand for finished goods must be regularly evaluated and improvement opportunities must be considered.

See Bottleneck resource; Contribution margin analysis; Cost per bottleneck minute (CPB) ratio; Drum-buffer-Rope; GAAP-based financial reporting guidelines; Operations cost; Primary ratio (T/TFC); Theory of constraints; Theory of constraints in manufacturing management; Throughput; Throughput accounting (TA) ratio.

References

- Constantinides, K., and J. K. Shank (1994). "Matching accounting to strategy: One mill's experience," Management Accounting (USA, September), 32-36.

- Dugdale, D., and D. Jones (1996). "Accounting for throughput: Part 1-the theory," Management Accounting (UK, April), 24-29.

- Galloway, D., and D. Waldron (1988). "Throughput Accounting: Part 2-ranking products profitably," Management Accounting (UK, December), 34-35.

- Galloway, D., and D. Waldron (1989a). "Throughput Accounting: Part 3-a better way to control labour costs," Management Accounting (UK, January), 34-35.

- Galloway, D., and D. Waldron (1989b). "Throughput Accounting: Part 4-moving on to complex products," Management Accounting (UK, February), 34-35.

- Goldratt, E. M. (1990). The Haystack Syndrome: Sifting Information out of the Data Ocean, North River Press, Croton-on-Hudson, New York.

Fig. 1. Examples for extracted blocks [25]

Initially, a set of features is generated by other machine learning modules in a first tier. The Stanford Named Entity Recognizer [17] is used to extract persons, dates, locations and organizations. This software in turn utilizes a Conditional Random Field to extract these entities. An example output for a reference block is given in Table 3(c). In this table the corresponding label to a word is appended by _LABEL. In comparison to blocks that do not belong to a reference section, the number of extracted entities is much higher. So the number of entities is a clear indicator for a reference entry.

Next the Stanford Log-linear Part-Of-Speech Tagger [18] is applied to each block. This uses a Cyclic Dependency Network for its labeling process. Table 3(d) and (e) show the output of this tagging process for a reference block and a random other block from our corpus. Almost all words in the reference string are tagged as NN (Noun, singular or mass) or NNP (Proper noun, singular). In the case of a random non-reference paragraph however, the number of conjunctions or verbs is much higher like VBD (Verb, past tense), VBZ (Verb, 3rd person singular present), VBP (Verb, non-3rd person singular present), VBN (Verb, past participle) or VBG (Verb, gerund or present participle). So the number of verb forms in a block is another indicator for classification. Other labels that appear in this table are: CC (Coordinating conjunction), DT (Determiner), IN (Preposition or subordinating conjunction), JJ (Adjective).

In addition to these, we use the output of our own reference parser as input for classification. While it is able to correctly label most reference sections, its output is very different for a non-reference section as shown in Table 3(f). All words are labeled as titles or notes. Not even one author is found in the string. So the output of the CRF reference tagger can also be of value to tell a reference and a non-reference section apart.

On top of the machine learning outputs, a number of other features are used for classification. All features generated by the reference parsing process are reused in this reference locating step as well. Meaning features for dictionaries, important keywords, regular expressions and so on.

Furthermore, we used the following features for each block:

- text length
- number of words
- average word length
- number of punctuation characters
- average number of punctuation character (normalized by the number of words)

The number of occurrences of each of the previously mentioned features is then used for training. To take the position of these features in the block into account, each of the features is also combined with a position number (from 1 to 6) that indicates where the feature appeared. It is for example expected to find many author names at the beginning of a reference entry and only a few at the end.

Next we use the Selenium web browser automation framework[6] to determine the y-coordinate of each block. This framework is able to render an HTML page in a browser and provides an interface to get all visual and DOM based information for this page. The blocks are then ordered according to their vertical position and the features of block neighbors are added to the current block. Because it is expected that reference blocks are found in groups, the classification results can be increased this way. The same is true for blocks that are not reference entries. We could retrieve additional layout information about each block from rendering it in a browser, but since we have multiple reference sections in some documents, additional visual information were not useful.

As Support Vector Machine implementation for our second tier we used the Weka Data Mining software [19]. It implements a SVM with Sequential Minimal Optimization and some additional improvements to increase training speed [20]. The SVM is then used to classify each block into the categories reference or non-reference. Since a CRF's purpose is to label an input sequence and not a single block, a whole document and all blocks in it are used as an input sequence for training. Unfortunately the Mallet CRF does not support continuous values

---

[6] seleniumhq.org.

for training, so first all features have to be discretized. This is done with the help of an algorithm by Fayyad and Irani [21], which is also implemented by the Weka framework.

4 Evaluation of Reference Locating

To evaluate the reference locating, we collected a random set of 1,000 articles from our SpringerReference corpus that had a reference section. These were then split into 500 articles for training and 500 for testing. The 500 testing articles contained 42,023 blocks of which 8,983 were reference blocks. From these 41,764 could be correctly classified by the SVM approach and 259 got mislabeled. This means 99.3837 % of the blocks got the correct classification, while only 0.6163 % blocks were classified wrong.

The CRF however did not perform this well. It only achieved 91.3 % accuracy. Perhaps the CRF is not able to extract the most important features out of such a variety of features. Additionally, since the CRF uses each document as only one training instance and treats all the blocks as an input sequence, 500 training instances might just not be enough training material. Another problem might be that the algorithm used for discretizing did not split the value ranges into usefull segments.

The results are slightly worse in comparison to those of Zou et al. [2]. Only 8 blocks out of 22,147 got misclassified by their approach using the MEDLINE 2006 database. As already mentioned before, their approach would not be as successful on our data. In contrast to the MEDLINE documents, many of our documents have zero or only one reference. Despite that, there are articles that have more than one reference section e.g. a further reading section. Using a 'repair step' to find the first and last entry of a reference section would not be possible here. Either because there is only one entry or all entries between two different reference section would be classified as a reference section through this step. So we concentrated our effort on generating better features for learning, instead of trying to increase results through an additional 'repair step' at the end.

5 Evaluation of Reference Parsing

As a reference parsing test domain we used the Cora reference extraction task [22] to compare our approach to previous ones. This set contains 500 labeled reference with 13 different labels like *AUTHOR, BOOKTITLE, DATE, EDITOR, INSTITUTION, JOURNAL, LOCATION, NOTE, PAGES, PUBLISHER, TECH, TITLE, VOLUME.*

5.1 Labeling Results

Generally, the same test approach as described in [23] was used. 500 reference instances were split into 300 for training, 100 for development and 100 for testing

Table 4. Comparison of token based accuracy for different supervised/semi-supervised training models for a varying number of labeled training examples N. Results are an average over 5 runs in percent.

N	Sup	PR	CODL	GE-KL	GE-L_2-Range
5	69.0	75.6	76.0	74.6	75.4
10	73.8	-	83.4	81.2	83.3
20	80.1	85.4	86.1	85.1	86.1
25	84.2	-	87.4	87.2	88.4
50	87.5	-	-	89.0	90.5
100	90.2	-	-	90.4	91.2
300	93.3	94.8	93.6	93.9	94.1

in the evaluation process. From this training set we take a varying number of samples from 5 to 300 for training. For semi-supervised learning we also use 1000 instances of unlabeled data, we took from the FLUX-CiM and CiteSeerX databases. These can be obtained from the ParsCit[7] website.

The labeling results are shown in Table 4. The results report the token based accuracy i.e. the percentage of correct labels and are calculated as averages over 5 runs. Column **Sup** contains the results for a CRF with the same features as previously described but with no constraints. Column **GE-KL** shows the results for our Generalized Expectation with Multiple Feature approach using the KL-divergence and the last column those for the L_2-Range distance metric **GE-L_2-Range**.

In this paragraph, our method is compared to other state-of-the-art semi-supervised approaches. Column **CODL** shows the results for the constraint-driven learning framework [23]. Here the top-k-inference results are iteratively used in a next learning step. In Posterior Regularization (column **PR**) the E-Step in the expectation maximization algorithm is modified to take constraints into account [24]. Dashes indicate that the referenced papers did not contain values for a comparison.

As one can see, our approaches can compete with other leading semi-supervised training methods. While our approaches perform slightly worse than the others with a very limited amount of training data, one of our approaches outperforms the other techniques with $N = 25$ and $N = 50$ training instances. One recognizes that the introduction of constraints greatly improves labeling results. For $N = 20$ the improvement of **GE-L_2-Range** in comparison to (column **Sup**) is 6 percentage points. In our experiments a CRF using GE constraints with multiple feature functions and $L_2 - Range$ as distance metric has the best labeling results.

The results also suggest that the positive influence of constraints decreases with an increasing number of training instances N. The traditional CRF (column **Sup**)

---

[7] http://aye.comp.nus.edu.sg/parsCit/.

Table 5. Precision, recall and F1 measure for label accuracy with $N = 15$ for a CRF with GE-L_2-Range.

Label	Precision	Recall	F_1
AUTHOR	98.5	98.6	98.6
DATE	95.0	82.5	88.3
EDITOR	92.3	52.8	67.2
TITLE	85.5	98.2	91.4
BOOKTITLE	84.3	84.1	84.2
PAGES	82.6	90.0	86.1
VOLUME	75.8	73.5	74.6
PUBLISHER	73.7	35.0	47.5
JOURNAL	71.9	66.6	69.1
TECH	67.5	25.7	37.2
INSTITUTION	62.4	43.4	51.2
LOCATION	51.4	60.0	55.4
NOTE	15.6	10.0	12.2

is then able to determine proper weights for features without user provided constraints. In the case of relatively many training instances $N = 300$ complex constraints even seem to have slightly negative effects in comparison to use of simpler constraints in other approaches (column **PR**).

Table 5 shows precision, recall and the F_1 measure for each separate label with 15 training instances using GE with multiple feature functions and L_2-Range as distance metric.

As Table 5 indicates, there is a big difference in the F_1 value for all labels. Because some labels like *AUTHOR* occur much more often in the training set and we set up more constraints for these labels, the performances in these cases are better. In contrast to that, labels like *NOTE* achieve a rather poor accuracy. However it is important to have a good performance for more common labels to achieve a good overall labeling performance.

It also has to be mentioned that in cases with only few reference instances the results can have a high standard deviation. The diversity of strings in this small portion of training material might simply be to high for proper training.

6   Conclusions and Future Work

We proposed a new method for locating and parsing bibliographic references in HTML documents with a two-tier machine learning approach. We pointed out how HTML content can be grouped into blocks based on the Document Object Model. Afterwards, we have shown how these then can be used for classification into the categories 'is a reference' or 'is not a reference'. Therefore, the features and output of the reference parsing step are used to locate the bibliographic reference section.

In addition to that, we used other machine learning techniques like Part Of Speech tagging and Named Entity Recognition to obtain further input for our reference location process and so improved classification results. Next, a Conditional Random Field and a Support Vector Machine for the classification task were compared against each other. In our setup the SVM approach achieved much better results than a similar approach using a CRF. Even though the reference location process by Zou et al. achieves slightly better results, we have shown that their approaches can not be transferred to our documents from the SpringerReference corpus. Their 'repair step' can not be applied to our document domain, because it assumes many references in one section and only one reference section per document. Despite the difficulties of our test domain, our approach yields very good locating results.

Furthermore, we introduced a new method of parsing references with constraints. The labeling results proof that the proposed semi-supervised machine learning algorithm can compete with other state-of-the-art approaches. It even outperforms other approaches with a certain amount of training instances.

Both the reference locating and the reference parsing approach can easily be adapted to other domains of data. For example, the CRF with constraints approach could also be used to build a custom Named Entity Recognizer for historical texts. Since many NER approaches already use CRFs in the background, constraints could improve results while only needing a few training instances. The proposed classification algorithms could for example be used to determine if a page contains a shopping cart and so used for data mining purposes.

In future we would like to evaluate how bibliographic databases can be used to correct citations and even augment them with missing information as proposed by Gao et al. [4]. We would also like to evaluate how content that was obtained through optical character recognition (OCR) can be improved by automatically replacing badly recognized reference entries with correct ones. To do this we are going to incorporate the Levenshtein distance into our dictionary feature. Afterwards the reference information can even be used to calculate a relatedness for documents. This information could for example be used in an automatic link generation process for disambiguation. Since a method for the automatic extraction of keywords was proposed for feature extraction, we would like to concentrate future effort in the automatic extraction of further features. On top of that, we are trying to not only include constraints in the learning phase of a Conditional Random Field, but also in the inference step. We believe that this could even improve labeling results.

References

1. Bollacker, K.D., Lawrence, S., Giles, C.L.: CiteSeer: An autonomous web agent for automatic retrieval and identification of interesting publications. In: Proceedings of the Second International Conference on Autonomous Agents, pp. 116–123. ACM (1998)
2. Zou, J., Le, D., Thoma, G.R.: Locating and parsing bibliographic references in HTML medical articles. Int. J. Doc. Anal. Recogn. **2**, 107–119 (2010)

3. Hetzner, E.: A simple method for citation metadata extraction using hidden markov models. In: Proceedings of the 8th ACM/IEEE-CS Joint Conference on Digital Libraries, pp. 280–284. ACM (2008)
4. Gao, L., Qi, X., Tang, Z., Lin, X., Liu, Y.: Web-based citation parsing, correction and augmentation. In: Proceedings of the 12th ACM/IEEE-CS Joint Conference on Digital Libraries, pp. 295–304. ACM (2012)
5. Park, S.H., Ehrich, R.W., Fox, E.A.: A hybrid two-stage approach for discipline-independent canonical representation extraction from references. In: Proceedings of the 12th ACM/IEEE-CS Joint Conference on Digital Libraries, JCDL 2012, pp. 285–294. ACM, New York (2012)
6. Sutton, C., McCallum, A.: Introduction to Conditional Random Fields for Relational Learning. MIT Press, Cambridge (2006)
7. Mann, G.S., McCallum, A.: Generalized expectation criteria for semi-supervised learning with weakly labeled data. J. Mach. Learn. Res. **11**, 955–984 (2010)
8. Lafferty, J., McCallum, A., Pereira, F.: Conditional random fields: Probablistic models for segmenting and labeling sequence data. In: Proceedings of the Eighteenth International Conference on Machine Learning (ICML-2001), pp. 282–289 (2001)
9. McCallum, A.: Mallet: A machine learning for language toolkit (2002). http://mallet.cs.umass.edu
10. Councill, I.G., Giles, C.L., Kan, M.Y.: ParsCit: An open-source CRF reference string parsing package. In: International Language Resources and Evaluation. European Language Resources Association (2008)
11. Sebastiani, F.: Machine learning in automated text categorization. ACM Comput. Surv. **34**(1), 1–47 (2002)
12. Lindner, S., Höhn, W.: Parsing and maintaining bibliographic references. In: International Conference on Knowledge Discovery and Information Retrieval (KDIR 2012) (2012)
13. Zhai, Y., Liu, B.: Structured data extraction from the web based on partial tree alignment. IEEE Trans. Knowl. Data Eng. **18**(12), 1614–1628 (2006)
14. Fontan, L., Lopez-Garcia, R., Alvarez, M., Pan, A.: Automatically extracting complex data structures from the web. In: International Conference on Knowledge Discovery and Information Retrieval (KDIR 2012) (2012)
15. Ha, J., Haralick, R.M., Phillips, I.T.: Recursive XY cut using bounding boxes of connected components. In: Proceedings of the Third International Conference on Document Analysis and Recognition, vol. 2, pp. 952–955. IEEE (1995)
16. Jain, A.K., Yu, B.: Document representation and its application to page decomposition. IEEE Trans. Pattern Anal. Mach. Intell. **20**(3), 294–308 (1998)
17. Finkel, J.R.: Named entity recognition and the stanford NER software (2007)
18. Toutanova, K., Klein, D., Manning, C.D., Singer, Y.: Feature-rich part-of-speech tagging with a cyclic dependency network. In: Proceedings of the 2003 Conference of the North American Chapter of the Association for Computational Linguistics on Human Language Technology, vol. 1, pp. 173–180. Association for Computational Linguistics (2003)
19. Hall, M., Frank, E., Holmes, G., Pfahringer, B., Reutemann, P., Witten, I.H.: The WEKA data mining software: an update. ACM SIGKDD Explor. Newsl. **11**(1), 10–18 (2009)
20. Keerthi, S.S., Shevade, S.K., Bhattacharyya, C., Murthy, K.R.K.: Improvements to platt's SMO algorithm for SVM classifier design. Neural Comput. **13**(3), 637–649 (2001)

21. Fayyad, U.M., Irani, K.B.: Multi-interval discretization of continuous-valued attributes for classification learning. In: Thirteenth International Joint Conference on Articial Intelligence, vol. 2, pp. 1022–1027. Morgan Kaufmann Publishers (1993)

22. McCallum, A., Nigam, K., Rennie, J., Seymore, K.: Automating the contruction of internet portals with machine learning. Inf. Retrieval J. **3**, 127–163 (2000)

23. Chang, M.W., Ratinov, L., Roth, D.: Guiding semi-supervision with constraint-driven learning. In: Proceedings of the 45th Annual Meeting of the Association of Computational Linguistics, pp. 280–287 (2007)

24. Ganchev, K., Graca, J., Gillenwater, J., Taskar, B.: Posterior regularization for structured latent variable models. J. Mach. Learn. Res. **11**, 2001–2049 (2010)

25. Swain, M., Fawcett, S.: Accounting system implications of TOC. In: Swamidass, P. (ed.) Encyclopedia of Production and Manufacturing Management. Springer, Heidelberg (2000). http://www.springerreference.com January 31 2011

Keyword Extraction from Company Websites for the Development of Regional Knowledge Maps

Christian Wartena[1]([✉]) and Montserrat Garcia-Alsina[2]

[1] University of Applied Sciences and Arts Hannover, Hannover, Germany
christian.wartena@hs-hannover.de
[2] Universitat Oberta de Catalunya, Barcelona, Spain
mgarciaals@uoc.edu

Abstract. Regional Innovation Systems describe the relations between actors, structures and infrastructures in a region in order to stimulate innovation and regional development. For these systems the collection and organization of information is crucial. In the present paper we investigate the possibilities to extract information from websites of companies. Especially we consider faceted classification of companies by keyword extraction using a specialized thesaurus. First we identify a number of challenges that arise when we want to extract information about companies from their websites. Then we describe a small scale experiment in which keywords related to economic sectors and commodities are extracted from the websites of over 200 companies. The experiment shows that the approach is at least feasible for the commodities facet. For the sectors facet the simple keyword extraction methods used do not perform well. We find that a good coverage of words in the text by the thesaurus is crucial and that hence the results can be improved by adding more alternative labels to the thesaurus terms. Furthermore, we find that weighting terms according to their relations to other terms on the website instead of using inverse document frequency gives better results than the classical tf.idf weighting of terms.

1 Introduction

A basic prerequisite for regional development, for stimulating regional innovation and for improved cooperation between companies in a region, is a high quality overview of all companies in that region, their activities and their strengths. However, hardly any such overviews exist. Information about companies is often incomplete, outdated or focuses only on one specific branch of industry. Official lists, if available at all, also suffer from the problem that officially registered companies might be inactive or just be administrative constructs.

When we try to build a regional overview of economic activities in a region, we face several challenges. In the first place, we have to identify which companies in a region are involved in innovation and regional development. The second challenge is the selection of information sources. In the third place, we should determine which information from each company is relevant to identify

© Springer-Verlag Berlin Heidelberg 2015
A. Fred et al. (Eds.): IC3K 2013, CCIS 454, pp. 96–111, 2015.
DOI: 10.1007/978-3-662-46549-3_7

its strengths, and to promote innovation and regional development. Finally, we have to describe the information about a company, its activities, products and customers in a uniform way. As with regard to the second challenge, it seems natural to extract information from the company's website to obtain basic information (formal data, products and activities) about a company, since virtually every company has a website. There are other sources of up to date information, like patent databases or commercial directories. However, in the present paper we will investigate the possibilities to use the websites of companies, leaving other sources for future work. We propose to use keyword extraction (KWE) techniques based on a restricted vocabulary and we identify what the main challenges are for extracting information about companies from their websites using this approach.

While we sketch the overall approach for the construction of comprehensive regional knowledge maps in [15], in this paper we focus on the possibilities of KWE. In a small empirical study we investigate the possibilities to extract knowledge from company websites for the construction of regional knowledge maps of innovation systems. We use simple thesaurus based KWE techniques, but show how these techniques can provide very rich information about companies when used in combination with a highly structured thesaurus. In contrast to previous work on KWE we do not obtain a flat list of keywords but we get a keyword based description of a company for different aspects. Thus the proposed keyword assignment can be seen as a kind of faceted classification.

We have implemented the thesaurus based keyword extraction method and evaluated it on over 200 manually classified websites from German companies from 10 economic sectors. The main conclusions of this initial study are:

1. The results of the simple thesaurus based approach can be improved strongly by adding more alternative labels that might be found in texts, to the thesaurus concepts.
2. The approach works quite well for keywords related to commodities, but not to keywords related to economic sectors.
3. The classical term weighting by the inverse document frequency has a negative effect on the results, probably because the data set is too small.

Furthermore we observe that the results can be improved using the thesaurus relations between concepts in the text. However, the improvement is not very convincing.

The remainder of the paper is organized as follows. In Sect. 2 we discuss the theoretical background and related work in territorial intelligence and text mining. In Sect. 3 we describe the material used in our experiment. Section 4 describes an experiment to access the feasibility of KWE as a tool to support the construction of regional knowledge maps. The results of the experiment are given and discussed in Sect. 5. We finish the paper with a conclusion and an outlook to future work.

2   Related Work

In this section we first discuss the information need for the construction of comprehensive regional knowledge maps and thus find that text mining might become an important tool for territorial intelligence. In Sect. 2.2 we then review the most important available techniques for keyword extraction.

2.1   Territorial Intelligence

Studies about the regional economics are done with methods coming from different disciplines: Economy, Geography, Sociology, Science of Education, Information and Communication Science [33]. More specifically, in the last 15 years a line of research has been developed, that identifies innovation processes as a key to the regional development [21,37].

Territorial Intelligence (TI) is a collective posture that explores the territorial possibilities by the collection of information and its treatment in order to anticipate risks and threats [23]. To Girardot [20] the territorial intelligence is the science whose object is the sustainable development of territories and whose subject is the territorial community. TI focuses not only on the economic efficacy and efficiency of the development model, but also includes all the dimensions that affect sustainable development (social, political, cultural, and environmental). It involves information about resources available in the territory, products, services (individual or collective, private or public) people needs, which activities are taking part in a territory and the territorial dynamics [20,21,23]. Hence, the collection of information about the territory plays an important role to planning strategic actions to put in touch different actors.

According to different authors [1,8,25,27] a Regional Innovation System (RIS) or National Innovation System (NIS) consists of the relation between actors, structures and infrastructures involved in a region, as well as the knowledge flow between them, that serve the goal of bringing innovative performance of companies in the region. Besides, an efficient system of distribution and access to knowledge contributes to increase the amount of innovative opportunities [5,29]. More specifically, RIS and NIS as research area have developed a framework to study the factors that enable innovation and regional economic development. Some of the features of this framework are: (a) institutions (rules, norms, or organizations) [26,35], (b) innovation processes, (c) knowledge's flows which take place within the regional learning process [28]; (d) the social capital and the regional context in which innovation happens, and that provides a set of rules, conventions and norms that prescribe behavioral roles and shape expectations [8,34]; (e) influence of national or local or social idiosyncrasies, which influence the social process [29,39], and (f) the inter-sectoral differences to explain innovation activities [7].

Considering these antecedents, we underline the relevance of the role of: (a) knowledge in the innovation process and consequently, knowledge management in firms and regions, where knowledge maps play an important role [2,9], (b) the role of companies as actors in a regional innovation system, and

(c) the information about the environment that the companies should achieve and (d) controlled vocabularies, like thesauri, with a common understanding of central terms to classify and retrieve knowledge [3,11,12,14]. Taking into account the amount of documents, it is difficult extract relevant information, and classify this information, so semantic technologies and data mining are needed [10,12].

2.2 Keyword Extraction

Our current goal is to classify companies and other organizations. A number of information sources can be used to do this. In the present study we focus however exclusively on the web sites as a source of information. Thus we can consider our task as a text classification problem. Text classification is a well studied field. A good overview of techniques and approaches is given by [41]. If there is a large number of categories, which is usual the case if each term of a thesaurus is considered as a potential document class, the standard classification approaches cannot be applied, as not enough training data will be available. In these cases a KWE approach is usually better suited. In its basic form, KWE is nothing else than selecting the most salient terms of a document. Determining such terms has been studied since the mid of the previous century [40]. In 1972 Spärck Jones (reprinted as [42]) proposed a weighting for specificity of a term based on $1 + log(\#\text{documents}/\#\text{term occurrences})$. This term weighting, which has become known as *tf.idf*, was refined in [38]. However, salience turns out not to be the only criterion for keywords. More features can be found, indicating whether a term is suited as a keyword or not. These features, together with the relevance weight of the term, can be used in a supervised machine learning setting to learn how to distinguish keywords from non-keywords. This approach to KWE was proposed by [13,44].

Alternatively, the set of possible keywords can be restricted by the terms of a thesaurus or some other restricted vocabulary. This approach is followed by [6] who use information from the thesaurus in combination with Bayesian statistics to suggest keywords. Wang et al. [45] use PageRank to determine the most central words in the graphs which is constructed with the WordNet relations between the potential keywords. The community structure formed by the potential keywords and their relations is used by [22]. In [18,30] words with a large number of relations get higher weights in order to promote central concepts.

In the following we will use the latter approach since we have at the one hand side the possibility to use a highly structured thesaurus, and at the other hand side we do not have enough annotated data to train a model.

3 Data

In order to investigate the potentials of KWE we crawl the pages from the websites of over 200 companies and store them in a local repository. We annotate the data manually for several facets in order to be able to evaluate KWE algorithms. In the following we will describe the thesaurus and the data sources in more detail.

3.1 Thesaurus

We use the STW Thesaurus for Economics[1] as a source for potential key-words [16]. The STW is organized in subthesauri. In order to describe a company we extract only words that are used as descriptors in the subthesauri *Commodities* and *Economic Sectors*. For classification of the economic sectors it might seem more natural to use the NACE (Nomenclature statistique des Activités économiques dans la Communauté Européenne) classification, that is an official standard in the European Union for the classification of economic activities. However, the STW Economic Thesaurus has the advantages that

1. It is available in SKOS format [24], and thus can be read easily by text analysis software [36];
2. It has a lot of different descriptors for each category, that might be found in texts;
3. It contains a classification of products and related areas as well.

Especially the second property is essential for our approach: since the descriptors are usually short terms, that are likely to be found in texts, we can use these descriptors as a list of potential keywords. In order to use the approach for the construction of a regional knowledge map, we could use the mapping between the STW and NACE [17].

The main disadvantage of the STW is, that it is designed to index documents about economics, not to classify companies and their products. We are not aware of any usage of the STW for this purpose. The difference between the two goals might mainly be one of focus and coverage. When we describe companies, we describe different aspects and use a different vocabulary than when we describe a scientific paper about economics.

The STW Thesaurus for Economics consists of 7 subthesauri. Each thesaurus has several subject categories, that can contain subcategories. Subject categories represent either a technical intermediate level of description (e.g. *Branches of Industry*) or classes of products, economic sectors, etc. Each category has a denotation and a number of descriptors. The descriptors are subclasses or examples of the category they belong to. E.g. *fashion* is a descriptor of the category *Textile and Clothing Industry* (W.06.01.11). Also the technical categories have descriptors, that usually refer to instances that cannot be classified into one of the subclasses.

A first inspection of the thesaurus terms found on websites and the classification based on these terms suggest that many problems arise from data sparsity in the thesaurus. It makes of course no sense to add ad-hoc categories for products or sectors that are completely missing, like medical equipment industry. We use a thesaurus because it is standardized and stable. However, we can add more descriptors, non-preferred labels, spelling variants and synonyms to the thesaurus. This would let the original thesaurus intact. The extended version of the thesaurus needs only to be used by the KWE algorithm, while the results are

---

[1] http://zbw.eu/stw/versions/latest/about.en.html.

still descriptors or categories from the official thesaurus. When more labels are available, especially synonyms and alternative labels, the final result is based on much more data, and the occurrence of an irrelevant word will have less impact. When we use the provided labels only, in many cases only two or three thesaurus terms are found on a page, and it is more or less by chance whether these are relevant or irrelevant terms. Enrichment of a thesaurus to enhance thesaurus based KWE is a common method used e.g. by [31,43].

The STW has a mapping[2] to the Integrated Authority File (GND)[3] of german subject headings that is maintained by the German national library. The GND has many alternative labels for each term. We have used this mapping to add more alternative labels to the descriptors of the STW: for each descriptor of the STW we have added all preferred and alternative labels form GND concepts related by either `skos:exactMatch` or `skos:narrowMatch`. The number of alternative labels for each descriptor term was almost doubled in this way. Details are given in Table 1. The number of preferred labels is of course the same as the number of descriptor terms. Note that the number of labels for each category does not some up to the total number of labels in the thesaurus. The reason is, that some descriptor terms are part of more than one subthesaurus.

Table 1. Number of (German) labels for each subthesaurus of the STW before and after enrichment with labels from the GND.

	#descriptors	#skos:altLabel	#skos:altLabel after enrichment
Commodities	785	2197	3429
Sectors	2267	5938	11988
Economics	2757	7234	12868
Related subject areas	1815	5312	10551
General descriptors	33	47	97
Business economics	1154	3806	8541
Geographic names	543	665	822
STW	6027	15426	29514

3.2 Data Sources

In order to investigate the possibilities of describing companies by extracting keywords from their websites we have compiled a set of 229 companies from 10 economic sectors. These companies are not situated in one region, as we do not yet aim at the construction of a regional knowledge map in this phase of the development. All companies are German and have German websites. The companies were selected by students, that had the task of selecting 20 websites of companies in a specific branch or sector as a source for keywords in the related domain.

---

[2] http://zbw.eu/stw/versions/latest/mapping/gnd/about.

[3] http://www.dnb.de/gnd.

We have used the crawler4j[4] to crawl the websites. For 21 companies the crawling was not successful and no pages could be retrieved. Thus the websites of 208 companies remain. Since a few companies have a site with a high number of pages, we limited the number of pages to be retrieved to 120. The limitation services the practical goal of keeping the size of the corpus moderate, but also has more fundamental reasons: A few companies have very large websites, which makes the amount of information per company very unbalanced. Moreover, we expect that even a large company should be described rather well on the first two levels of a web site. If we crawl in a breadth first way, as we do, we might expect that at some point we have seen the core information of a company. If more pages follow, we might get more and more specific information on detailed topics, that even could obscure the more important and central information. The limit of 120 is rather arbitrary and turned out to be a size that allows us for almost all companies in our list to crawl the complete site. In total 14 673 pages were retrieved, which averages to 70.5 pages per company, with a total amount of about 4.6 million words.

We did not do any boiler plate removal since it turned out that in many cases essential information is removed. E.g. a list of products or departments is often given as a menu, that might be removed. For companies with very limited websites, or in cases where the products or departments mentioned in the list point to other domains (we crawl only pages from one domain for each company) the most important information then would be removed.

3.3 Establishment of a Ground Truth

In order to create a base for evaluation of the automatic KWE, we classified all companies manually with the subject categories from the subthesauri *Commodities* and *Economic Sectors*. The categories from the STW are quite broad and can be assigned with a low rate of error and subjectivity and do not require very deep analysis of the information available about a company. In many cases several economic sectors and products had to be assigned. In the first place this was necessary for a number of large companies or business groups that are active in several (related sectors), like machinery construction and electrical industry. The second reason for multiple assignments are businesses that can be viewed from different perspectives. E.g. there is a number of goat farms in our data set. These farms can be classified as *Animal Husbandry* (W.01.02), but since they usually produce goat cheese, they also can be classified as *Food and Tobacco Industry* (W.06.01.12), but as they usually also have a shop where they sell the cheese, *Retail Trade* (W.10.05) is also not completely wrong. Moreover, such farms often offer possibilities to view the animals to the public and they have a small restaurant or even offer cottages for rent. Thus a number of further subject categories apply. These problems partially might stem from the fact that the thesaurus was developed to classify texts about economics, not to classify companies.

---

[4] http://code.google.com/p/crawler4j/.

4   Extracting Company Keywords from Websites

In the past decades a lot of experience has been collected and reported in literature on the automatic extraction of keywords from texts. However, we are faced
here to a unusual situation: we are not looking for keywords that describe the
text they are extracted from, but the keywords should describe the companies
behind the texts. This is not trivial, since the websites of a company usually
are not designed to describe all aspects of a company, but are at least partially
written to present the products to possible customers. Thus at the one hand
side only some aspects are represented, and moreover the sites might give more
information about potential applications of the products than about the internal
processes of the company. Moreover, we note that most work on keyword extraction considers relatively short texts, like news paper articles, scientific papers or
even only abstract thereof, whereas we try to extract a single set of keywords
from about 70 pages on average.

In a first experiment we identified a number of challenges that are specific
for our situation, that we describe in Sect. 4.1. In Sect. 4.2 we than describe our
KWE method.

4.1   Challenges

For most cases, in which no keywords from the right class have been assigned
the underlying problem is data sparseness. Sparseness might have led either to
unprecise manual classification, or to lack of thesaurus terms that could be used
by the analysis of the internet sites.

A typical case is a set of over 10 companies that are specialized in library
(software) systems. The closest branch in the STW is *Information Services*
(W.19.05). The KWE finds mainly words related to libraries. Since a software
company writing software for libraries is not a library, these keywords are evaluated as being incorrect.

If we analyze the wrong keywords, we find five brought classes of errors:

Minor Aspects. In a number of cases, the keyword found is in fact correct,
   but its category was not used to classify the company, usually because the
   category does not reflect the main business of that company. This is especially
   the case for the products category, where we usually assigned only the main
   category in the manual classification, while many other product categories
   might apply.

Random Classes. In some cases the extracted keywords seem completely random. This is the result of the idf-weighting scheme. If some seldom term is
   mentioned two or three times on a small website it will get a very high score.
   This effect is reinforced by the fact that the correct terms usually do not
   have a very high tf.idf value, since we started collecting several companies
   from the same sector. This problem is observed frequently in thesaurus based
   KWE. A method to exclude completely irrelevant terms, is to analyze the
   (thesaurus) relations between the keywords initially found: if a keyword has

no relation (or only few relations) to other keywords, it is likely that it is not related to the main topic of the document. Various proposals have been made to operationalize this idea [18, 22, 30]. In the following we will test exactly this approach in order to overcome this problem.

Customer Classes. Web pages are usually designed for the customers of a company. This might result in texts that contain much more terms describing the branch of the customers of a company than the company itself. E.g. for most pharmaceutical companies descriptors for the branch *Health care systems* (W.25), which is reserved for hospitals and physician practices, are found. Another typical example are the shipyards, classified as *Vehicle Construction* (W.06.01.03), but for which many terms are found from the categories *Port Management* (W.12.01.03.03) and *Shipping* (W.12.01.03). In a number of cases customer classes are chosen, since no appropriate class for the company itself is available. A typical case are the companies specialized in library (software) systems mentioned before.

Ambiguous Terms. Some terms are in some way ambiguous and might refer to two different branches of industry, but are only used for one of both in the STW. E.g. the word *radio* is used a descriptor for *Broadcasting Industry* (W.19.03), but is in our data set found for a company producing radio sets. In a number of cases, the structure of the thesaurus is also problematic. E.g. the term *sail*, sometimes found on the pages of shipyards, is an alternative term for *off-the-peg textiles* that belongs to the category *Clothing* (P.19).

Broader Terms. A typical error is the selection of a term that is too broad, but in fact not incorrect. This is also a common problem for thesaurus based KWE and alternative evaluation methods have been proposed to deal with this [19, 32].

Besides the issues arising from data sparsity another class of problems arises from our decision to use the subject categories to classify companies and to evaluate the descriptors found as keywords. In most cases this works quite well. E.g. the category *Fishery* (W.03) has descriptors like *Aquaculture, Fishery, Fishermen, Fishery fleet*, etc. These terms both are likely to occur in texts and are good descriptors for companies in this sector. However, a category like *Branches of Industry* (W.06.01) has subcategories, representing specific branches, that have descriptors describing them, but it has also descriptors itself. These descriptors represent branches, like *Toy industry* that do not have other descriptors themselves. Consequently we have to annotate a toy manufacturer as *Branches of Industry* (W.06.01), which is not adequate.

4.2 Keyword Extraction

As a first step for KWE all texts are analyzed using a GATE pipeline [4] that consists of a language guesser, a tokenizer, a sentence splitter, a part of speech tagger and lemmatizer an ontology lookup component and several JAPE grammars. The language guesser is used to ensure that only German texts are analyzed. The ontology lookup component Apolda [46] finds all labels of thesaurus terms in the texts, including multiword terms and inflected forms. The JAPE grammar formalism is part of the GATE software, and enables the definition of patterns over words and previous annotations. The Jape grammars are used to extract addresses, phone numbers and names of companies, that we do not evaluate in the present paper.

Since advanced algorithms for KWE usually have been reported to give only limited benefit over a simple tf.idf weighting scheme, we decided to use keyword extraction by selection thesaurus terms with the highest tf.idf score as a baseline. As well for the computation of the document frequencies as for the computation of the term frequencies we consider the whole of all web pages of a company as one document. The intuition behind this is, that a term like *Geschäftsführer* (CEO) that is found on 348 pages of 111 companies in our data set is much less indicative for the business of a company than the word *Schiff* (ship) that occurs on approximately the same number of pages (330) but only for 29 companies.

As an alternative to the tf.idf weighting we also tested a weighting scheme scheme based on the number of relations between thesaurus terms found on a page that was proposed in [18]. The advantage of this weighting scheme is, that it does not depend on the selection of a corpus but uses explicit knowledge from a thesaurus instead. According to this weighting scheme a term becomes more important, if many related terms are found in the same text as well. Now, concepts that are central in a text get high weights whereas rare concepts (with a high idf-value!) that don't play an important role are suppressed. However, If there are concepts in a thesaurus that function as a hub, these might get too high weights. We also tested the combination of relation based and idf-weighting.

In order to compute the tf.idf value of a term w in a document d w.r.t. some document collection D we denote the number of occurrence of w in d as $n_d(w)$ and we let

$$\mathrm{idf}(w, d, D) = \log\left(\frac{|D|}{\mathrm{df}(w)}\right) \tag{1}$$

where $\mathrm{df}(w)$ is the number of documents in which w occurs. Now we define the tf.idf value as

$$\mathrm{tf.idf}(w, d, D) = n_d(w) \cdot \mathrm{idf}(w, d, D). \tag{2}$$

In order to define the weight of a term based on its connections to other terms, first we let $r_n(w, d)$ be the number of thesaurus terms found in d related to w by at most $n - 1$ intermediate concepts. To be precise we first define the set of concepts that can be reached with n steps, given a SKOS-Thesaurus T:

$$R_n(w, d, T) = R_{n-1}(w, d) \cup \{v \mid u\,p\,v \text{ and } u \in R_{n-1}(w, d) \text{ and } v \in C_T(d)\} \tag{3}$$
$$R_0(w, d, T) = \{w\} \text{ if } w \in d \text{ and } \emptyset \text{ otherwise,} \tag{4}$$

where $u\,p\,v$ represents an rdf triple in T and $C_T(d)$ is the set of all concepts from T in d. For the present experiment we restrict p to `skos:exactMatch` and `skos:narrowMatch`. We let

$$r_n(w, d) = |R_n(w, d)| \tag{5}$$

Now the relation strength of a word w in a document d is defined as

$$rr(w,d) = 1 + \mu r_1(w,d) + \mu r_2(w,d) \tag{6}$$

where $\mu = \alpha/\text{avNumLinks}$ for some damping factor α and with avNumLinks the average number of links for each concept. Like in [18] we take $\alpha = 0.5$ and we estimate the average number of relations to be 2.

$$\text{tf.rr}(w,d,T) = n_d(w) \cdot rr(w,d,D) \tag{7}$$
$$\text{tf.rr.idf}(w,d,T,D) = n_d(w) \cdot rr(w,d,T) \cdot \text{idf}(w,d,D) \tag{8}$$

5   Results and Discussion

As mentioned above, we classified the companies according to the main categories in the subthesauri *Commodities* or *Economic Sectors*. We consider a keyword correct if it is a descriptor of a category assigned to the company. Thus we can compute the precision, i.e. the fraction of correct keywords, for each top n elements of the ranked keyword list. The precision of the top n keywords is referred to as prec@n. Since we have no exhaustive list of keywords that should be assigned to a company, we cannot compute a recall value. The precision for economic sectors and commodities is given in Table 2 for the baseline using tf.idf weighting and the original thesaurus without additional labels.

Table 2. Precision of extracted keywords for economic sectors and commodities using tf.idf weighting and original thesaurus.

	Sectors	Commodities
prec@1	0.076	0.27
prec@2	0.088	0.27
prec@5	0.063	0.23

The results show immediately that the extraction of product categories is much easier than the extraction of economic sectors. This is not very surprising since companies will tell about their products on their website, not about the sector they operate in. Even worse, they might write about the use of their products in the sectors their customers are from. In the further conditions the results for the sectors are each time very low, and no significant differences between the conditions are found. Thus we conclude that the approach does not work for sectors and do not report on the results for sectors in the following.

The first column in Table 3 gives the results for the keyword extraction using the enriched thesaurus and keeping all other parameters the same as in the first condition. We can observe a clear improvement. In the following runs reported in Table 3 we use different weighting schemes for the terms found in the texts. We see that a strong improvement is obtained by using the term frequency without idf weighting. Also the use of the rr-measure in isolation gives quite good results.

Table 3. Precision of extracted keywords for commodities using various weighting schemes and enriched thesaurus.

	tf.idf	tf	rr	tf.rr	tf.rr.idf
prec@1	0.35	0.42	0.38	0.44	0.37
prec@2	0.32	0.35	0.36	0.37	0.36
prec@5	0.27	0.28	0.29	0.28	0.29

Best results are obtained by the tf.rr weighting. Overall we observe that inverse document frequency has a negative effect on the results.

Though we did not use any advanced technique for KWE the results are in the same order of magnitude as reported in literature. E.g. [32] find a prec@5 of 0.21 for assignment of terms from the Agrovoc thesaurus to documents, [19] find a prec@5 of 0.23 for assignment of terms from the Dutch GTAA thesaurus for audio-visual archives. However, in our case a high precision is also favored by our evaluation method. In fact we evaluate the results as if each descriptor from the right subject category is a correct one, which would not have been the case when descriptors would have been assigned manually.

Keywords that do not match our self defined gold standard do not have to be wrong. Exactly these keywords might be very interesting and informative. The keywords that do not match with the obvious economic sector, might give hints to hidden expertise. E.g., a company from the sector *Vehicle construction* (W.06.01.03), which main product is ships construction (*Watercrafts*; P.09.03), but with an extracted keyword *Textiles* (P.18), might indeed have expertise on textiles that could also be useful for other companies. Nevertheless, first the main classification for each company has to be established.

5.1   Future Work

For future work, we have to extract more types of information. Obviously, there are other sources of information and a challenge for the future will be to integrate the knowledge about the regional economy that comes from different sources, uses different vocabularies, is written for different target audiences and might even be contradictory.

The other line of research we have to follow is the improvement of the information extraction methods using machine learning techniques. E.g. we could first classify companies into the main thesaurus categories (using a supervised approach) and then restrict the keyword extraction to descriptors from the assigned categories.

Besides keywords we extract other data such as addresses, phone numbers and names of (other) companies as well. These entities are detected using patterns that were constructed by hand. Addresses and phone numbers can help to construct a database of companies. The names of other companies that are mentioned on the website might offer interesting possibilities to exploit network analysis techniques, especially when a large number of companies and other organizations from one region are analyzed. This information then can serve as a further step to map regional knowledge.

6   Conclusion

We have argued that automatic methods to find and aggregate information on the internet are deserved for the construction of concise regional knowledge maps in which economic activities and potentials of companies and other organizations in a region are described. The abundance of information rich websites of companies seems to offer good possibilities to do so.

In an experimental study we have shown that websites of companies indeed can be used as a source of information. In the first place, factual information like addresses and phone numbers can be extracted relatively easily, which we did not discuss in more detail in the paper. Secondly, if an appropriately structured thesaurus is available, the description of various aspects of a company can be described separately by automatically extracting keywords related to that aspect. In a small scale experiment we analyzed over 14 500 web pages from over 200 companies, and extracted company keywords for two facets: economic sectors and commodities.

In our study the keyword extraction worked quite well for keywords related to commodities, but not to keywords related to economic sectors. Furthermore, we observed that the results of the simple thesaurus based approach can be improved strongly by adding more alternative labels that might be found in texts to the thesaurus concepts. Finally, we found that the classical term weighting by the inverse document frequency has a negative effect on the results, probably because the data set is too small. As well ranking without term weighting as ranking using a weighting scheme based on the thesaurus structure gave better results.

Overall we have shown that analysis of web information and KWE constitutes an interesting source of information for territorial intelligence and the construction of concise knowledge maps, that is worth to be explored further.

Acknowledgements. The research presented in this paper was partially funded by the Spanish Ministry of Education, Culture and Sport (Ref. CAS 12/00155).

References

1. Asheim, B., Gertler, M.: The geography of innovation: regional innovation systems. In: Fagerberg, J., Mowery, D., Nelson, R. (eds.) The Oxford Handbook of Innovation, pp. 291–317. Oxford University Press, Oxford (2005)
2. Barinani, A., Agard, B., Beaudry, C.: Competence maps using agglomerative hierarchical clustering. J. Intell. Manuf. **24**(2), 1–12 (2011)

3. Canongia, C.: Synergy between competitive intelligence (CI), knowledge management (KM) and technological foresight (TF) as a strategic model of prospecting — the use of biotechnology in the development of drugs against breast cancer. Biotechnol. Adv. **25**(1), 57–74 (2007)
4. Cunningham, H., Maynard, D., Bontcheva, K., Tablan, V.: A framework and graphical development environment for robust nlp tools and applications. In: Proceedings of the 40th Annual Meeting of the Association for Computational Linguistics, 6–12 July, pp. 168–175. ACL, Philadelphia (2002)
5. David, P., Foray, D.: Assessing and expanding the science and technology knowledge base. STI Rev. **14**, 13–68 (1995)
6. De Campos, L.M., Fernández-Luna, J.M., Huete, J.F., Romero, A.E.: Automatic indexing from a thesaurus using bayesian networks: application to the classification of parliamentary initiatives. In: Mellouli, K. (ed.) ECSQARU 2007. LNCS (LNAI), vol. 4724, pp. 865–877. Springer, Heidelberg (2007)
7. Doloreux, D., Nabil, A., Landry, R.: Mapping regional and sectoral characteristics of knowledge-intensive business services: Evidence from the province of Quebec (Canada). Growth Change **39**(3), 464–496 (2008)
8. Doloreux, D., Parto, S.: Regional innovation systems: Current discourse and unresolved issues. Technol. Soc. **27**, 133–153 (2005)
9. Driessen, S., Huijsen, W., Grootveld, M.: A framework for evaluating knowledge-mapping tools. J. Knowl. Manage. **11**(2), 109–117 (2007)
10. Eckert, K., Stuckenschmidt, H., Pfeffer, M.: Interactive thesaurus assessment for automatic document annotation. In: Proceedings of the 4th International Conference on Knowledge Capture, pp. 103–110. ACM (2007)
11. Escorsa, P., Rodriguez, M., Maspons, R.: Technology mapping, business strategy and market opportunities. Compet. Intell. Rev. **11**(1), 46–57 (2000)
12. Färber, M., Rettinger, A.: A semantic wiki for novelty search on documents. In: Proceedings of the 13th Dutch-Belgian Workshop on Information Retrieval, Delft, pp. 60–61 (2013)
13. Frank, E., Paynter, G.W., Witten, I.H., Gutwin, C., Nevill-Manning, C.G.: Domain-specific keyphrase extraction. In: Proceedings of the Sixteenth International Joint Conference on Artificial Intelligence, IJCAI 1999, Stockholm, Sweden, July 31–August 6, pp. 668–673 (1999)
14. Garcia-Alsina, M., Ortoll, E.: La Inteligencia Competitiva: evolución histórica y fundamentos teóricos. Trea, Gijón (2012)
15. Garcia-Alsina, M., Wartena, C., Lieberam-Schmidt, S.: Regional knowledge maps: potentials and challenges. In: Fifth International Conference on Knowledge Management and Information Sharing (KMIS 2013) (2013)
16. Gastmeyer, M.: Standard-thesaurus wirtschaft. Technical report Deutsch Zentralbibliothek für Wirtschaftswissenschaften, Kiel (1998)
17. Gastmeyer, M., Weskamp, W.: Nace-konkordanz. In: Standard-Thesaurus Wirtschaft, vol. 2, Kiel (1998)
18. Gazendam, L., Wartena, C., Brussee, R.: Thesaurus based term ranking for keyword extraction. In: Tjoa, A.M., Wagner, R. (eds.) Database and Expert Systems Applications, DEXA, 10th International Workshop on Text-based Information Retrieval, TIR, pp. 49–53. IEEE (2010)
19. Gazendam, L., Wartena, C., Malaisé, V., Schreiber, G., De Jong, A., Brugman, H.: Automatic annotation suggestions for audiovisual archives: Evaluation aspects. Interdis. Sci. Rev. **34**(2–3), 172–188 (2009)

20. Girardot, J.J.: Evolution of the concept of territorial intelligence within the coordination action of the european network of territorial intelligence. Ricerca e Sviluppo per le politiche sociali **1**(1–2), 11–29 (2008)
21. Girardot, J.J., Brunau, É.: Territorial intelligence and innovation for the socioecological transition. In: 9th International conference of Territorial Intelligence, ENTI, Strasbourg (2010)
22. Grineva, M.P., Grinev, M.N., Lizorkin, D.: Extracting key terms from noisy and multitheme documents. In: Proceedings of the 18th International Conference on World Wide Web, WWW 2009, Madrid, Spain, 20–24 April, pp. 661–670 (2009)
23. Herbaux, P.: Tools for territorial intelligence and generic scientific methods. In: Internationa Annual Conference on Territorial Intelligence. Besançon: 16–17 October 2008
24. Isaac, A., Summers, E.: Skos simple knowledge organization system primer. W3C Working Group Note (August 2009). http://www.w3.org/TR/skos-primer/
25. Jimenez, F., Fernández, I., Menéndez, A.: Los sistemas regionales de innovación: revisión conceptual e implicaciones en américa latina. In: Los Sistemas Regionales de Innovación en América Latina. Banco Interamericano de Desarrollo, Washington (2011)
26. Lundvall, B.A., Christensen, J.L.: Broadening the analysis of innovation systems-competition, organisational change and employment dynamics in the danish system. In: Conceição, P., Heitor, M., Lundvall, B.-A. (eds.) Innovation, Competence Building and Social Cohesion in Europe: Towards a Learning Society, pp. 144–179. Edward Elgar, Cheltenham (2003)
27. Lundvall, B.A., Johnson, B.: The learning economy. J. Ind. Stud. **1**(2), 23–42 (1994)
28. Lundvall, B. (ed.): National Systems of Innovation: Towards a Theory of Innovation and Interactive Learning. Pinter, London (1992)
29. Lundvall, B.A.: Why study national systems and national styles of innovations? Technol. Anal. Strateg. Manag. **10**(4), 407–421 (1998)
30. Malaisé, V., Gazendam, L., Brugman, H.: Disambiguating automatic semantic annotation based on a thesaurus structure. In: Hathout, N., Muller, P. (eds.) Actes de la 14e conférence sur le Traitement Automatique des Langues Naturelles (communications orales), pp. 197–206. Association pour le Traitement Automatique des Langues, Toulouse (2007)
31. Malaisé, V., Isaac, A., Gazendam, L., Brugman, H.: Anchoring dutch cultural heritage thesauri to wordnet: two case studies. In: ACL 2007, pp. 57–63 (2007)
32. Medelyan, O., Witten, I.H.: Thesaurus-based index term extraction for agricultural documents. In: Proceedings of the 6th Agricultural Ontology Service Workshop (2005)
33. Mollo, M.: The survey on territory research in europe, In: International Conference of Territorial Intelligence, Papers on Tools and methods of Territorial Intelligence (MSHE). Besançon (2009)
34. Nahapiet, J., Ghoshal, S.: Social capital, intellectual capital, and the organizational advantage. Acad. Manage. Rev. **23**(2), 242–266 (1998)
35. Nelson, R.R. (ed.): National Innovation Systems: A Comparative Study. Oxford University Press, Oxford (1993)
36. Neubert, J.: Bringing the "thesaurus for economics" on the web of linked data. In: Proceedings of the Linked Data on the Web Workshop (LDOW 2009) (2009)
37. OECD, EUROSTAT: Oslo Manual: Guidelines for collecting and interpreting innovation data. OECD Publising and European Commission. 3rd edn. (2005)
38. Robertson, S., Jones, K.: Relevance weighting of search terms. J. Am. Soc. Inform. Sci. **27**(3), 129–146 (1976)

39. Salavisa, I., Vali, M.: Social Networks, Innovation and the Knowledge Economy. Routledge, London (2012)
40. Salton, G., Buckley, C.: Term weighting approaches in automatic text retrieval. Technical report Cornell University (1987). http://hdl.handle.net/1813/6721
41. Sebastiani, F.: Machine learning in automated text categorization. ACM Comput. Surv. (CSUR) **34**(1), 1–47 (2002)
42. Jones, K.S.: A statistical interpretation of term specificity and its application in retrieval. J. Doc. **60**, 493–502 (2004)
43. Tiun, S., Abdullah, R., Kong, T.E.: Automatic topic identification using ontology hierarchy. In: Gelbukh, A. (ed.) CICLing 2001. LNCS, vol. 2004, pp. 444–453. Springer, Heidelberg (2001)
44. Turney, P.D.: Learning algorithms for keyphrase extraction. Inf. Retr. **2**(4), 303–336 (2000)
45. Wang, J., Liu, J., Wang, C.: Keyword extraction based on PageRank. In: Zhou, Z.-H., Li, H., Yang, Q. (eds.) PAKDD 2007. LNCS (LNAI), vol. 4426, pp. 857–864. Springer, Heidelberg (2007)
46. Wartena, C., Brussee, R., Gazendam, L., Huijsen, W.: Apolda: A practical tool for semantic annotation. In: Database and Expert Systems Applications, DEXA, 7th International Workshop on Text-based Information Retrieval, TIR, pp. 288–292. IEEE (2007)

Knowledge Engineering and Ontology Development

Pitfalls in Ontologies and TIPS to Prevent Them

C. Maria Keet[1]([✉]), Mari Carmen Suárez-Figueroa[2],
and María Poveda-Villalón[2]

[1] School of Mathematics, Statistics, and Computer Science,
UKZN/CSIR-Meraka Centre for Artificial Intelligence Research,
University of KwaZulu-Natal, Durban, South Africa
keet@ukzn.ac.za
[2] Ontology Engineering Group, Departamento de Inteligencia Artificial,
Facultad de Informática, Universidad Politécnica de Madrid, Madrid, Spain
{mcsuarez,mpoveda}@fi.upm.es

Abstract. A growing number of ontologies are already available thanks
to development initiatives in many different fields. In such ontology developments, developers must tackle a wide range of difficulties and hand-
icaps, which can result in the appearance of anomalies in the resulting
ontologies. Therefore, ontology evaluation plays a key role in ontology
development. OOPS! is an on-line tool that automatically detects pitfalls,
considered as potential errors or problems—and thus may help ontology
developers to improve their ontologies. To gain insight in the existence
of pitfalls and to assess whether there are differences among ontologies
developed by novices, a random set of already scanned ontologies, and
existing well-known ones, data of 406 OWL ontologies were analysed on
OOPS!'s 21 pitfalls, of which 24 ontologies were also examined manually
on the detected pitfalls. The various analyses performed show only minor
differences between the three sets of ontologies, therewith providing a
general landscape of pitfalls in ontologies. We also propose guidelines to
avoid the inclusion of such common pitfalls in new ontologies, the Typical
pItfalls Prevention Scheme (TIPS), so as to increase the baseline quality
of OWL ontologies.

1 Introduction

A growing number of ontologies are already available in different domains thanks
to ontology development initiatives and projects. However, the development of
ontologies is not trivial. Early ontology authoring suggestions were made by [1],
and Rector et al. [2] present the most common problems, errors, and misconcep-
tions of understanding OWL DL based on their experiences teaching OWL. OWL
2 DL contains more features and there is a much wider uptake of ontology development by a more diverse group of modellers since. This situation increases the
need for training, for converting past mistakes into useful knowledge for ontology
authoring, to prevent common flaws and it requires a clear notion of ontology
quality both in the negative sense (what are the mistakes?) and in the positive
(when is some representation good?). Several steps have been taken with respect

© Springer-Verlag Berlin Heidelberg 2015
A. Fred et al. (Eds.): IC3K 2013, CCIS 454, pp. 115–131, 2015.
DOI: 10.1007/978-3-662-46549-3_8

to quality in the negative sense, such as to identify antipatterns [3] and to create a catalogue of common pitfalls—understood as potential errors, modelling flaws, and missing good-practices in ontology development—in OWL ontologies [4,5], and in the positive sense by defining good and 'safe' object property expressions [6] and taxonomies [7]. The catalogue of common pitfalls included 29 types of pitfalls at the time of evaluation and 21 of them are detected automatically by the online OntOlogy Pitfall Scanner! (OOPS! http://www.oeg-upm.net/oops). With the automation of scanning pitfalls as well as advances in ontology metrics, this now provides the opportunity to obtain quantitative results, which has been identified as a gap in the understanding of ontology quality before [8]. Here, we are interested in answering two general questions, being:

A. What is the prevalence of each of those pitfalls in existing ontologies?
B. To what extent do the pitfalls say something about quality of an ontology?

The second question can be broken down into several more detailed questions and hypotheses, which one will be able to answer and validate or falsify through a predominantly quantitative analysis of the ontologies:

1. Which anomalies that appear in OWL ontologies are the most common?
2. Are the ontologies developed by experienced developers and/or well-known or mature ontologies 'better' in some modelling quality sense than the ontologies developed by novices? This is refined into the following hypotheses:
    (i) The prevalence and average of pitfalls is significantly higher in ontologies developed by novices compared to ontologies deemed established/mature.
    (ii) The kind of pitfalls observed in novices' ontologies differs significantly from those in well-known or mature ontologies.
    (iii) The statistics on observed pitfalls of a random set of ontologies is closer to those of novices' ontologies than the well-known or mature ones.
    (iv) There exists a positive correlation between the detected pitfalls and the size or number of particular elements of the ontology.
    (v) There exists a positive correlation between the detected pitfalls and the DL fragment of the OWL ontology.

To answer these questions, we used the 362 ontologies scanned by OOPS! over the past year, 23 novices ontologies, and 21 ontologies that are generally considered to be well-known, where the latter two sets were also scanned by OOPS! and evaluated manually. Although all 21 types of pitfalls have been detected, the most common pitfalls concern lack of annotations and domain and range axioms, and issues with inverses, and to some extent creating unconnected ontology elements and using a recursive definition. The results falsify hypotheses (i), (ii), and (v), partially validate (iv)—for novices, the number of pitfalls/ontology does relate to the size and complexity of the ontology—and validate (iii); i.e., there are no striking differences between the three sets of ontologies, therewith providing a general landscape of pitfalls in ontologies. Taking the pitfall results into account, we propose the **T**ypical p**I**tfall **P**revention **S**cheme, TIPS. The TIPS differ from earlier suggestions [1,2], as the suggestions are applicable to OWL 2 instead of its predecessor languages, they contain an order of importance, and the TIPS embeds emphases with respect to occurrence of the pitfall so that common pitfalls can be prevented first compare to treating any possible pitfall as equally relevant in the overall ontology authoring activity.

In the remainder of this paper, we describe the state of the art in Sect. 2, report on the experimental evaluation of the ontologies in Sect. 3, present the proposed TIPS (Typical pItfalls Prevention Scheme) in Sect. 4 and conclude in Sect. 5.

2   State of the Art

When developing ontologies, developers must tackle a wide range of difficulties, which are related to the inclusion of anomalies in the modelling. Thus, ontology evaluation, which checks the technical quality of an ontology against a frame of reference, plays a key role when developing ontologies. To help developers during the ontology modelling, early ontology authoring guidelines to avoid typical errors were provided in [1]. Such guidelines help developers to prevent errors related to the definition of classes, class hierarchies, and properties during frame-based ontology developments, but they are becoming outdated due to increased expressiveness of ontology languages for which guidance is needed, and advances in ontology authoring guidelines have been made over the past 13 years. Rector and colleagues [2] help with the precise meaning of OWL DL and provide some guidelines on how to avoid diverse pitfalls when building OWL DL ontologies. These pitfalls were mainly related to (a) the failure to make information explicit, (b) the mistaken use of universal and existential restrictions, (c) the open world reasoning, and (d) the effects of domain and range constraints. A classification of errors was identified during the evaluation of consistency, completeness, and conciseness of ontology taxonomies [9]. First steps towards a catalogue of common pitfalls started in 2009 [4] leading to a first stable version in [10]. This catalogue is being maintained and is accessible on-line as part of the OOPS! portal. OOPS! [5] is a web-based tool for detecting potential pitfalls, currently providing mechanisms to automatically detect a subset of 21 pitfalls of those included in the catalogue and therewith helping developers during the ontology validation activity. Related to the aforementioned catalogue of pitfalls, is the identification of a set of antipatterns [3]. Theory-based methods to help developers to increase ontology quality include defining good and 'safe' object property expressions [6] and ontologically sound taxonomies [11]. To help developers during the ontology evaluation activity, there are different approaches: (a) comparison of the ontology to a "gold standard", (b) use of the ontology in an application and evaluation of the results, (c) comparison of the ontology with a source of data about the domain to be covered, and (d) evaluation by human experts who assess how the ontology meets the requirements [12]. A summary of generic guidelines and specific techniques for ontology evaluation can be found in [13]. A three-layered approach to ontology evaluation is presented in [14]: (1) O2 (a meta-ontology), (2) oQual (a pattern based on O2 for Ontology

Quality), and qood (for Quality-Oriented Ontology Description). This allows one to measure the quality of an ontology relative to structural, functional, and usability-related dimensions. A compendium of criteria describing good ontologies is reported in [8] (including accuracy, adaptability, clarity, completeness, computational efficiency, conciseness, consistency/coherence and organizational fitness) and it presents a review of domain and task-independent evaluation methods related to vocabulary, syntax, structure, semantics, representation and context aspects.

A separate strand of suggestions for good practices of representations within Semantic Web are for data and SKOS [15,16] that may complement good practices for ontology development once those connections are better established.

To the best of our knowledge, what is missing at present in the ontology and evaluation field is a quantitative analysis of the most common pitfalls developers include in the ontologies. Based on this study, one then may create a relevant set of guidelines to help developers in the task of developing ontologies and refine ontology quality criteria.

3 Experimental Evaluation of Pitfalls in Ontologies

The experimental evaluation is described in the standard order in this section: first, materials & methods regarding data collection and analysis, then the quantitative and qualitative results, and, finally, the discussion of the results.

3.1 Materials and Methods

Data Collection. With the aim of identifying the most common pitfalls typically made when developing ontologies in different contexts and domains, we have collected and analyzed 44 ontologies (Set1 and Set2) and used the data stored in OOPS! for a random set (Set3):

Set1: 23 ontologies in different domains (a.o., furniture, tennis, bakery, cars, soccer, poker, birds, and plants) developed by novices. These ontologies were developed as a practical assignment by Computer Science honours (4th year) students attending the course "Ontologies & Knowledge bases (OKB718)" in 2011 and 2012 at the University of KwaZulu-Natal.

Set2: 21 existing well-known ontologies that may be deemed 'mature' in the sense of being a stable release, well-known, a real OWL ontology (i.e., no toy ontology nor a tutorial ontology, nor an automated thesaurus-to-OWL file), the ontology is used in multiple projects including in ontology-driven information systems, and whose developers have ample experiences in and knowledge of ontologies, and the selected ontologies are in different subject domains; a.o., DOLCE, BioTop, and GoodRelations.

Set3: 362 ontologies analyzed with OOPS! They were selected from the 614 times that ontologies were submitted between 14-11-2011 and 19-10-2012. The full set was filtered as follows: maintain those repeated ontologies for which OOPS! obtained different results in each evaluation, eliminate those repeated ontologies for which OOPS! obtained the same results in every evaluation, and eliminate those ontologies whose namespace is deferenceable but it does not refer to an ontology.

OOPS! output for the three sets, including calculations, manual analyses of OOPS! detected pitfalls for ontologies in Set1 and Set2, and the names and URIs of the ontologies of Set2 and the names of the ontologies in Set1, are available at http://www.oeg-upm.net/oops/material/KEOD2013/pitfallsAnalysis.xlsx.

All ontologies are evaluated by being scanned through OOPS!, which checks the ontology on most pitfalls that have been collected in the pitfall catalogue that has been presented and discussed in earlier works [4,5] and are taken at face value for this first quantitative evaluation: Creating synonyms as classes (P2); Creating the relationship "is" instead of using rdfs:subClassOf, rdf:type or owl:sameAs (P3); Creating unconnected ontology elements (P4); Defining wrong inverse relationships (P5); Including cycles in the hierarchy (P6); Merging different concepts in the same class (P7); Missing annotations (P8); Missing disjointness (P10); Missing domain or range in properties (P11); Missing equivalent properties (P12); Missing inverse relationships (P13); Swapping intersection and union (P19); Misusing ontology annotations (P20); Using a miscellaneous class (P21); Using different naming criteria in the ontology (P22); Using recursive definition (P24); Defining a relationship inverse to itself (P25); Defining inverse relationships for a symmetric one (P26); Defining wrong equivalent relationships (P27); Defining wrong symmetric relationships (P28); and Defining wrong transitive relationships (P29). Detailed descriptions are available online from the pitfall catalogue at http://www.oeg-upm.net/oops/catalogue.jsp. Note that OOPS! analyses also properly imported OWL ontologies, i.e., when they are available and dereferencable online at the URI specified in the import axiom.

In addition, we collected from the ontologies of Set1 and Set2: DL sublanguage as detected in Protégé 4.1, number of classes, object and data properties, individuals, subclass and equivalence axioms.

Analyses. The data was analysed by computing the following aggregates and statistics. The basic aggregates for the three sets are: (a) percentage of the incidence of a pitfall; (b) comparison of the percentages of incidence of a pitfall among the three sets; (c) average, median, and standard deviation of the pitfalls per ontology and compared among the three sets; and (d) average, median, and standard deviation of the pitfall/ontology.

For Set1 and Set2 ontologies, additional characteristics were calculated, similar to some of the ontology metrics proposed elsewhere [8,14]. Let $|C|$ denote the number of classes, $|OP|$ the number of object properties, $|DP|$ the number of data properties, $|I|$ the number of individuals, $|Sax|$ the number of subclass axioms, and $|Eax|$ the number of equivalences in an ontology. The number of *Ontology Elements* (OE) is computed by Eq. 1, and an approximation of the *Ontology Size* (OS) by Eq. 2.

$$OE = |C| + |OP| + |DP| + |I| \tag{1}$$

$$OS = |C| + |OP| + |DP| + |I| + |Sax| + |Eax| \qquad (2)$$

We use two measures for quantifying the 'complexity' of the ontology. First, an *Indirect Modelling Complexity* (IMC) is computed based on the axioms present (Eq. 3), where a lower value indicates a more complex ontology with relatively more axioms declaring properties of the classes compared to a lightweight ontology or bare taxonomy.

$$IMC = |C| : (|Sax| + |Eax|) \qquad (3)$$

Second, the OWL features used are analysed twofold: (i) by calculating the overall percentage of use of $\mathcal{S}$, $\mathcal{R}$, $\mathcal{O}$, $\mathcal{I}$, $\mathcal{Q}$ and (D), i.e., a rough measure of the OWL 2 DL features used; (ii) by converting the DL fragment into a numerical value, where $\mathcal{AL}$ is given the lowest value of 0 and $\mathcal{SROIQ}$ the highest value of 10, to be used in correlation calculations (see below). The DL fragment and IMC are compared as well, for they need not be similar (e.g., a bare taxonomy with one object property declared reflexive already 'merits' detection of an $\mathcal{R}$, but actually is still a simple ontology with respect to the subject domain represented, and, vv., an ontology can be comprehensive with respect to the subject domain, but originally developed in OWL DL but not updated since OWL 2).

Basic correlations are computed for the ontology sizes and complexities with respect to the pitfalls, and detailed correlations are computed for certain individual pitfalls: P5, P11, P13, P25, P26, P27, P28, and P29 are pitfalls specific to object properties, hence, the amount of properties in the ontologies may be correlated to the amount of pitfalls detected, and likewise for P3, P6, P7, P10, P21, and P24 for classes, and P8 for classes and ontology elements.

Finally, manual qualitative analyses with ontologies in Set1 and Set2 were conducted on possible false positives and additional pitfalls.

3.2   Results

We first present the calculations and statistics, and subsequently a representative selection of the qualitative evaluation of the ontologies in Set1 and Set2.

Aggregated and Analysed Data. The raw data of the ontologies evaluated with OOPS! are available online at http://www.oeg-upm.net/oops/material/KEOD2013/pitfallsAnalysis.xlsx. The type of mistakes made by novice ontology developers are: P4, P5, P7, P8, P10, P11, P13, P19, P22, P24, P25, P26, P27, and P29. The percentages of occurrence of a pitfall over the total set of 23 ontologies in Set1 is included in Fig. 1, the average amount of pitfalls is shown in Fig. 2, and aggregate data also with minimum, maximum, median and standard deviation is listed in Table 1. The analogous results for Set3 are shown in Figs. 1 and 2, and in Table 1, noting that all OOPS! pitfalls have been detected in Set3 and that the median amount of pitfalls/ontology is similar to that of Set1. The high aggregate values are caused by a few ontologies each with around 5000 or more detected pitfalls; without P8 (missing annotations), there are three ontologies

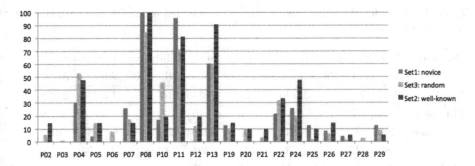

Fig. 1. Percentage of occurrence of a pitfall in the three sets of ontologies.

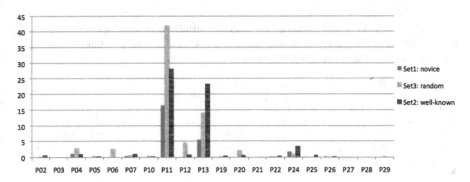

Fig. 2. Average number of pitfall/ontology, by set; for P8, the averages are 62, 297, and 303, respectively.

that have more than 1000 detected pitfalls at the time of scanning the ontology. The results obtained with the 21 well-known ontologies (Set2) can be found in the same table and figures, and include pitfalls P2, P4, P5, P7, P8, P10, P11, P12, P13, P19, P20 (0 upon manual assessment), P21, P22, P24, P25, P26, P27, and P29, noting that the percentages and averages differ little from those of the novices and random ones. The high aggregate values for Set2 is largely due to OBI with a pitfall count of 3771 for P8 (annotations) and DMOP with a pitfall count of 866 for P8; without P8, OBI, DMOP, and the Government Ontology exceeded 100 pitfalls due to P11 (missing domain and range axioms) and P13 (missing inverses—but see also below). P8 is an outlier both in prevalence and in quantity for all three sets of ontologies and only some of the ontologies have very many missing annotations, which skews the average, as can be observed from the large standard deviations.

For Set1 and Set2, we collected data about the content of the ontologies and analysed them against the pitfalls, as described in Sect. 3.1. The usage of the OWL 2 DL features in Set1 are: $\mathcal{S}$ 44 %, $\mathcal{R}$ 26 %, $\mathcal{I}$ 83 %, $\mathcal{O}$ 26 %, $\mathcal{Q}$ 52 %, and $\mathcal{D}$ 17 %, whereas for Set2, the percentages are 62 %, 19 %, 81 %, 24 %, 5 %, and 86 %, respectively; the difference is largely due to the difference in timing of

Table 1. Totals for the three sets of ontologies (rounded off), with and without the annotation pitfall (P8).

Ontology	Set1: Novices		Set3: Random		Set2: Well-known		Combined	
Pitfalls	All	All – P8	All	All – P8	All	All – P8	All	All – P8
Total	2046	626	133746	26330	7639	1277	143436	28238
Minimum	23	3	0	0	15	2	0	0
Maximum	366	95	7948	1999	3920	207	7948	1999
Average	89	27	735	145	364	61	353	70
Median	65	19	50	14	137	48	54	16
St. dev.	74	26	1147	244	846	53	1101	231

Table 2. Correlations and p-values for specific pitfalls and ontology size and complexity, with the relatively interesting values in boldface; where $p < 0.0001$, only 0 is written in the cell due to width limitations.

Set	Set1: Novices				Set2: Well-known				Both			
	All		All – P8		All		All – P8		All		All – P8	
pitfalls/onto.	Corr.	p	Corr.	p	Corr.	p	Corr.	p	Corr.	p	Corr.	p
DL fragment	0.33	0	0.18	0.0002	0.49	0.066	0.52	0	0.38	0.020	0.38	0
IMC	0.06	0	−0.14	0	−0.21	0.056	−0.36	0	−0.14	0.017	−0.2	0
OE	0.998	0.47	0.70	0.0003	0.993	0.84	0.57	0.068	0.990	0.79	0.58	0.025
OS	0.58	0.0072	0.67	0	0.998	0.34	0.52	0.10	0.995	0.24	0.52	0.044

the development of the ontology, with some of the well-known ontologies having been developed before the OWL2 standard, and the use of data properties was discouraged in the lectures for the ontologies in Set1. In order to include the DL fragment in the analyses, we assigned values to the fragments prior to analysis, ranging from a value of 0 for an ontology in $\mathcal{AL}(D)$ to 10 for an ontology in $\mathcal{SROIQ}(D)$, and intermediate values for others (e.g., $\mathcal{ALCHI}(D)$ with a value 3 and $\mathcal{SHIF}$ with value 6—see supplementary data). With the calculated IMC (recall Eq. 3), the correlation between DL fragment and the IMC is -0.18 for the ontologies in Set1 and -0.74 for the ontologies in Set2. This possibly may change a little by tweaking the values assigned to the DL fragments, but not such as to obtain a strong, meaningful correlation between detected DL fragment and the IMC.

Correlations for several measures are included in Table 2. The only substantial correlations found are between all pitfalls per ontology elements and size (in boldface), although with all pitfalls minus P8, there is no obvious correlation anymore. p-values were computed with the 1-tailed unpaired Student t-test, which are also included in Table 2. Using a generous $p < 0.05$ for no difference between the number of pitfalls per ontology and DL fragment, IMC, OE, or OS as measures for ontology size, then the hypotheses have to be rejected mainly for novices (boldface in Table 2). Correlations were also computed for certain pitfalls and relevant ontology elements, as shown in Table 3; e.g., P5 is

Table 3. Correlations by pitfalls and ontology metric, with the most interesting values in boldface, and potentials in italics; "–": no pitfall detected, hence, no correlation.

Ontology correlation	Set1	Set2	Both
P5 (wrong inverses) – number of object properties	*0.71*	0.52	0.58
P11 (missing domain/range) – number of object properties	0.41	0.34	0.40
P13 (missing inverses) – number of object properties	0.54	*0.78*	*0.77*
P25 (inverse to itself) – number of object properties	0.36	0.30	0.32
P26 (inverse for symmetric) – number of object properties	*0.72*	−0.25	0.25
P27 (wrong equivalence) – number of object properties	*0.71*	0.61	0.59
P28 (wrong symmetric) – number of object properties	–	–	–
P29 (wrong transitive) – number of object properties	−0.20	0.15	−0.03
P3 (adding is-a) – number of classes	–	–	–
P6 (cycles) – number of classes	–	–	–
P7 (merging classes) – number of classes	0.17	0.04	0.06
P10 (missing disjointness) – number of classes	−0.08	−0.13	−0.09
P21 (miscellaneous classes) – number of classes	–	−0.10	−0.06
P24 (recursive definition) – number of classes	0.15	−0.04	0.01
P8 (missing annotation) – number of classes	0.22	**0.9975**	**0.9909**
P8 (missing annotation) – number of ontology elements	0.51	**0.9899**	**0.9848**

about inverse relationships, hence, one might conjecture it is correlated with the amount of object properties in the ontology. This only holds strongly for P8 and the elements in the Set2 ontologies, which explains why there are significant correlations for all pitfalls but not all minus P8 in Table 2. A weakly possibly interesting correlation exists for P5, P26, P27 in the Set1 ontologies, and for P13 in the well-known ontologies.

Comparing pitfalls among Set1, Set2, and Set3 with the 1-tailed unpaired Student t-test, then the null hypothesis—no difference—has to be rejected for novice vs. mature if one ignores pitfall P8 ($p = 0.0096$), i.e., one can observe a difference, but this does not hold anymore for all pitfalls ($p = 0.13$). The results are inconclusive for the other combinations: all pitfalls novice vs. random $p = 0.15$, all mature vs. random $p = 0.98$, all minus P8 novice vs. random $p = 0.37$, and all minus P8 mature vs. random $p = 0.82$.

Qualitative Analysis of the Detected Pitfalls. As the pitfalls in the catalogue (and thus OOPS!) are relatively coarse-grained, we examined the OOPS!-detected pitfalls of the ontologies in Set1 and Set2 on the correctness of detection. That is, although the algorithms in OOPS! are implemented correctly, they may detect more pitfalls than what an ontology developer may see as a problem, and such insights may, on the one hand, help refining a pitfall and, on the other hand, downgrade a pitfall to being irrelevant practically. Of the analyses carried out (included in the supplementary data file), we highlight four types of pitfalls that illustrate well an aspect of ontology development practices (P4), subject domain peculiarities (P7), language features (P13), and modelling (P24).

P4: Unconnected Ontology Elements. OOPS! correctly identifies 'orphan' classes and properties, but they are debatable in some cases. For instance, an orphan's subclasses are used in a class expression, i.e., the orphan class is used merely as a way of grouping similar things alike a so-called 'abstract class' in UML. The Deprecated and Obsolete orphans are typically present in bio-ontologies, which is deemed a feature in that field. A recurring type of orphan class was to add a class directly subsumed by `owl:Thing` to indicate the subject domain (e.g., a Bakery class for an ontology about bakery things), which might be defensible in a distributed ontology, but not in a single domain ontology. Overall, each of these practices require a more substantive argument whether they deserve to be a false positive or not.

P7: Merging Different Concepts in the Same Class. OOPS! detects a few occurrences that are false positives, besides the many correctly identified ones. For instance, a RumAndRaisinFlavour of ice cream does not constitute merging different classes, but a composite flavour and would not have been a false positive if that flavour had obtained its own name (e.g., RummyRaisin). From a computational perspective, there is no easy way to detect these false positives.

P13: Missing Inverse Relationships. The issues with inverses are contentious and hard to detect, especially since OWL and OWL 2 differ in their fundamental approach. Unlike OWL, OWL 2 has a feature `ObjectInverseOf`, so that for some object property hasOP in an OWL 2 ontology, one does not have to extend the vocabulary with an OPof property and declare it as the inverse of hasOP with `InverseObjectProperties`, but instead one can use the meaning of OPof with the axiom `ObjectInverseOf`(hasOP). In addition, GFO's exists_at and BioTop's abstractlyRelatedTo do not readily have an inverse name, and a modeller likely will not introduce a new property for the sake of having a named inverse property when it is not needed in a class axiom. Overall, P13 is detected more often than warranted from a modeller's viewpoint, and it could be refined to only those cases where the declaration of `InverseObjectProperties` is missing; e.g., both manufacturedBy and hasManufacturer are in the car ontology, but they are not declared inverse though they clearly are, which OOPS! detects already.

P24: Using Recursive Definition. This pitfall is tricky to define and detect. In general, recursive definitions are wrong, such as the pattern $X \equiv X \sqcap R.Y$, which should be detected, and likewise detecting unintended assertions, such as CarrotFilling $\sqsubseteq \exists$ hasFillingsAndToppimg.CarrotFilling (in the bakery (novice's) ontology). However, P24 currently detects whether the class on the left-hand side of the subsumption or equivalence occurs also on the right-hand side, which is not always a problem; e.g., DM-Process $\sqsubseteq \exists$hassubprocess.DM-Process in DMOP is fine. These subtle differences are difficult to detect automatically, and require manual inspection before changing or ignoring the pitfall.

Removal of the false positives reduces the observed minor differences between the three sets of ontologies, i.e., roughly equalize the percentages per pitfall. Put differently, this supports the observation that there is a *general landscape of pitfalls*.

New and More Detailed Pitfalls. The novices' ontologies had been analysed manually on modelling mistakes before OOPS! and before consulting the catalogue. In addition to detecting the kind of pitfalls already in the catalogue, new ones were detected, which typically occurred in more than one ontology. We refer to them here as new *candidate pitfalls* (Cs) which are currently being added to the catalogue:

C1. *Including some form of negation in ontology element names.* For example, DrugAbusePrevention (discussed in [17]), and NotAdults or ImpossibleHand (in the poker ontology). This pitfall refers to an anomaly in the element naming.

C2. *Distinguishing between file name and URI.* This is related to naming issues where the .owl file has a meaningful name, but the ontology URI has a different name (also observed in [18]).

C3. *Confusing part-of relation with subclass-of relation.* This pitfall is a special and very common case of pitfall P23 (using incorrectly ontology elements) (see [19]). As part of this pitfall, there is also the case in which the most appropriate part-whole relation in general is not selected (see also [20]).

C4. *Misusing* min 1 *and* some. This pitfall affects especially ontology feature usage due to the OWL restrictions (note: Protégé 4.x already includes a feature to change all such instances).

C5. *Embedding possibility/modality in the ontology element's name.* This pitfall refers to encapsulating a modality ("can", "may", "should") in an element's name (e.g., canCook).

3.3   Discussion

Whilst giving valuable insight in the prevalence of pitfalls in existing ontologies, the results obtained falsify hypotheses (i) (except for novice vs. mature when discounting P8), (ii), and (v), partially validate (iv) (for all pitfalls and mature ontologies), and validate (iii), which is not exactly as one may have expected, and it raises several possible interpretations.

First, the set of pitfalls currently implemented in OOPS! is limited and with more and more refined checks, substantial differences may be found. Perhaps this is the case, but it does not negate the fact that it is not the case for the 21 already examined and therefore not likely once extended. In addition, recently, the notion of good and safe object property expressions has been introduced [6], where manual evaluation with a random set of ontologies—including some of the ones in Set2—revealed advanced modelling issues concerning basic and complex object property expressions. This further supports the notion that, for the time being, there is a general landscape compared to saliant differences among levels of maturity.

Second, the well-known ontologies are possibly not really mature and exemplary after all (the converse—that the novices' ontologies in Set1 are 'as good as the well-known ones'—certainly does not hold), for they are quite close to the ones in Set3; i.e., that some ontology is widely known does not imply it is 'good'—or, at least: has fewer pitfalls than an—ontology being developed by a novice ontologist. This makes it more difficult to use them in ontology engineering courses, where one would like to point students to 'good' or 'exemplary' ontologies: if well-known ontologies have those pitfalls, they are more likely to be propagated by the students "because ontology x does it that way". This attitude was observed among the novices with respect to P11, because the popular Protégé OWL Pizza tutorial (http://www.co-ode.org) advises against declaring domain and range of properties (page 37), which may explain why P11 was detected often.

Third, it may be reasonable to argue that 'maturity' cannot be characterised by absence of pitfalls at all, but instead is defined by something else. Such a 'something else' may include its usefulness for its purpose—or at least meeting the requirements—or, more abstract, the precision and coverage as introduced by Guarino (see also Fig. 2 in [11]). Concerning the latter, this means both a high precision and maximum coverage of the subject domain one aims to represent in the ontology. It is known one can improve on one's low precision— i.e., the ontology admits more models than it should—by using a more expressive language and adding more class expressions, and recently the approach of letting users choose among possible models is used to add more disjointness axioms [21] and in ontology-driven conceptual modelling [22]. For domain ontologies, another option that influences to notion of being well-known and mature is its linking to a foundational ontology and that therewith less modelling issues occur [18], but this has to do with the knowledge that is represented, not with, e.g., language feature misunderstandings. We leave a more detailed investigation in this direction for future works.

4   Ontology Authoring Guidelines: TIPS

Given the pervasiveness of the pitfalls regardless the level of maturity and usage of an ontology, there is a need to increase the average quality of ontologies at least to the extent that pitfalls are avoided upfront or fixed once detected. While one can take the same approach here as with using an automated reasoner—model first, then check the deductions—and design one's ontology first and then test with OOPS!, from an academic viewpoint, prevention is better, because it means the modeller has a better grasp of ontology development.

The easiest way is to devise a checklist by simply turning around the pitfalls. The pitfall catalogue is a *list*, however, and it is possible categorise the pitfall into different groups; e.g., grouping pitfalls by a structural, functional, and usability profiling dimension or by 'consistency' (not necessarily resulting in logical inconsistencies), completeness, and conciseness [5,10]. Importantly for useful guidelines, is to recognise that first structural and modelling issues—like

P15 and P29—have to be addressed before making an ontology 'neat' with a consistently applied naming pattern or evaluating the somewhat vague notion of meeting the requirements. In addition, in a similar fashion to a "conceptual schema design procedure" [23] or some other ordering of activities in formalization of the subject domain (e.g., DiDOn [24]), one can structure the guidelines accordingly; here, we choose the order of addressing pitfalls as: classes, taxonomy, properties, constraints, documentation. Taking these notions into account, we propose the Typical pItfall Prevention Scheme, TIPS. While there are indeed earlier suggestions [1,2], these TIPS are applicable to the latest OWL 2, contain an order of importance, and embeds emphases with respect to occurrence of the pitfall so that common pitfalls can be prevented first. The descriptions of the tips are written in the imperative indicating what a developer should be checking.

T1: Class Naming and Identification (includes P1, P2, P7, C2, and C5): When identifying and naming classes in ontologies, avoid synonymy and polysemy: distinguish the concept itself from the different names such a concept can have (the synonyms) and create just one class for the concept and provide, if needed, different names for such a class using `rdfs:label` annotations. Regarding polysemy, where the same name has different meanings, try to disambiguate the term, use extension mechanisms and/or axioms. Other important cases regarding class naming and identification are (a) creating a class whose name refers to two or more different concepts by including "and" or "or" in the name (e.g., StyleAndPeriod or ProductOrService) and (b) using modality ("can", "may", "should") in the ontology element's names. In situation (a) consider dividing the class into different subclasses, and in case (b) consider a more appropriate name avoiding the use of modality or change to a logic language that can express it. Take care about providing proper names for both the ontology file and the URI.

T2: Class Hierarchy (includes P3, P6, P17, and P21): The class taxonomy is based on is-a relations, where a class A is a subclass of class B, if and only if every instance of A is also instance of B, and the is-a in the hierarchy is transitive. So, do not introduce such a relation as an object property, but use primitives provided by the ontology language: e.g., `subclassOf` for representing the subclass of relationship, and `instanceOf` for representing membership of an individual in a class. Another issue when creating class taxonomies is to avoid cycles in the hierarchy, i.e., to avoid defining a class as a specialization or generalization of itself either directly or indirectly, because if a cycle is included between two classes in a taxonomy, the implication is like defining such classes and the ones involved in the cycle as equivalent. Also avoid the temptation of creating a class named Unknown, Other or Miscellaneous in a class hierarchy just because the set of sibling classes defined is incomplete. Finally, consider the leaf elements of the hierarchy, and ask yourself whether they are still classes (entities that can have instances) or individuals (entities that cannot be instantiated anymore), if the latter, then convert them into instances.

T3: Domain and Range (includes P11 and P18): When creating an object or data property, answer the question "What is the most general class in the ontology for which this property holds?" and set the answer as domain of the property. If the answer is `owl:Thing` consider using several subclasses joined by "or" operators (i.e., `owl:unionOf`). Repeat the process to set the range for object properties. For a data property's range, answer the question "What would be the format of data (strings of characters, positive numbers, dates, floats, etc.) used to fill in this information?".

T4: Equivalent Relations (includes P12 and P27): Is any object or data property declared equivalent to another one? To have 'safe' equivalent properties that will not generate unexpected deductions, assess the following: the domains of both properties should be the same class (e.g., Country) and also the ranges should be the same class (e.g., LanguageCode) or datatype. Also check that both properties refer to the same meaning (e.g., hasLanguageCode and has_language_code) between classes or to the same attributes in case of datatype properties. Finally, check if both relations are really needed: if they are defined in the same namespace they could be either redundant or refer to different real-world relations and require disambiguation instead.

T5: Inverse Relations (includes P5, P13, P25, and P26): If there is an object property declared inverse to another one, then, to avoid unexpected deductions, be certain to check that the domain class of one is the same class as the range of the other one, and vv. (e.g., Country and LanguageCode for hasLanguageCode and isCodeOf) and try to create sentences from domain to range using the property name for both (potential) inverse properties to double-check it is possible. Note that this TIPS applies for object property pairs; if only a single object property is involved, consider T6.

T6: Object Property Characteristics (includes P28 and P29): Go through the object properties and check their characteristics, such as symmetry, functional, and transitivity. To have 'safe' object property characteristics declared that will not have unexpected deductions, examine: for transitive properties, the domain and range of the object property should be the same class (e.g., both Process) so that a chain can be formed; for symmetric relations, verbalise the assertion both from domain to range and from range to domain to double-check it is possible and has no hidden second property; reflexivity: if the relation holds for all objects in your ontology, declare it reflexive, if only for a particular relation, then use the `Self` construct.

T7: Intended Formalization (includes P14, P15, P16, P19, C1, and C4): A property's domain (resp., range) may consist of more than one class, which is usually a union of the classes (an `or`), not the intersection of them. Considering the property's participation in axioms, the `AllValuesFrom`/only/∀ can be used to 'close' the relation, i.e., that no object can relate with that relation to the class other than the one specified. If you want to say there is at least one such relation (more common), then use `SomeValuesFrom`/some/∃ instead. To state there is *no* such relation in which the class on the left-hand side participates, put the negation before the quantifier ($\neg\forall$ or $\neg\exists$), whereas stating that there is a relation but just not with some particular class, then the negation goes in front of the class on the right-hand side; e.g., a vegetarian pizza does not have meat as ingredient ($\neg\exists$hasIngredient.Meat), not that it can have all kinds of ingredients—cucumber, marsh mellow, etc.—as long as it is not meat ($\exists$hasIngredient.$\neg$Meat). To avoid the latter (the unintended pizza ingredients), one ought not to introduce a class with negation, like NotMeat, but use negation properly in the axiom. Finally, when convinced *all* relevant properties for a class are represented, consider making it a defined class, if not already done so.

T8: Modelling Aspects (includes P4, P23, and C3): Even though it is too difficult to check whether every ontology element is defined in the most appropriate way, we attempt to provide some general rules for checking basic modelling issues. First, add only elements to the ontology that will be used somewhere in the ontology. Second, in case ontology population is part of the project, one could check whether all the classes will contain instances and, following that, whether the (potential) individual will have property assertions to other individuals, and that those properties indeed do exist. Also, check that the necessary data properties exist for those individuals that would need to be linked to values (e.g., dates, booleans, floats, strings, etc.). Finally, remember that the primitive subclassOf is used to define taxonomies and membership for individuals (see T2), and they are different from part-whole relations that do need a separate object property (e.g., Province is part of (or: located in) Country, not a subclass of Country).

T9: Domain Coverage and Requirements (includes P9 and P10): Check if functional ontology requirements—referring to the particular knowledge to be represented by the ontology and the particular terminology to be included in the ontology—are covered by the ontology. One important check in this regard is to inspect whether disjointness has been declared explicitly in the ontology (two ontology elements are disjoint if they cannot share instances or are already different individuals). It is crucial to remember that in OWL, elements are not disjoint unless disjointness statements are stated. Disjointness knowledge is important both as a means to describe the world as it is and to obtain the expected results from ontology inferences.

T10: Documentation and Understandability (includes P8, P20, and P22): Because of multi-authored ontologies, understanding of the ontology, and eyeing long-term and broad use and reuse, it is good practice to include human readable descriptions, such as comments and labels, in the ontology. Consider providing names as labels (with rdfs:label) and definitions as comments (with rdfs:comment) for all ontology elements. Remember not to confuse the name for an ontology element with its definition.

5 Conclusions

We performed a quantitative analysis of the pitfalls developers included in ontologies by analyzing different sets of data obtained after using OOPS!. All implemented types of pitfalls have been detected in the ontologies scanned with OOPS!, but the most common ones are lack of annotations, absence of domain and range axioms, and issues with inverses, and to a lesser extent creating unconnected ontology elements and using a recursive definition. Five new pitfalls have been identified upon closer inspection of the novices' ontologies. Analysis showed that there is no clear evidence of noteworthy differences between ontologies developed by novices, well-known ones, and the random set of ontologies, except for novice vs. mature when disregarding pitfall P8, and for novices, the pitfalls per ontology is related to the size of the ontology complexity of the ontology. Thus, the analysis provides a data-driven general landscape of pitfalls in current ontologies. Taking advantage of the results of our study, we proposed the Typical pItfalls Prevention Scheme (TIPS) in order to facilitate avoiding the inclusion of such common pitfalls in ontologies and thus to benefit the ontology quality.

We are extending the pitfall catalogue, and are working on a better characterization of 'maturity' in ontologies and how such a characterization is related to the set of most common pitfalls.

Acknowledgements. This work has been partially supported by the Spanish projects BabelData (TIN2010-17550) and BuscaMedia (CENIT 2009-1026).

References

1. Noy, N., McGuinness, D.: Ontology Development 101: A guide to creating your first ontology. Number KSL-01-05, and Stanford Medical Informatics Technical Report SMI-2001-0880, March 2001
2. Rector, A., Drummond, N., Horridge, M., Rogers, J., Knublauch, H., Stevens, R., Wang, H., Wroe, C.: OWL pizzas: practical experience of teaching OWL-DL: common errors & common patterns. In: Motta, E., Shadbolt, N.R., Stutt, A., Gibbins, N. (eds.) EKAW 2004. LNCS, vol. 3257, pp. 63–81. Springer, Heidelberg (2004)
3. Roussey, C., Corcho, O., Vilches-Blázquez, L.: A catalogue of OWL ontology anti-patterns. In: Proceedings of K-CAP 2009, pp. 205–206 (2009)
4. Poveda, M., Suárez-Figueroa, M.C., Gómez-Pérez, A.: Common pitfalls in ontology development. In: Meseguer, P., Mandow, L., Gasca, R.M. (eds.) CAEPIA 2009. LNCS, vol. 5988, pp. 91–100. Springer, Heidelberg (2010)
5. Poveda-Villalón, M., Suárez-Figueroa, M.C., Gómez-Pérez, A.: Validating ontologies with OOPS!. In: ten Teije, A., Völker, J., Handschuh, S., Stuckenschmidt, H., d'Acquin, M., Nikolov, A., Aussenac-Gilles, N., Hernandez, N. (eds.) EKAW 2012. LNCS, vol. 7603, pp. 267–281. Springer, Heidelberg (2012)
6. Keet, C.M.: Detecting and revising flaws in OWL object property expressions. In: ten Teije, A., Völker, J., Handschuh, S., Stuckenschmidt, H., d'Acquin, M., Nikolov, A., Aussenac-Gilles, N., Hernandez, N. (eds.) EKAW 2012. LNCS, vol. 7603, pp. 252–266. Springer, Heidelberg (2012)

7. Guarino, N., Welty, C.: An overview of ontoclean. In: Staab, S., Studer, R. (eds.) Handbook on Ontologies, pp. 201–220. Springer, Heidelberg (2009)

8. Vrandečić, D.: Ontology evaluation. In: Staab, S., Studer, R. (eds.) Handbook on Ontologies, 2nd edn, pp. 293–313. Springer, Heidelberg (2009)

9. Gómez-Pérez, A.: Ontology evaluation. In: Staab, S., Studer, R. (eds.) Handbook on Ontologies. International Handbooks on Information Systems, pp. 251–274. Springer, Heidelberg (2004)

10. Poveda-Villalón, M., Suárez-Figueroa, M.C., Gómez-Pérez, A.: A double classification of common pitfalls in ontologies. In: Proceedings of Workshop on Ontology Quality (OntoQual 2010). CEUR-WS (2010) C-located with EKAW 2010

11. Guarino, N., Oberle, D., Staab, S.: What is an ontology? In: Staab, S., Studer, R. (eds.) Handbook on Ontologies, pp. 1–17. Springer, Heidelberg (2009)

12. Brank, J., Grobelnik, M., Mladenic, D.: A survey of ontology evaluation techniques. In: Proceedings of SiKDD 2005, Ljubljana, Slovenia (2005)

13. Sabou, M., Fernandez, M.: Ontology (network) evaluation. In: Suárez-Figueroa, M.C., Gómez-Pérez, A., Motta, E., Gangemi, A. (eds.) Ontology Engineering in a Networked World, pp. 193–212. Springer, Heidelberg (2012)

14. Gangemi, A., Catenacci, C., Ciaramita, M., Lehmann, J.: Modelling ontology evaluation and validation. In: Sure, Y., Domingue, J. (eds.) ESWC 2006. LNCS, vol. 4011, pp. 140–154. Springer, Heidelberg (2006)

15. Poveda-Villalón, M., Vatant, B., Suárez-Figueroa, M.C., Gomez-Perez, A.: Detecting good practices and pitfalls when publishing vocabularies on the web, Sydney, Australia, 21 October 2013

16. Suominen, O., Mader, C.: Assessing and improving the quality of SKOS vocabularies. J. Data Semant. **2**(2), 1–27 (2013)

17. Schulz, S., Stenzhorn, H., Boekers, M., Smith, B.: Strengths and limitations of formal ontologies in the biomedical domain. Electron. J. Commun. Info. Innov. Health (Special Issue on Ontologies, Semantic Web and Health) **3**(1), 31–45 (2009)

18. Keet, C.M.: The use of foundational ontologies in ontology development: an empirical assessment. In: Antoniou, G., Grobelnik, M., Simperl, E., Parsia, B., Plexousakis, D., De Leenheer, P., Pan, J. (eds.) ESWC 2011, Part I. LNCS, vol. 6643, pp. 321–335. Springer, Heidelberg (2011)

19. Aguado de Cea, G., Gómez-Pérez, A., Montiel-Ponsoda, E., Suárez-Figueroa, M.C.: Natural language-based approach for helping in the reuse of ontology design patterns. In: Gangemi, A., Euzenat, J. (eds.) EKAW 2008. LNCS (LNAI), vol. 5268, pp. 32–47. Springer, Heidelberg (2008)

20. Keet, C.M., Fernández-Reyes, F.C., Morales-González, A.: Representing mereotopological relations in OWL ontologies with ONTOPARTS. In: Simperl, E., Cimiano, P., Polleres, A., Corcho, O., Presutti, V. (eds.) ESWC 2012. LNCS, vol. 7295, pp. 240–254. Springer, Heidelberg (2012)

21. Curé, O., Prié, Y., Champin, P.-A.: A knowledge-based approach to augment applications with interaction traces. In: ten Teije, A., Völker, J., Handschuh, S., Stuckenschmidt, H., d'Acquin, M., Nikolov, A., Aussenac-Gilles, N., Hernandez, N. (eds.) EKAW 2012. LNCS, vol. 7603, pp. 317–326. Springer, Heidelberg (2012)

22. Braga, B.F.B., Almeida, J.P.A., Guizzardi, G., Benevides, A.B.: Transforming ontoUML into alloy: towards conceptual model validation using a lightweight formal methods. Innov. Syst. Softw. Eng. **6**(1–2), 55–63 (2010)

23. Halpin, T.: Information Modeling and Relational Databases. Morgan Kaufmann Publishers, San Francisco (2001)

24. Keet, C.M.: Transforming semi-structured life science diagrams into meaningful domain ontologies with DiDOn. J. Biomed. Inform. **45**, 482–494 (2012)

Foundational Ontology Mediation in ROMULUS

Zubeida C. Khan[1] and C. Maria Keet[1,2](✉)

[1] School of Mathematics, Statistics, and Computer Science,
UKZN/CSIR-Meraka Centre for Artificial Intelligence Research,
University of KwaZulu-Natal, Durban, South Africa
zkhan@csir.co.za, keet@ukzn.ac.za
[2] University of Cape Town, Cape Town, South Africa
mkeet@cs.uct.ac.za

Abstract. An approach for semantic interoperability among heterogeneous systems is to assist with the integration of foundational ontologies. In order to achieve this, we have selected three popular foundational ontologies DOLCE, BFO, and GFO, and their related modules. We perform ontology mediation (alignment, mapping, and merging) on these ontologies by aligning their ontology entities using tools, documentation, and our manual alignments, and comparing their effectiveness. Thereafter, based on the alignments, we created mappings in the ontology files resulting and merged ontologies. However, during the mapping process, it was found that structural differences in foundational ontologies, caused by conflicting axioms due to complement and disjointness, and incompatible domain and range restriction, cause logical inconsistencies in foundational ontology alignments, thereby reducing the number of mappings. In this paper, we present each phase of the mediation process, including the mediation issues we encountered with solutions where available.

Keywords: Foundational ontology · Ontology mediation · Semantic interoperability · Ontology alignment · Ontology mapping · Ontology matching · Ontology merging

1 Introduction

There has been an exponential growth in ontology development for the Semantic Web, including a move toward modular and networked ontologies that require coordination among ontologies. Foundational ontologies are commonly used to facilitate semantic interoperability, where Semantic Web system developers choose a preferred foundational ontology among several available ones for their domain ontologies; these include, among others DOLCE, BFO, GFO, OCHRE, UFO, YAMATO, SUMO, and GIST. The semantics and underlying Ontology of each foundational ontology differs, however, causing a problem in semantic interoperability even when a foundational ontology is used. Heterogeneous systems on the Semantic Web are restricted to committing to a single foundational ontology in order to promote interoperability. However, no single foundational ontology is

© Springer-Verlag Berlin Heidelberg 2015
A. Fred et al. (Eds.): IC3K 2013, CCIS 454, pp. 132–152, 2015.
DOI: 10.1007/978-3-662-46549-3_9

used across all systems, therewith preventing interoperability. In order for these applications to share and process information correctly, there is a need for foundational ontology interoperability, so that ontology developers committing to a preferred foundational ontology will achieve seamless linking to other domain ontologies linked to another foundational ontology. An infrastructure to support such a scenario was envisioned as the "WonderWeb Foundational Ontologies Library" (WFOL) [1], but this infrastructure still does not exist. The main preconditions for a WFOL are content comparisons and ontology mediation. Ontology mediation refers to identifying and solving differences between heterogeneous ontologies, in order to allow reuse and interoperability. Its three main processes are alignment, mapping, and merging [2]. There are only few paper-based alignments of foundational ontologies, being between GFO and DOLCE [3] and between DOLCE and BFO [4,5], which, however, are partial, with older versions of the ontologies, informal, and/or aligned but not mapped. To the best of our knowledge, no systematic comparison of the contents of foundational ontologies has been done, nor full alignments, let alone consistent mappings.

We aim to contribute to fill this gap of semantic interoperability by selecting three well-known foundational ontologies, DOLCE [1], BFO (http://www.ifomis. org/bfo) with RO [6], and GFO [3] with which we perform a rigorous foundational ontology content comparison and mediation to aid in achieving foundational ontology interchangeability. The alignment process is carried out by using the manual alignment as a gold standard and (semi-)automated alignment with seven alignment tools to examine them on their capabilities to align foundational ontologies. The accuracy and percentage of alignments that were found vary greatly among the tools due to their diverse alignment algorithms, ranging from 18 to 94 % and 17 to 31 %, respectively. Further alignment issues appear in the transitivity of alignments across the three foundational ontologies due to absence of some entity or conflicting parthood theories, whilst some may be resolved by asserting them as sibling classes. Mapping the aligned entities whilst keeping a consistent ontology reduces the feasible set from 85 alignments to 43 successful mappings due to disjointness and complement axioms elsewhere in the ontology, and due to incompatible domain and range axioms, which in some cases can be solved from a logic viewpoint by asserting subsumption instead. For each mediation process (alignment, mapping and mediation), we present the issues encountered for foundational ontology mediation and how some of them may be solved.

In the remainder of the paper, we provide a literature review in Sect. 2. A content comparison of the foundational ontologies is described in Sect. 3, which is followed by an analysis of alignments in Sect. 4, and of the mappings in Sect. 5. We discuss the results in Sect. 6 and conclude in Sect. 7.

2   Literature Review

Few results are available on comparisons among the foundational ontologies of its classes and relationships. Seyed compared the primitive relations of BFO (i.e., the Relation Ontology (RO)) and DOLCE, who observed that the philosophies behind the foundational ontologies affect the way the relations are modelled. For instance, BFO is based on realist principles and has no abstract entities while GFO is both descriptive and realist in nature and allows abstract entities in an ontology, and BFO's parthood relation has_part does not consider abstract entities, while GFO has a parthood relation abstract_has_part that considers abstract entities at a higher-level than its has_part relation.

Temal et al. [5] created a BFO-DOLCE mapping in order to integrate medical information. The classes (universals or categories) are mapped with equivalence and subsumption relations. Based on the older so-called SNAP and SPAN version of BFO, they found that all BFO universals were successfully mapped to DOLCE, but not all DOLCE entities could be mapped to BFO. These alignments were not checked on consistency of the mappings and were done on some First Order Logic version of the ontologies, where the SNAP-BFO has, e.g., Boundary, that BFO v1.1 in OWL does not have, and DOLCE is claimed to have Collection, which appears neither in the principal documentation [1] nor in the OWLized version of DOLCE. Some of their alignments are useful, however, which we will return to in Sect. 4.

Broadening the scope toward general ontology mediation and matching and from a computational viewpoint, some principles and definitions are useful also for the foundational ontology setting. Ontology mediation [2] is divided into three operations: mapping, alignment, and merging. To be precise in the terminology we use throughout the paper, we provide several definitions on ontology matching in this section, which are taken from [7]. First, there is the matching process:

Definition 1 (Matching Process [7]). *The matching process can be seen as a function f which, from a pair of ontologies to match o and o', an input alignment A, a set of parameters p and a set of oracles and resources r, returns an alignment A' between these ontologies: $A' = f(o, o', A, p, r)$.*

To be able to talk about an actual alignment or mapping, the notion of "entity language" has to be introduced, which is used to express precisely those entities that will be matched.

Definition 2 (Entity Language [7]). *Given an ontology language L, an entity language Q_L is a function from any ontology $o \subseteq L$ which defines the matchable entities of ontology o.*

Then, a correspondence consists of a relation between two entities in different ontologies, which is uniquely identified and has some confidence value assigned to it.

Definition 3 (Correspondence [7]). *Given two ontologies o and o' with associated entity languages Q_L and $Q_{L'}$, a set of alignment relations θ and a confidence structure over Ξ, a correspondence is a 5-tuple: $\langle id, e, e', r, n \rangle$, such that id is a unique identifier of the given correspondence, $e \subseteq Q_L(o)$ and $e' \subseteq Q'_{L'}(o')$, $r \subseteq \theta$, and $n \subseteq \Xi$.*

Ontology alignment, then, is the process of specifying correspondences between entities, by using a particular alignment relation, such as equivalence, subsumption, or a predefined similarity relation.

Definition 4 (Alignment [7]). *Given two ontologies o and o', an alignment is made up of a set of correspondences between pairs of entities belonging to $Q_L(o)$ and $Q_{L'}(o')$ respectively.*

Ontology mapping deals with creating correspondences between ontologies based on alignments such that the resultant ontology is still consistent and does not have unsatisfiable classes or relations. Euzenat and Shvaiko do consider this with respect to models of aligned ontologies, which is too lengthy to repeat here, and De Bruijn et al. does not provide a definition of their idea of mapping as a 'consistent alignment in the context of the whole ontology' either. Therefore, we capture the gist in the following definition, using Euzenat and Shvaiko's notational conventions.

Definition 5 (Mapping). *Given two ontologies o and o', a mapping is made up of a set of correspondences between pairs of entities belonging to $Q_L(o)$ and $Q_{L'}(o')$, respectively, and this mapping is satisfiable and does not lead to an unsatisfiable entity in either o or o'.*

In merging, a new merged ontology is created from the original ontologies.

Definition 6 (Merging). *Given two ontologies o and o', a merging is the creation of a new ontology o'' containing o and o' and all mappings between entities belonging to $Q_L(o)$ and $Q_{L'}(o')$ such that o'' does not have unsatisfiable entities and is consistent.*

Overviews of approaches, frameworks, and technologies used to perform ontology mapping, alignment and merging are discussed elsewhere (e.g., [2]), and more detail about algorithms and issues can be found in [7].

As mentioned earlier, there are many foundational ontologies, in whole and modularised modules, and foundational ontologies are regularly being updated. This makes it rather time-consuming to explore each foundational ontology time and again, especially when there are differences in hierarchy and structure. Therefore, it makes sense to use matching tools to align foundational ontologies, which thereby also is an opportunity to determine which tools are better suited for foundational ontologies and to the type of alignments that are misaligned or not discovered by those tools. We summarize the alignment tools that are used in the experimental evaluation, of which we note that LogMap [8], YAM++ [9], HotMatch [10], Hertuda [11] and Optima [12] have been evaluated with positive results by the Ontology Alignment Evaluation Initiative (OAEI) in terms of their precision, recall and other performance measures.

H-Match [13] is an algorithm for matching ontologies at different depth levels, with different accuracies. The algorithm takes into account linguistic and semantic features of ontologies to perform matching and uses one of four matching models: surface, shallow, deep or intensive. The surface model considers linguistic affinity between entity names to measure similarity. In shallow, deep and intensive models, context is also considered to determine entity similarity.

PROMPT [14] is an ontology matching plug-in for Protégé that allows for comparison, mappings, and merging between ontologies. It is a semi-automatic method that invokes algorithms based on a combination of concept-representation structure, the relations between entities and user's actions. PROMPT offers the user four different algorithms to use for initial comparison: lexical matching, FOAM plugin, lexical matching with synonyms and using UMLS concept identifiers for matching. It is only supported in older versions of Protégé, which makes it unstable.

LogMap [8] automatically generates mappings between ontologies using logic-based semantics of the input ontologies. It offers an improvement to other mapping tools in that it addresses scalability and logical inconsistencies. LogMap allows a user to upload ontologies in a number of formats and implements existing reasoners to check the satisfiability of the ontologies.

YAM++ [9] aligns entities by information retrieval or machine learning if training data is available. Three matchers are implemented in YAM++: an element level matcher, a structural matcher and a semantic matcher. The element level and structural mapper discover alignments while the semantic matcher revises these alignments to remove inconsistencies and ensure logical mappings.

HotMatch [10] is a tool based on a combination of many matching algorithms. The two types of algorithms are element level and structural matching. However, there is more than one of each implemented. There are also filters in HotMatch, used to remove duplicate mappings found by the matchers. Upon input of a source and target ontology, HotMatch deploys its matchers and filters sequentially resulting in mappings between the two.

Hertuda [11] is an entity matcher that applies element level matching with a string comparison. The alignments generated by Hertuda are only satisfiable in OWL Lite/DL. As a result, object properties in the ontologies are handled separately. This may cause some difficulties in aligning object properties in the foundational ontologies because their domains and ranges affect the alignments.

Optima [12] is a fully automatic tool which iteratively improves alignments. It is aimed at aligning large ontologies but may also be used for smaller ontologies. Its similarity measure is based on both syntactic and semantic similarity.

3   Foundational Ontology Content Comparison

In this section, we provide an informal content comparison between the foundational ontology pairs by identifying differences and similarities between the them. A content comparison is beneficial in that it forms the basis for performing ontology mediation operations. It does not include abstract comparisons such as those based on philosophical choices, ontological alignments and software engineering properties, which has been addressed elsewhere [15], but rather a high-level comparison of the structure, organisation, and entities of the ontologies.

DOLCE, BFO and GFO contain both 3D and 4D entities. Both BFO and GFO name these entities Continuant and Occurrent while DOLCE names them endurant and perdurant. Some syntactic variants exist between DOLCE, BFO, and GFO, e.g., DOLCE's space-region vs. BFO's SpatialRegion vs. GFO's Spatial_region. In DOLCE, BFO, and GFO, classes that share the same name and idea are process, function and role.

DOLCE entities are of type particular, BFO's entities are Universals while GFO contains a combination of the two, both Individual and Universal entities. DOLCE and BFO have similar structures at a high-level only in that both have separate branches of 3D and 4D entities. GFO's high-level structure is different as it offers a distinction between Category and Individual entities. DOLCE's endurant and perdurant branches are linked by participation relations; BFO's and GFO's 3D and 4D entity branches are completely independent of each other.

The three foundational ontologies have entities and axioms that represent quality, temporal and spatial entities in different ways. DOLCE and GFO have advanced support for representing entity properties (e.g., colour) and their values (e.g., blue) while BFO has limited support for this. However, similar entities within the ontologies do exist e.g., DOLCE's quality, BFO's Quality and GFO's Property. Similarly, for temporal and spatial entities, the treatment differs in the three foundational ontologies but there are some similar entities. GFO subsumes them in a Space-time entity, while in DOLCE and GFO, the spatial and temporal entities are subsumed by different classes.

DOLCE and GFO contain relational properties. BFO does not have relational properties included in the ontology, but rather as a separate ontology, the Relational Ontology (RO) [6]. BFO 2.0 is currently being developed, where BFO is integrated with RO. DOLCE's relational properties are all based on either of its six primitive relations: parthood, temporary parthood, constitution, participation, quality, and quale. For mereology, DOLCE adopts the axioms of General Extensional Mereology (GEM), which includes parthood, proper part, overlap, strong supplementation, and unrestricted fusion. BFO core is a comprehensive mereology represented in first-order logic and contains collections, sums and universal axioms. GFO's mereology contains the following axioms: antisymmetry, transitivity, set inclusion, proper parthood, and other GFO-specific axioms based on these.

Thus, the organisation of entities within the three ontologies differ. In some cases, entities that seem similar fall in contradicting or disjoint classes. These differences in structure and organisation may cause inconsistencies when performing mapping, as we shall see later in detail.

4   Alignment

For foundational ontology alignment, i.e., aligning on an entity-by-entity basis, certain aspects of the underlying philosophies of each foundational ontology have been ignored, because else it would result in few or no alignments and for practical usage of their OWL files, they are less pressing issues. In particular, DOLCE is descriptive and contains particulars, while BFO is realist and contains universals (but OWL treats them all as classes either way). We align classes and object properties with equivalence relations first, and use subsumption relations afterward to resolve some mapping inconsistencies.

We create alignments for 20 pairs of ontologies. These ontologies include DOLCE-Lite, BFO, GFO, FunctionalParticipation, SpatialRelations, and TemporalRelations (which are more-detailed modules of DOLCE), BFORO and GFO-Basic. BFORO refers to the merged ontology of BFO with the RO, and GFO-Basic is a less-detailed module of GFO. We perform ontology alignment by using existing tools, documentation and manually using the content comparison, with its axioms and annotations. Further, for each resource (tool, documentation or manual alignment), we measure its *accuracy* by firstly examining each of its output alignments to determine whether or not the equivalence relation is correct. Accuracy is defined as the number of 'correct' alignments over the total alignments given by the resource (Eq. 1), where 'correct' denotes the alignment is also in the set of alignments found manually, i.e., what is typically considered as the 'gold standard'. We define the *found* measure of the resources as the number of correct alignments over the total possible correct alignments, after manual intervention (Eq. 2).

$$Accuracy = \frac{|correct\ alignments|}{|total\ alignments_{resource}|} \times 100 \qquad (1)$$

$$Found = \frac{|correct\ alignments|}{|total\ alignments_{gold}|} \times 100 \qquad (2)$$

4.1   Alignment Results

We describe the results of the manual alignments first, and then the results obtained with the matching tools.

Table 1. Equivalence alignments between DOLCE-Lite and BFO; the alignments numbered in bold font can also be mapped.

Entity			Relational property		
	DOLCE-Lite	BFORO		DOLCE-Lite	BFORO
1.	endurant	Independent Continuant	1	generic-location	located_in
2.	physical-endurant	MaterialEntity	2	generic-location-of	location_of
3.	physical-object	Object	**3.**	part	has_part
4.	perdurant	Occurrent	**4.**	part-of	part_of
5.	process	Process	**5.**	proper-part	has_proper_part
6.	quality	Quality	**6.**	proper-part-of	proper_part_of
7.	spatio-temporal-region	SpatioTemporal region	7	participant	has_participant
8.	temporal-region	TemporalRegion	8	participant-in	participates_in
9.	space-region	SpatialRegion			

Table 2. Equivalence alignments between DOLCE-Lite and GFO; the alignments numbered in bold font can also be mapped.

Entity			Relational property		
	DOLCE-Lite	GFO		DOLCE-Lite	GFO
1.	particular	Individual	1.	generic-constituent	has_constituent_ part
2.	endurant	Presential	2.	generic-constituent-of	constituent_part _of
3.	physical-endurant	Material_persistant	3.	generically-dependant-on	depends_on
4.	physical-object	Material_object	4.	generic-dependant	necessary_for
5.	amount-of-matter	Amount_of_ substrate	5.	has-quale	has_value
6.	perdurant	Occurrent	6.	quale-of	value_of
7.	process	Process	7.	boundary	has_boundary
8.	state	State	8.	boundary-of	boundary_of
9.	abstract	Abstract	9.	q-present-at	exists_at
10.	set	Set	**10.**	temporary-participant-in	agent_in
11.	quality	Property	**11.**	temporary-participant	has_agent
12.	quale	Property_value	12.	generic-location	occupies
13.	quality-space	Value_space	13.	generic-location-of	occupied_by
14.	time-interval	Chronoid	14.	part	abstract_has_part
15.	space-region	Spatial_Region	15.	part-of	abstract_part_of
16.	temporal-region	Temporal_Region	16.	proper-part	has_proper_part
			17.	proper-part-of	proper_part_of
			18.	participant	has_participant
			19.	participant-in	participates_in

Manual Alignments. The yield of the manual alignments between the main foundational ontologies (DOLCE-Lite, BFO and GFO) resulted in 35 alignments for GFO ↔ DOLCE-Lite, 17 alignments for DOLCE-Lite ↔ BFO and 23 alignments for BFO ↔ GFO; hence, 75 in total which are listed in Tables 1, 2 and 3. When we consider entity alignments including the related modules of the foundational ontologies (e.g., GFO-Basic), there is a total of 85 alignments. Naturally, there are many more than 85 alignments if we consider identical alignments that occur among the same entities in related modules; e.g., DOLCE-Lite:particular ↔ GFO:Individual and FunctionalParticipation:particular ↔ GFO:Individual. There are 14 alignments common between these three ontologies, based on the alignments of the ontologies in Tables 1, 2 and 3 which is displayed in Table 4.

The manual alignments were aided by the GFO documentation [3] and checked against the alignments proposed by [4,5]. The GFO documentation [3] contains a list of similarities between GFO and DOLCE which helped with the alignment process. Some of the alignments could not be used, however, due to changes in the two foundational ontologies in the meantime. We were able to use 42 % of the alignments from the documentation. We discuss four equivalence alignments from [5]. We changed the alignment bfo:ProcessualEntities ↔ dolce:perdurant to bfo:Occurrent ↔ dolce:perdurant, because by definition occurrents and perdurants

Table 3. Equivalence alignments between BFO and GFO; the alignments in bold are also mapped.

Entity			Relational property		
	BFORO	GFO		BFORO	GFO
1.	Entity	Entity	**1.**	has_part	has_part
2.	Independent Continuant	Presential	**2.**	part_of	part_of
3.	Dependent Continuant	Dependent	**3.**	has_proper-part	has_proper_part
4.	MaterialEntity	Material_persistant	**4.**	proper_part_of	proper_part_of
5.	Object	Material_object	**5.**	has_participant	has_participant
6.	ObjectBoundary	Material_boundary	**6.**	participant_in	participates
7.	Function	Function	**7.**	located_in	occupies
8.	Role	Role	**8.**	location_of	occupied_by
9.	Occurrent	Occurrent	**9.**	has_agent	has_agent
10.	Process	Process	**10.**	agent_in	agent_in
11.	Quality	Property			
12.	SpatialRegion	Spatial_region			
13.	TemporalRegion	Temporal_region			

Table 4. Common alignments between DOLCE-Lite, BFO and GFO.

	DOLCE-Lite	BFORO	GFO
Classes			
1	endurant	Independent Continuant	Presential
2	physical-object	Object	Material_object
3	perdurant	Occurrent	Occurrent
4	process	Process	Process
5	quality	Quality	Property
6	space-region	SpatialRegion	Spatial_region
7	temporal-region	Temporal-Region	Temporal_region
Relational properties			
1	proper-part	has_proper_part	has_proper_part
2	proper-part-of	proper_part_of	proper_part_of
3	participant	has_participant	has_participant
4	participant-in	participates_in	participates_in
5	generic-location	located_in	occupies
6	generic-location-of	location_of	occupied_by

both represent entities that have temporal parts and unfold in time. Temal et al.'s alignment of bfo:Quality with dolce:physical-quality is more precise than ours, because, as mentioned above, we chose to ignore the some philosophies (the realist debate) with the hope of achieving a higher number of alignments. That is, our mapping has bfo:Quality ↔ dolce:quality, thereby ignoring the fact that BFO does not consider abstract entities. We agree with bfo:SpatialRegion ↔ dolce:

Table 5. Comparison of manually performed alignment accuracies of the GFO documentation [3], related works, and ours, and aggregates for mappings.

	Seyed	Herre	Temal et al.	Ours
Class alignments				
DOLCE − Lite ↔ BFO	-	-	2/7	9/9
BFO ↔ GFO	-	-	-	13/13
GFO ↔ DOLCE − Lite	-	13/31	-	16/16
Object property alignments				
DOLCE − Lite ↔ BFO	0	-	-	8/8
BFO ↔ GFO	-	-	-	10/10
GFO ↔ DOLCE − Lite	-	0	-	19/19
Overall alignments				
Total	0/0	13/31	2/7	75/75
Accuracy	0 %	42 %	29 %	100 %
Found	0 %	37 %	12 %	100 %
Overall mappings				
Total	0/0	8/31	1/7	40/40
Accuracy	0 %	26 %	14 %	100 %
Found	0 %	61 %	9 %	100 %

space-region and bfo:TemporalRegion ↔ dolce:temporal-region, and use this equivalence, too. Seyed [4] examined only three relations—dependency, quality, and constitution—and found that they are different in DOLCE and BFO. The basic numbers of the alignments are included in Table 5.

Automated Alignments. Table 6 lists the numbers of alignments found by the selected tools. We describe some further data in the remainder of this section.

H-Match generated many alignments, but most of the output was not accurate. Many entity pairs that were matched using H-Match were found to be incorrectly aligned; e.g., DOLCE-Lite:quale ↔ bfo:Role. This resulted in us being able to use only 18 % of these alignments, with the rest being false positives. PROMPT was generally unstable resulting in force closure of the application. We could use 56 % of the suggestions it generated, with the rest being false positives; e.g., bfo:Site ↔ gfo:Situoid.

While LogMap provided few alignments between the foundational ontologies (less than ten in all cases), most alignments were accurate. The one false positive in LogMap was the alignment of bfo:IndependentContinuant ↔ gfo:Independent. YAM++ generated many alignments. However, while most of the alignments for DOLCE ↔ BFO and BFO ↔ GFO were accurate, only about half were accurate for GFO ↔ DOLCE. Overall we were able to use almost 64 % of its alignments. Like LogMap, YAM++ also incorrectly aligned bfo:IndependentContinuant ↔ gfo:Independent. Some of YAM++'s other false positive alignments include dolce:

Table 6. Comparison of alignment accuracies of the matching tools and aggregates for mappings.

	H-Match	PROMPT	LogMap	YAM++	Hot Match	Hertuda	Optima
Class alignments							
DOLCE – Lite ↔ BFO	4/16	3/8	2/2	4/4	3/3	3/3	4/12
BFO ↔ GFO	5/31	7/12	7/8	6/7	7/7	7/7	8/14
GFO ↔ DOLCE – Lite	4/25	4/8	3/3	8/11	5/5	5/5	5/16
Object property alignments							
DOLCE – Lite ↔ BFO	0	0	0	0	0	0	0/1
BFO ↔ GFO	0	0	4/4	0	0	0	1/3
GFO ↔ DOLCE – Lite	0	4/4	0	5/14	5/7	6/8	2/23
Overall alignments							
Total	13/72	18/32	16/17	23/36	20/22	21/23	20/69
Accuracy	18 %	56 %	94 %	64 %	91 %	91 %	29 %
Found	17 %	24 %	21 %	31 %	27 %	28 %	27 %
Overall mappings							
Total	10/72	11/32	16/17	15/36	11/22	12/23	13/69
Accuracy	14 %	34 %	94 %	42 %	50 %	52 %	19 %
Found	25 %	28 %	40 %	38 %	28 %	30 %	33 %

generic-constituent ↔ gfo:has_sequence_constituent, dolce:quality-space ↔ gfo: Space and dolce:temporary-proper-part ↔ gfo:has_constituent_part.

HotMatch generated a fair amount of alignments between the ontologies. Overall, we were able to use 91 % of HotMatch's alignments, with just 2 alignments out of all 22 being false positives. Hertuda's output was surprisingly similar to HotMatch's output, with just one more alignment than HotMatch. We were able to use 91 % of Hertuda's alignments, with just 2 alignments out of all 23 being false positives. Common false positives in YAM++, Hertuda and HotMatch were the alignments between dolce:part ↔ gfo:has_part and dolce:part-of ↔ gfo:part_of, which is discussed in Sect. 4.2. Optima generated many alignments for each pair. However, there were many false positives, consequently we could use only 29 % of its alignments overall. Optima incorrectly aligned gfo:Continuous ↔ bfo: Continuant, dolce:Region ↔ bfo:SpatialRegion and dolce:dependent-place ↔ bfo: Dependent.

4.2   Alignment Issues

We have encountered two types of issues in alignment: transitivity, where there was no 'full circle' alignment between some entities of the three ontologies, and approximate alignments, where there is no clear relationship to describe the match.

Transitivity. Transitivity in entity alignments works as follows: if the equivalence relation holds between entities from the first and second ontology and it holds between entities from the second and third ontology; it necessarily holds between entities from the first and third ontology. Applying transitivity to entity alignments assists in detecting errors. For instance, if one were to align dolce:endurant ↔ gfo:Persistant, and gfo:Persistant ↔ bfo:Continuant, then by transitivity this means that dolce:endurant is equivalent to bfo:Continuant, which is incorrect, because in most cases, the foundational ontology alignments are transitive. There were two types of exceptions, being the absence of an entity and what can be termed consequences of conflicting philosophies.

Absence of an Entity. An alignment cannot be a candidate for transitivity if there is an equivalence between only two out of the three ontologies. From the three main ontology alignments, the following ones were not transitive due to the absence of an entity:

- *Absence of a DOLCE entity (7 cases):* bfo:Entity ↔ gfo:Entity, bfo: Dependent-Continuant ↔ gfo:Dependent, bfo:ObjectBoundary↔gfo:Material_boundary, bfo: Function ↔ gfo:Function, bfo:Role ↔ gfo:Role, bfo:has_agent ↔ gfo:has_agent, bfo: agent_in ↔ gfo:agent_in.
- *Absence of a GFO Entity (1 Case):* dolce:spatio-temporal-region ↔ bfo: SpatioTemporalRegion.
- *Absence of a BFO Entity (17 Cases):* gfo:Individual ↔ dolce:particular, gfo: Amount_of_substrate ↔ dolce:amount-of-matter, gfo:State ↔ dolce:state, gfo: Abstract ↔ dolce:abstract, gfo:Set ↔ dolce: set, gfo:Property_value ↔ dolce: quale, gfo:Value_space ↔ dolce:quality-space, gfo:Chronoid ↔ dolce:time_ interval, gfo: has_constituant_part ↔ dolce:generic-consitituant, gfo: constituant part_of ↔ dolce: generic-constituant-of, gfo: necessary_for ↔ dolce: generic-dependent, gfo: depends_on ↔ dolce: generically-dependent-on, gfo: has_value ↔ dolce: has-quale, gfo: value_of ↔ dolce:quale-of, gfo: has_boundary ↔ dolce: boundary, gfo: boundary_of ↔ dolce:boundary-of, and gfo:exists_at ↔ dolce:q-present-at.

From this type of transitivity issue, we see that for the three main ontology alignments, in most cases BFO entities are absent. There are a few cases of absent DOLCE entities and one case of an absent GFO entity.

Conflicting Philosophies. The philosophies of foundational ontologies affect their entities to a certain extent, despite already having been lenient. In some cases, two entities that are aligned to each other may not be aligned to the same entity of a third ontology.

- *dolce:physical-endurant ↔ bfo:MaterialEntity, dolce:physical-endurant ↔ gfo: Discrete_ presential and bfo:MaterialEntity ↔ gfo:Material_ persistant.* Let us align bfo:MaterialEntity ↔ dolce:physical-endurant, ignore their underlying philosophies (i.e., that BFO is an ontology of universals and DOLCE of particulars). However, in GFO, there are two entities for representing this type of entity, based on distinct philosophical notions: gfo: Discrete_presential, being subsumed gfo:Individual, is suited for dolce:physical-endurant while gfo: Material_persistant, being subsumed by gfo:Universal, is suited for bfo:MaterialEntity.

- *dolce:*part ↔ bfo:has_part, dolce:part ↔ gfo:has_abstract_part and bfo: has_part ↔ gfo:has_part (idem for their inverses). In DOLCE, both the domain and range of part is particular. In BFORO, there is no domain and range for has_part. In GFO, both the domain and range of abstract_has_part is Item, while both the domain and range for has_part is Concrete. The former relational property may be better suited for DOLCE because it is a descriptive ontology and contains abstract entities. The latter is better suited for BFORO as it is a realist ontology, representing the world as is, thereby containing concrete entities only.

The ontology matching tools discussed in Sect. 4.1 misaligned dolce:part ↔ gfo: has_part and their inverses. This is because object property inconsistencies are not fully recognised by reasoners [16], hence their conflicting domains and ranges did not affect the satisfiability of the ontology.

Approximate Alignments. There are a number of approximate alignments between foundational ontology entities. By this we mean that they are not equivalent to each other or subsumed by one another, but share some common characteristics. By identifying these relations between these entities, foundational ontology developers could possibly relate them as sibling classes by grouping them both under a common superclass. We mention three of them.

- *dolce:arbitrary-sum, bfo:ObjectAggregate and gfo:Configuration*: All three of these entities describe a collection of something. dolce:arbitrary-sum, however, has no unity criterion e.g., a pencil and laundry basket are together a dolce:arbitrary-sum, and it can contain both dolce:physical-endurant and dolce: non-physical-endurant entities. dolce:physical-endurant is not restricted just to instances of dolce:physical-object but can possibly include dolce:feature and dolce:amount-of-matter. bfo:ObjectAggregate, on the other hand, has overall unity and can be considered as a whole. It is restricted to bfo:Object only, and in the case of BFO, all objects are physical. gfo:Configuration is simply a collection of gfo:Presential facts. gfo:Presentials are not restricted to whole physical objects and can include other gfo:Presential entities. For this reason, it cannot equate to bfo:ObjectAggregate. Furthermore, it holds a restriction that it must contain at least one material entity. dolce:arbitrary-sum could contain physical, non-physical or both entities, with no restrictions.
- *dolce:state and bfo:SpatioTemporalInstant*: DOLCE describes dolce:state by using an example of a rock erosion describing state as a time interval of the erosion is collapsed into a time point. Similarly BFO defines bfo: SpatioTemporalInstant as a "connected spatiotemporal region at a specific moment". The difference between the two lies in the fact that dolce:state is homeomeric while bfo:SpatioTemporalInstant is not.
- *dolce:relevant-part and bfo:FiatObjectPart*: DOLCE describes dolce:relevant-part as a feature that is a relevant part of their host; e.g., the edge of a cube. BFO defines bfo:FiatObjectPart as a material entity that is part of an object but not demarcated by physical discontinuities; e.g., the lower portion of the leg. In this sense they are both part objects that are physical entities. However,

it is unclear whether dolce:relevant-part is demarcated by physical discontinuities or not and whether BFO's fiat object parts are 'relevant' somehow. This requires further investigation.

4.3 Evaluating Alignments

The alignments identified may be open to further investigation by ontologists for some time, but interoperability is becoming a pressing matter. Therefore, we chose to evaluate the alignments with end-users, who are ultimately the ones who would be using the foundational ontology library for practical ontology development purposes. To this end, we set up an experiment via a web-based survey with a time-limit of two weeks to complete the evaluation. The survey presents the participant with a set of alignments, where every alignment has the following options as answer: Agree, Partially Agree, Disagree, Unsure (i.e., 'I though about it, and I still do not know') and Skip. The participants for this evaluation were members of the Digital Enterprise Research Institute (DERI).

Each alignment set received a different number of responses: DOLCE↔BFO had 18 responses, BFO↔GFO had 10 responses and GFO↔DOLCE had 13 responses; Table 7 provides a summary of the responses received for each option. For each alignment set, the highest percentage of the responses were Agree, although not more than for half of the cases: on average, 44.9 % of all responses were for the Agree option, and then Partially Agree. Thereafter, 11.0 % and 17.7 % of the responses were for the Unsure and Skip options, respectively. The smallest portion of responses, 7.1 %, were from the Disagree option. For what the participants agreed upon, in most cases they agreed on the same alignments. An alignment that many participants agreed on is the equivalence of DOLCE:spatio-temporal-region and BFO:SpatioTemporalRegion. In most cases, the Agree option received few or no responses when ontology entity annotations were not clearly defined; e.g., in aligning bfo:DependentContinuant and gfo:Dependent, the latter was annotated with only "Dependent entities.". The few Disagree options were for different alignments; one that received some Disagree responses is the equivalence of DOLCE:perdurant and GFO:Occurrent. Participants were not united in their Unsure and Skip responses. Most of the general comments received from the participants indicated that the annotations from the foundational ontologies were difficult to understand, not properly defined, and missing in some cases. Perhaps if the annotations were better defined, the number of Unsure and Skip responses would decrease. Also, a more in-depth investigation into the motivations for the participant's choices may reveal useful results for examining the alignments further.

5 Mapping and Merging

Ontology Mapping uses the alignments from the alignment process to create correspondences between entities in the ontologies. The output from the alignment process is broader, while the output from the mapping process is narrower as inconsistencies affect the mapping process. Merging is performed by

Table 7. Comparison of alignment evaluation responses.

Ontologies	DOLCE ↔ BFO	BFO ↔ GFO	GFO ↔ DOLCE	Average
Number of responses	18	10	13	13.7
Agree	49.4 %	47.1 %	38.1 %	44.9 %
Partially agree	21.7 %	20.0 %	16.3 %	19.3 %
Disagree	7.8 %	6.4 %	7.1 %	7.1 %
Unsure	8.3 %	9.3 %	15.4 %	11.0 %
Skip	12.8 %	17.1 %	23.1 %	17.7 %

creating a new ontology of the source ontologies with their mappings between each other. Ontology mapping and merging was performed by relating classes and object properties in Protégé v4.2 using the Hermit v1.3.6 reasoner. Entities were mapped in the order of their level in the hierarchy, from higher to lower level, because foundational ontologies by definition are general high-level ontologies. Therefore, in mapping, preference must be first given to high-level entities to have agreement among general entities and avoid inconsistencies at that level.

Alignments that cannot be mapped due to logical inconsistencies result in unsuccessful mappings. The inconsistencies were identified by using the following method. For each candidate class mapping:

1. Assert the equivalence for the found alignment.
2. Run the automated reasoner.
3. Check if there are any unsatisfiable classes.
4. If there are unsatisfiable classes, use the reasoner explanation feature to generate an explanation.
5. Analyse explanations.
6. Remove inconsistent mapping, if applicable.

For each candidate object property mapping, since object property inconsistencies and flaws are not properly recognised by reasoners [16], we identified inconsistencies by checking if an object property pair's domain and range restrictions are satisfiable by using the above method.

The numbers in bold face in Table 4 represent the alignments that resulted in successful mappings between the common entities of the three main ontologies based on the mappings of Tables 1, 2 and 3. From the 14 alignments in Table 4, six successful mappings exist. Recall from the previous section on alignment, there was a total of 85 distinct alignments between all foundational ontologies and related modules, and 75 alignments between the main foundational ontologies. Performing the method to identify inconsistencies in alignments resulted in 42 distinct logical inconsistencies of which 35 alone were from the main ontologies. From all the distinct equivalence alignments, only half were satisfiable and resulted in successful mappings. Comparing these mappings to the alignments found by the tools, LogMap doubled its percentage found to 40 % and performed best compared to the six others evaluated (see Table 6, bottom three rows).

To solve inconsistencies in the mapping attempts, we analysed each alignment on the logical explanation for the inconsistency and the description of the entity provided by the foundational ontology developers, and checked whether it was possible to change the alignment from equivalence to subsumption. However, there are still many unsolvable inconsistencies, mainly due to hierarchical and structural differences in the foundational ontologies. Due to space limitations, we describe only a representative selection of the logical inconsistencies and (logically satisfiable) possible solutions; the full list of inconsistencies is available at http://www.thezfiles.co.za/ROMULUS/.

Inconsistencies Due to Disjoint Classes. For this type of inconsistency, the entities to be aligned are disjoint to each other, either directly, through higher-level equivalence relations or through their subclasses. If entities are disjoint, they cannot overlap, hence cannot be equivalent.

- *dolce:temporal-region - gfo:Temporal_region - bfo:TemporalRegion*: The issue with incompatible temporal regions between BFO, GFO, and DOLCE is depicted in Fig. 1 and is a result of the OWL DisjointClasses class axiom between gfo:Concrete, gfo:Space_Time and gfo:Abstract, and between dolce: Abstract and dolce:Perdurant, or, from the other viewpoint: because BFO made TemporalRegion an Occurrent, DOLCE made it Abstract, and GFO neither. This does not seem to be resolvable.

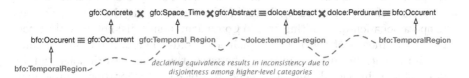

Fig. 1. Graphical depiction of why the aligned gfo:Temporal_Region, bfo:Temporal Region, and dolce:temporal-region cannot be mapped in any way without causing an inconsistency; ≡: aligned entities, ×: disjoint entities.

- *bfo:Role - gfo:Role:* This is due to disjointness between classes and multiple inheritance. The essential aspects of the situation is depicted in Fig. 2: gfo: Processual_role becomes inconsistent if an equivalence were to be declared between gfo: Role and bfo:Role. <u>Solution:</u> Logically, bfo:Role cannot be equivalent to gfo:Role, but bfo:Role can be subsumed by gfo:Role, or one can have gfo: Relational_role and gfo:Social_role subsumed by bfo:Role.
- *gfo:necessary_for - dolce:generic-dependent:* If we were to equate these object properties, we would have to assume that their domains and ranges are equivalent, which is not the case; the situation is depicted in Fig. 3. <u>Solution:</u> Logically, gfo:necessary_for cannot be equivalent to dolce:generic-dependent, because equating their domains and ranges causes inconsistencies. However, dolce:generic-dependent's domain and range, dolce:particular can logically be subsumed by gfo:necessary_for's domain and range, gfo:Item. Therefore the relation can be changed to gfo: necessary_for subsumes bfo:generic-dependent.

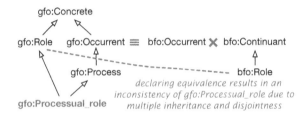

Fig. 2. Visualisation of the root cause of the non-mappable gfo:Role and bfo:role; ×: disjointness, ≡: equivalence mapping.

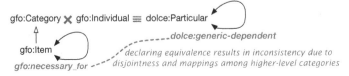

Fig. 3. Visualisation of the root cause of the non-mappable gfo:necessary_for and DOLCE-Lite:generic-dependent; ×: disjointness, ≡: equivalence mapping.

- *dolce:generic-location - bfo:located_in:* This issue is due to disjointness among domain/range. dolce:generic-location's range is dolce:particular and bfo: located_in's range is bfo:Continuant. bfo:Continuant is disjoint to bfo:Occurrent and bfo: Occurrent ≡ dolce:perdurant. In DOLCE, perdurant ⊑ ∃has-Quality. temporal-location-q and the domain of dolce:has-Quality is dolce:particular (the superclass of dolce:perdurant). Thus, bfo:Continuant is disjoint to ∃has-Quality. temporal-location-q, resulting in bfo:Continuant being disjoint to dolce: particular, by means of the above explained axioms. The two relations cannot be equivalent, because equivalence between the range restrictions will be unsatisfiable in the alignment. Therefore dolce:generic-location cannot map to bfo:located_in.

Another unresolvable case is dolce:set - gfo:Set.

Inconsistencies Due to Complement Classes. For this type of inconsistency, the entities to be aligned were found to be complements of each other, either directly, through higher-level equivalences or through subsumption. We describe here one such case.

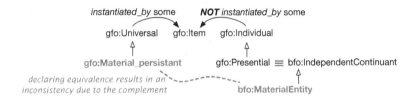

Fig. 4. Visualisation of the root cause of the non-mappable bfo:MaterialEntity and gfo:Material_persistant; ≡: equivalence mapping.

- *bfo:MaterialEntity* - *gfo:Material_persistant*, which is visualised in Fig. 4. The crucial aspect in GFO is the class axiom Universal ⊑ ∃instantiated_by.Item, and the complement for individuals. Concerning mappings, bfo: IndependentContinuant ≡ gfo:Presential. However, in GFO, Presential ⊑ Individual and Individual⊑ ¬∃instantiated_by.Item. Thus, gfo:Material_persistant is a subclass of gfo: instantiated_by some gfo:Item while bfo:MaterialEntity is a subclass of the *complement* of that class, hence bfo:MaterialEntity cannot be equivalent to gfo: Material_persistant. Solution: The alignment can be changed into bfo: MaterialEntity - gfo: Discrete_presential, which avoids the complement issue but it is not free of argument (recall the "conflicting philosophies" item in Sect. 4.2).

6   Discussion

Given the size of the ontologies and our high tolerance by ignoring underlying philosophies, the amount of alignments, and, even more so, the amount of mappings is less than one may have expected; or: once investigated in detail, the foundational ontologies are, at present, not particularly interchangeable even at the logical level. Only six pairwise mappings exist, i.e., they being, essentially, equivalent throughout all three examined foundational ontologies.

Concerning feasibility to carry out automated alignments, in most cases, the tools evaluated with the OAEI performed better than the others, with the exception of Optima. LogMap had the highest accuracy, because it also considers the logic-based semantics of the ontologies and uses automated reasoning services throughout the process, therewith eliminating those false positives that would have led to a logical inconsistency. However, LogMap generated very few alignments compared to other accurate tools (YAM++, Hertuda and HotMatch),

Table 8. False positives caused by syntactic matching generated by the alignment tools; the terms in italics represent the strings that are common between aligned entities.

DOLCE-Lite	BFO
physical-*region*	Connected Spatio Temporal *Region*
non-physical-*object*	*Object*
region	Spatio Temporal *Region*
BFO	GFO
*Independent*Continuant	*Independent*
Site	*Sit*uoid
*Continu*ant	*Continu*ous
GFO	DOLCE-Lite
has_sequence_*constituent*	generic-*constituent*
has-*part*	*part*
Space	quality-*space*

indicating that the additional heuristics implemented are too strict at least for foundational ontology alignment.

Most false positive alignments generated by the tools, such as bfo:Independent Continuant ↔ gfo:Independent, indicate that the algorithms implement syntactic matching, which, based on the results we obtained, is not sufficient or suitable for foundational ontology matching because many entities have a common syntax e.g., dolce:quality-space ↔ gfo:Space both have the string 'space' in common but are entirely different entities; Table 8 includes a selection of such false positives that are caused by syntactic matching in the tools when aligning the three foundational ontologies. The tools failed to recognise simple alignments such as dolce:perdurant ↔ gfo:Occurrent, bfo:Quality ↔ gfo:Property. In this sense, semantic matching is not considered, or if it is, it fails to recognise synonyms of the philosophical scope on which foundational ontologies are built upon. Structural matching is not an effective method either, due to the fact that the hierarchies and structures of the foundational ontologies differ greatly which causes the root distances of mappable entities to differ. For aligning foundational ontologies, it will be useful if existing semantic matchers would include something alike a 'philosophy WordNet' that specialises in philosophical terms, synonyms, and definitions used in foundational ontologies.

The results of the tool analysis is a good indication of which tools to experiment with for foundational ontology alignment in general. However, they found less than a third of the actual alignments at this stage, and therefore it is still vital to perform manual alignment for foundational ontologies. The tools also did not generate subsumption relations for any of the alignments, but this could perhaps be an extension to the basic idea of LogMap by means of another call to the reasoner. One could investigate whether Optima is useful to identify accurate alignments among the larger foundational ontologies SUMO [17] and YAMATO [18].

On a positive note, the systematised list of issues now can be taken up by ontologists. While some of the inconsistencies found are quite elaborate, others should be easier to resolve both ontologically (philosophically) and where in the ontology the entity is positioned; e.g., the notion of a mathematical Set is fairly well investigated already, and likewise the different theories of parthood. As such, the results presented here provide a solid foundation for ample ontological investigations. From an engineering viewpoint and in case of urgent need for interoperability, one could take a quite different strategy: OWL 2 EL does not have negation, and therefore it should be possible to assert more mappings between the OWL 2 EL modules of the foundational ontologies. Whether that is the best strategy is a different matter, and it does not take away the substantial list for which there was no transitivity due to 'missing' entities. In any case, we now know that some mappings are possible, hence, also some foundational ontology interoperability.

7   Conclusion

The foundational ontologies DOLCE, BFO, and GFO were pairwise aligned and mapped. They were aligned manually, which served as the 'gold standard', and with the aid of seven alignment tools. The accuracy and percentage of alignment found were compared, where LogMap had the highest accuracy with 94 % and HotMatch and Hertuda as close second, and YAM++ found the most correct alignments (31 % of the total manual alignments among the three main ontologies (75)). The evaluation of the tools indicated that the algorithms currently implemented by the tools are not well-suited for foundational ontology mediation. Declaring the correspondences in all ontology files based on its 85 alignments resulted in only 43 mappings, with the remaining 42 causing logical inconsistencies. The inconsistencies are due primarily to differences in their respective hierarchical structure with conflicting axioms, such as complement and disjointness, and incompatible domain and range restriction. On closer inspection, some inconsistencies may be resolved using subsumption or making them sibling classes.

Future research includes mapping other foundational ontologies, adding subsumption mappings, and evaluating the current alignments with the foundational ontology developers. We also aim to implement a facility for community input on the alignments and mappings, which could to be facilitated via the foundational ontology library that is available online at http://www.thezfiles.co.za/ROMULUS/. ROMULUS [19] is the first online repository of machine- processable, modularised, aligned, and logic-based merged foundational ontologies. It encompasses the typical repository functions e.g., online browsing, metadata, downloadable resources as well as specific tools for foundational ontologies such as a foundational ontology recommender, ontology modules for easier reuse, and a ontology mediation outputs (alignment, mapping, merging) among the BFO, GFO and DOLCE foundational ontologies.

References

1. Masolo, C., Borgo, S., Gangemi, A., Guarino, N., Oltramari, A.: Ontology library. WonderWeb Deliverable D18 (ver. 1.0, 31–12-2003) (2003). http://wonderweb. semanticweb.org
2. de Bruijn, J., Ehrig, M., Feier, C., Martíns-Recuerda, F., Scharffe, F., Weiten, M.: Ontology mediation, merging, and aligning. In: Davies, J., Studer, R., Warren, P. (eds.) Semantic Web Technologies, pp. 1–20. Wiley Online Library, Chichester (2006)
3. Herre, H.: General Formal Ontology (GFO): A foundational ontology for conceptual modelling. In: Poli, R., Healy, M., Kameas, A. (eds.) Theory and Applications of Ontology: Computer Applications, pp. 297–345. Springer, Netherlands (2010)
4. Seyed, A.P.: BFO/DOLCE primitive relation comparison. In: The 12th Annual Bio-Ontologies Meeting Colocated with Intelligent Systems for Molecular Biology (ISMB'09). Stockholm, Sweden, 28 June 2009
5. Temal, L., Rosier, A., Dameron, O., Burgun, A.: Mapping BFO and DOLCE. Stud. Health Technol. Inform. **160**(Pt 2), 1065–1069 (2010)
6. Smith, B., Ceusters, W., Klagges, B., Kohler, J., Kumar, A., Lomax, J., Mungall, C., Neuhaus, F., Rector, A., Rosse, C.: Relations in biomedical ontologies. Genome Biol. **6**(5), 46 (2005)
7. Euzenat, J., Shvaiko, P.: Ontology Matching. Springer, Heidelberg (2007)

8. Jiménez-Ruiz, E., Cuenca Grau, B.: LogMap: Logic-based and scalable ontology matching. In: Aroyo, L., Welty, C., Alani, H., Taylor, J., Bernstein, A., Kagal, L., Noy, N., Blomqvist, E. (eds.) ISWC 2011, Part I. LNCS, vol. 7031, pp. 273–288. Springer, Heidelberg (2011)
9. Ngo, D.H., Bellahsene, Z.: YAM++: A multi-strategy based approach for ontology matching task. In: ten Teije, A., Völker, J., Handschuh, S., Stuckenschmidt, H., d'Acquin, M., Nikolov, A., Aussenac-Gilles, N., Hernandez, N. (eds.) EKAW 2012. LNCS, vol. 7603, pp. 421–425. Springer, Heidelberg (2012)
10. Dang, T.T., Gabriel, A., Hertling, S., Roskosch, P., Wlotzka, M., Zilke, J.R., Janssen, F., Paulheim, H.: HotMatch results for OEAI 2012. In: Seventh International Workshop on Ontology Matching (OM'12). CEUR Workshop Proceedings, vol. 946. http://CEUR-WS.org (2012)
11. Hertling, S.: Hertuda results for OEAI 2012. In: Seventh International Workshop on Ontology Matching (OM'12). CEUR Workshop Proceedings, vol. 946. http://CEUR-WS.org (2012)
12. Kolli, R., Doshi, P.: OPTIMA: tool for ontology alignment with application to semantic reconciliation of sensor metadata for publication in SensorMap. In: IEEE International Conference on Semantic Computing (ICSC'08), pp. 484–485. IEEE, SantaClara, 4–7 August 2008
13. Castano, S., Ferrara, A., Montanelli, S.: H-MATCH: an algorithm for dynamically matching ontologies in peer-based systems. In: The First International Workshop on Semantic Web and Databases (SWDB'03), pp. 231–250. Humboldt-Universitt, Berlin (2003)
14. Noy, N.F., Musen, M.A.: PROMPT: algorithm and tool for automated ontology merging and alignment. In: Seventeenth National Conference on Artificial Intelligence and Twelfth Conference on on Innovative Applications of Artificial Intelligence (AAAI/IAAI), pp. 450–455. AAAI Press (2000)
15. Khan, Z., Keet, C.M.: ONSET: automated foundational ontology selection and explanation. In: ten Teije, A., Völker, J., Handschuh, S., Stuckenschmidt, H., d'Acquin, M., Nikolov, A., Aussenac-Gilles, N., Hernandez, N. (eds.) EKAW 2012. LNCS, vol. 7603, pp. 237–251. Springer, Heidelberg (2012)
16. Keet, C.M.: Detecting and revising flaws in OWL object property expressions. In: ten Teije, A., Völker, J., Handschuh, S., Stuckenschmidt, H., d'Acquin, M., Nikolov, A., Aussenac-Gilles, N., Hernandez, N. (eds.) EKAW 2012. LNCS, vol. 7603, pp. 252–266. Springer, Heidelberg (2012)
17. Niles, I., Pease, A.: Towards a standard upper ontology. In: Second International Conference on Formal Ontology in Information Systems (FOIS'01), pp.17–19. IOS Press Ogunquit, October 2001
18. Mizoguchi, R.: YAMATO: yet another more advanced top-level ontology. In: Proceedings of the Sixth Australasian Ontology Workshop. Conferences in Research and Practice in Information, pp. 1–16. ACS, Sydney (2010)
19. Khan, Z.C., Keet, C.M.: The foundational ontology library ROMULUS. In: Cuzzocrea, A., Maabout, S. (eds.) MEDI 2013. LNCS, vol. 8216, pp. 200–211. Springer, Heidelberg (2013)

PRONTOE: An Ontology Editor for Domain Experts

Scott Bell[1]([✉]), Pete Bonasso[1], Mark Boddy[2],
David Kortenkamp[1], and Debra Schreckenghost[1]

[1] TRACLabs Inc., 16969 N Texas Ave, Suite 300,
Webster, TX 77598, USA
scott@traclabs.com
[2] Adventium Labs, 111 Third Ave South, Suite 100,
Minneapolis, MN 55401, USA
mark.boddy@adventiumlabs.com

Abstract. In this paper, we describe a set of software tools called the PRIDE ONTOlogy Editor (PRONTOE) and a methodology that allows system operators and domain experts to build and maintain ontologies of their systems with no explicit understanding of the underlying ontology representation. We present three case studies: one using NASA flight controllers, one using the DARPA Robotic Challenge, and one using unmanned vehicles.

1 Motivation

Ontologies provide a structural framework for system knowledge that is useful for many applications. One particular application is to provide the knowledge necessary for software tools that assist operators in monitoring and controlling complex and dynamic systems. Using an ontology to model system information has advantages. The ontology models provide monitoring and control concepts as objects that can be used in multiple control domains. These models define properties that can be used to automatically populate object data fields and derive relations between objects to improve search of system information.

The representational power of ontologies, however, introduces a number of challenges. One such challenge is developing and using ontologies for operations that the operators and other domain experts do not have any experience in developing or maintaining ontologies of their systems. Using ontology experts to build and maintain these ontologies is prohibitively expensive, especially because the knowledge necessary to build the ontologies exists in a variety of documents and in the operator's or domain expert's heads. Another challenge is that the states and configurations of the specific objects in the domain are both voluminous and dynamic, making manual entry and maintenance prohibitive. A final challenge is that the data required, especially state updates, need to be extracted or imported from other disparate systems. In this paper, we describe a set of software tools called the PRIDE ONTOlogy Editor (PRONTOE) and a methodology that allows system operators and domain experts to build and maintain

© Springer-Verlag Berlin Heidelberg 2015
A. Fred et al. (Eds.): IC3K 2013, CCIS 454, pp. 153–167, 2015.
DOI: 10.1007/978-3-662-46549-3_10

ontologies of their systems with no explicit understanding of the underlying ontology representation.

PRONTOE consists of a graphical editing tool, that allows users to define and edit objects and their properties and relationships and to view those properties and relationships in a variety of ways and in the context of their particular domain. PRONTOE supports the integration of different ontology *kernels* that divide complex systems into interacting components. PRONTOE includes reasoners for assisting in object definition and consistency. PRONTOE has software tools for importing (and exporting) data to domain-specific databases. PRONTOE also allows for viewing real-time system data in the context of defined objects and their relationships. PRONTOE is being evaluated in three domains. The first is operation of the International Space Station (ISS) by NASA flight controllers. In this case study, the domain experts are the flight controllers who have engineering degrees and years of experience in operating ISS. The ontology is used by software tools such as task planners, procedure editing and execution systems, and diagnosis systems. The second domain is a humanoid robot being developed by DARPA for disaster relief operations. In this case study, the domain experts are robotics engineers. The ontology is used to develop operator interfaces for using robot capabilities as well as robot scripts for automating common activities. The third is an ontology-driven decision support tool that can assist an operator in controlling and directing several unmanned vehicles at the same time

2   Approach

Our approach consists of several interacting components. At the core of PRONTOE is a graphical user interface (GUI) that allows an operator to visually inspect and edit an OWL ontology. The ontology itself is divided into a set of kernels that correspond to the different operational aspects of the system. The kernel approach makes it easier for an operator to focus on the specifics of their system and not the requirements of the underlying ontology. OWL reasoners provide consistency and fill in required ontology information automatically.

2.1   Graphical User Interface

The PRONTOE Graphical User Interface (GUI) can be used to inspect a system ontology, search for specific system information in that ontology, modify information already in the ontology, and add new information to the ontology. The PRONTOE GUI is an Eclipse based Rich Client Platform (RCP) application. Developing PRONTOE as an RCP application allowed us to use a variety well developed libraries for both Java and Eclipse. For ontology data manipulation and reasoner interaction, we used the OWL-API Java library. To render the ontology graph, the Eclipse Zest Toolkit was used. For rendering system schematics, we used the Batik SVG library. Other libraries include integration with version control, workflows, and user authentication.

The RCP approach also allows us to easily add or remove features for different domains. For example, the space domains have different schematics, bundled ontologies, and editing widgets than the robot domain. By specifying these differences in a product definition, we can simultaneously release PRONTOE for different domains for multiple platforms. We also provide extension points into PRONTOE to help other developers create new domain specific features.

PRONTOE's editing environment is divided into several different editing panes. Figure 1 shows a typical environment. The left window shows the current open ontology along with available ontologies to edit. This particular project has been checked out using the Subversion plugin, so the user can right click to add, commit, or update the ontology from version control within the PRONTOE application.

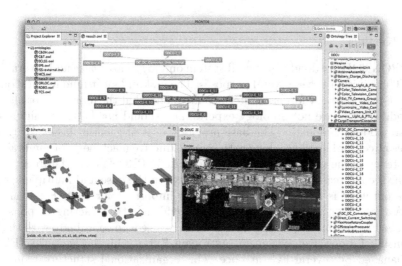

Fig. 1. The PRONTOE editing environment.

The central pane is a graph of a selection of the ontology. The graph shows class and subclass relationships. At the individual level, object properties are displayed with the property name labeled on the edge. Nodes can by right clicked on to expand or collapse their children. Double clicking on a node opens it for editing its name, object properties, and data properties.

The lower right hand pane is a schematic of the physical layout of the system represented by the ontology rendered as an SVG. The different colors in the schematic show the locations of the class and individuals shown in the center graph. Clicking on an area in the schematic will show all the classes and individuals associated with that location. Different SVGs may be used for different domains by creating a PRONTOE plugin. Figure 1 shows an SVG of the International Space Station (ISS).

The upper right hand pane is a tree showing the ontology's subclasses and individual. From here, classes and individuals may be added, edited, deleted, and reasoned upon. A search bar on the top provides incremental searching for classes and individual names.

All the editing panes can be rearranged or even detached according to user preference. The layout of the windows can be named and saved by the user using perspectives. These perspectives are available on the upper right of the PRONTOE window. The user can rapidly flip through different perspectives depending on their current work or role. By default, selecting an item in one pane automatically syncs the information in the remaining panes to reflect the selection. For example, clicking on a class in the ontology tree changes color the schematic with the locations of all the individuals under that class, and changes the graph to display the subclasses and individuals. This synching feature can be disabled per editing pane by deselecting the sync button in the upper right hand corner of each editing pane.

2.2 Kernels

For PRONTOE, we divide an overall system ontology into a *base* and several *kernels*. This division serves several purposes. The first and most fundamental one is that the base ontology forms an intermediate model that the users are not allowed to modify. Users are not ontology experts, and should not be left to figure out how to model the domain. The base provides a set of generic classes for things like resources, or locations, along with some deductive machinery for maintaining consistency in the description of the world state in PRONTOE's database.

The second reason for this separation between the base and multiple kernels in the ontology is bureaucratic, but no less important for that. For example, NASA's ISS flight controllers (FCs) have divided but interacting responsibilities. The FC responsible for orbital maintenance needs to coordinate with the FC responsible for power management. Astronaut Extra-vehicular activities (EVAs) are managed by several FCs at once. This organizational structure requires a supporting structure in PRONTOE. One flight control discipline must have complete control over their own part of the domain, with visibility into parts of the domain under the purview of other disciplines.

As a result, PRONTOE's database has several features that complicate maintenance. First, it is subject to asynchronous access by multiple parties, who may be making interacting or conflicting changes. Second, it is *deductive*: for many changes to the database describing the world state, there are rules that will fire, making additional changes. All of these changes need to be included in the definition of a database *transaction* that allows us to keep PRONTOE's database in a consistent state.

Distributed authority adds more complication, because the separation between disciplines is only partial. For example, there is a defined relationship between two different components of the ISS, such as a pump and the power-channel that energizes it. Disconnecting the pump is a change that must be reflected in the relations for both the pump and the power-channel, which may be in different kernels in the ontology. This requires us to come up with some way to make the change in one kernel and keep the database consistent, which requires an unauthorized change in another kernel.

For now, the solution we have implemented is to use a version-tracking system. All changes, authorized or not, will be made as required to keep the database consistent. Changes made in other kernels will result in the "owners" of those kernels being informed so that they can either approve or reject those changes. Rejection by any flight discipline will then result in the entire set of changes being rolled back.

2.3   Reasoners

For editing-time classification, consistency checking, property inference, and Semantic Web Rule Language (SWRL) reasoning, we use Pellet. The inferred axioms are added back to the currently open ontology allowing the user to save them if they wish. As we are using OWL-API, it is easy to plug in different reasoners (e.g., HermiT) for evaluation. For example, in the case of the ISS if the operator entered the property that a certain computer controls a piece of equipment, then the reasoner will assert a property that the piece of equipment is controlled by that computer. This reduces the burden on the operator of having to specify completely the relationships in the ontology when many of them can be inferred.

3   Case Studies

We have used PRONTOE to develop ontologies for two different complex systems. The first system is the International Space Station (ISS) being operated by NASA and the Johnson Space Center in Houston Texas. The second system is the Atlas humanoid robot being developed by the Defense Advanced Research Projects Agency (DARPA). In this section, we describe these two case studies and how we used PRONTOE to simplify ontology development and maintenance.

3.1   International Space Station

Our principle development activity for PRONTOE is an ontology for the International Space Station (ISS). We have worked for the past two years with NASA flight controllers to develop and design an ontology that partially models the ISS. The focus of the effort is on planning for Extravehicular Activities (EVA), basically space walks, so our ontological concentration is on ISS objects that are located on the outside (or external to) ISS, such as power module and antennas. PRONTOE, as mentioned in Sect. 2.2, comes with a base ontology, a domain base we call ISS-base, and kernel extensions for EVA and for each flight discipline that supports a given EVA, such as electrical power and motion-control systems. The users can then use PRONTOE to extend these kernels, incrementally as new ISS activities arise. To prepare for an upcoming EVA, the flight controllers start with a current configuration of the ontology, and use PRONTOE to develop and save a snapshot of the configuration of equipment, power and control that is anticipated to be true at the time of the activity. In an extension to PRONTOE, we are developing a capability to generate change forms concerning location and configuration changes that resulted from the EVA to be distributed to other ISS parties such as mass properties analysis teams and ISS guidance and navigation teams.

We have developed interfaces that allow PRONTOE to automatically import from two large NASA databases of ISS equipment: the External Configuration Analysis and Tracking Tool (ExCATT) and the Inventory Management System (IMS). By connecting to existing databases, we reduce the upfront editing time necessary to build the ontology. The ontologies created by these systems are large, with 4855 axioms containing 283 classes and 897 individuals. We have thus broken the ontology into kernels to ease editing. As shown in Fig. 2, PRONTOE has a merge tool to assist in updates from the external databases. We can also export to these databases, so any changes that operators make using PRONTOE can be pushed back into the official databases of record.

In addition to databases, PRONTOE has been integrated with a three dimensional model of ISS called the Dynamic Onboard Ubiquitous Graphics (DOUG) as shown in Fig. 3. By clicking on any ontological entity within PRONTOE with a physical location on ISS, PRONTOE algorithmically directs DOUG's camera to an appropriate location to view the entity and flashes it if possible. If the exact location of the entity on ISS isn't specified, PRONTOE tries to infer a physical location using SWRL rules. For example, a power switch on ISS is attached to a mount, which is in turn attached to a cold plate. Because the cold plate has a specified location on ISS, PRONTOE can infer the mount's position, and then the power switch's location. PRONTOE can also receive updates from DOUG, so if a user clicks on a piece of equipment in DOUG, PRONTOE will display ontological information associated with it.

The end goal of PRONTOE is to have operators add equipment to the ontology. As an example of editing in PRONTOE, we will walk through a user adding a new type of gas tank assembly for the ISS. The existing gas tank assemblies are show in Fig. 4 by selecting the GasTankAssemblies class in the ontology tree. The classes are all colored to mark their different locations on ISS. The user will create a new class by clicking on the "Add Class" button in the ontology tree toolbar. The resultant dialog is show in Fig. 5. Inherited object and data properties that can be bound are show in cyan. New properties for the class can be added by clicking the appropriate plus button. In this case, the user wanted to create a new class called OxygenTankAssembly. Clicking OK on the dialog creates the appropriate axioms and marks the ontology as dirty. In Fig. 6, the user is in the process of creating and specifying a new individual of OxygenTankAssembly. Object properties are specified in the top table and data properties specified in the bottom table. If object or data property hasn't been specified, we mark the field yellow.

The first user trials of PRONTOE involved EVA flight controllers, and to a lesser extent the robotics flight controllers (known as ROBOs). The EVA flight controllers generally approved of our current development, but asked if we might build a tighter interface to the 3D graphics engine they use known as DOUG (Dynamic Onboard Ubiquitous Graphics). But they also indicated that knowledge of how EVA serviced equipment was related to the information in the others kernels, e.g., power and computer control, would be useful to them for setting up preconditions on their EVA tasks. Our ROBO flight controller was skeptical that the ROBO team would use PRONTOE for procedures, but she saw a number of potential uses for the tool, such as providing support for operational planning meetings, for collaboration among disciplines, for troubleshooting training, and for use in simulation scripting meetings.

Later in the project, we demonstrated our current version of PRONTOE to core systems flight controllers. They all were impressed with how we had pulled together data from disparate sources into one integrated view and suggested that we have a series of one-on-one knowledge engineering meetings with each of them to see if our kernels had enough key concepts modeled for the users to extend them without our help. We began those sessions with the vehicle motion control flight controller who spent an afternoon with us investigating the ontology and pointing out what was missing if he were to use it in his day-to-day operations. The resulting additions included being able to model internal items that connect to external items, allowing multiple remote power controller modules (RPCMs) in our power channel models, and adding computer control channels to augment our relations, controllerFor and controlledBy.

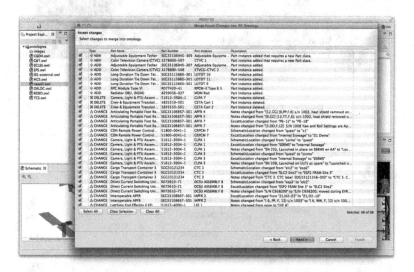

Fig. 2. PRONTOE's merge tool handling updates to the ontology from the ExCATT database.

3.2 DARPA Atlas

In addition to using PRONTOE for space systems, we have also been developing ontologies in the robot domain. TRACLabs is developing automation and control software for the simulated Atlas robot used for the DARPA Virtual Robotic

Challenge (VRC). We have defined an ontology of robot affordances of the Atlas robot to improve user understanding of the capabilities of a robot. These affordances define the perceived and actual capabilities of the robot based on Norman's definition of affordances [1]. This robot-centric ontology can be used to ground human-robot interaction about what the robot can know about itself and its environment based on what it can sense, and what the robot can do based on encoded behaviors. The ontology of robot affordances models the capabilities of a robot as Behaviors to change the robot's Stance. A Stance is a meaningful configuration of a robot's components and/or systems. For example, sit is

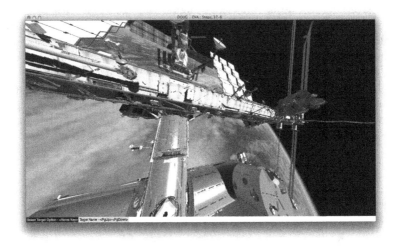

Fig. 3. DOUG, the 3D visualization of ISS.

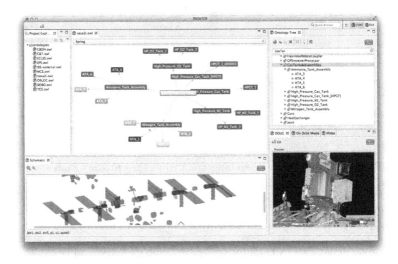

Fig. 4. The gas tank assemblies on ISS.

a Behavior of a bipedal robot that produces the Stance of sitting. A Behavior is accomplished by executing command sequences modeled as CommandLists. Each Command in the CommandList is associated with a BodyPart of the robot. The execution of a Command changes the State of the robot defined for components associated with that BodyPart. This ordered sequence of transient States is captured in a StateList. For example, a sequence of Move Commands in a Motion Behavior produces a corresponding sequence of Pose States that change the JointStates associated with the Command. The end state resulting from the

Fig. 5. Adding a new class called OxygenTankAssembly to the ontology.

Fig. 6. Adding a new instance of OxygenTankAssembly.

execution of a CommandList is a Posture Stance. Figure 7, summarizes the key concepts in the ontology of robot affordances and the properties relating these concepts.

As shown in Fig. 8, we use the PRONTOE ontology editor to visualize and inspect this ontology of robot affordances for Atlas. We automatically generate an ontology from the Unified Robot Description Format (URDF), which is an XML standard for representing a robot model. A schematic of the Atlas robot illustrates the robot components in the ontology, such as Joints and BodyParts. The user interacts with this schematic to identify concepts corresponding a particular component. For example, the user can search the ontology for all Behavior

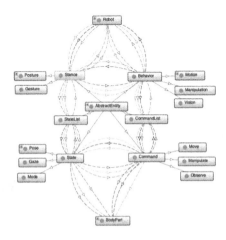

Fig. 7. Key concepts and properties in the ontology of robot affordances.

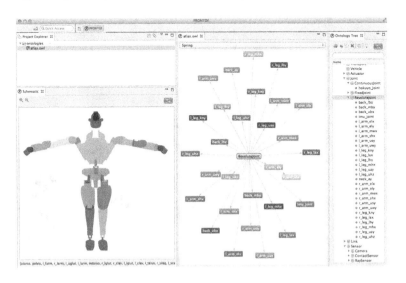

Fig. 8. The Atlas ontology created from a URDF.

instances defined for the Atlas robot. We are currently investigating the use of the ontology of robot affordances to build operator interfaces to the robot.

3.3   Autonomy Management Platform

The Autonomy Management Platform (AMP) is a project to enable one operator to control many unmanned vehicles simultaneously. This is primarily done by increasing the autonomy of each unmanned vehicle, freeing the operator to focus on more complex decision making. Achieving this objective requires significant advances in decision support, supervisory control, and human-autonomy interface concepts. Current operator interfaces require continuous attention from the operator to a single unmanned vehicle. In fact, many require the attention of several operators. In addition, designing and building the decision support and human interaction components often requires significant expertise in computer science, artificial intelligence, and software engineering. This makes the cost of developing, maintaining, and modifying these systems exorbitant. AMP's approach is to design reusable, ontology-driven decision support. As shown in Fig. 9, PRONTOE lets subject matter experts (SMEs) create and update the assets, capabilities, tasks, and environment of the various unmanned vehicles. These ontologies are used by the decision support and human interface tools, which allows for flexibility and extensibility of AMP. The benefits will be reduced cost of operations and increased productivity of the operators while saving time and money in the design, deployment, and maintenance of decision support and human interface systems for multi-asset coordination.

To control the multiple UAVs, a a tiered control architecture was implemented as shown in Fig. 10. At the top level is a task planner that reasons about

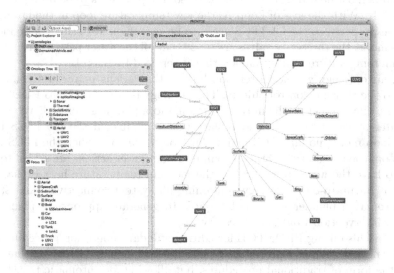

Fig. 9. AMP's ontology open for editing in PRONTOE.

resources, agent capabilities, timelines, and activities necessary to achieve high-level goals. At the next level is a task executive that runs conditionally executed scripts for individual agents to perform specific actions. Both the task planner and the task executive can read in and reason about OWL ontologies. For example, a weapon is added to an existing UAV. Those model updates would then affect both scripts and their associated planning actions. PRONTOE allows a subject matter expert who does not have any ontology background to maintain, modify, and extend the base ontology provided with the AMP system.

4   Applications

The ontology developed in PRONTOE is not an end in itself. It is designed to support a set of core autonomy capabilities. For example, we are able to translate the OWL ontology into a planning language called Planning Domain Description Language (PDDL) [2]. We then have a task planning tool, called AP [3], that can read in PDDL and use those models to schedule tasks that need to be performed. For example, let's take the overall task of replacing one of the ISS power modules. This requires subtasks of shutting down certain systems that are on that power module. The ontology describes the connectivity between the power modules and various subsystems. However, that is not enough information for the planner. For example, the internal thermal control system (ITCS) of the US Lab on the ISS. The ITCS is controlled by ISS computers S01 for primary control and S11 for backup control for a power pump of Loop A (the low temperature loop) of an external thermal control system (ETCS). Basically, the pump moves heat from inside of the ISS to outside of the ISS. Similarly, Loop B performs the same function for the medium temperature loop. When the ISS power module is shut down, backup computers must be brought up to run it. Also, Loop A needs to be brought down completely. With the loss of Loop A, the low temperature loop in the lab won't function properly, so the lab ITCS needs to be reconfigured to single loop mode. In this mode, the three-way valves are set so that all the water passes through the medium temperature heat exchanger that is serviced by Loop B.

In our ontology we model relationships of the various power and computer units to the thermal control objects. The ontology is translated into PDDL and read into the planner as the initial situation and a plan is generated. When AP generates a plan to remove and replace power module, if the lab ITCS is in single loop mode, no lab ITCS action must be taken. But when the ITCS is in dual loop, the necessary additional computer and power actions are added to the plan to have the water pass through the medium temperature heat exchanger. Without an up-to-date ontology containing all system connectivity and state, the automated planner would not be able to generate appropriate plans and manual intervention would be necessary.

We are also using PRONTOE for editing standard operating procedures for ISS. We have developed a procedure representation language (PRL) [4], that captures the form of traditional procedures, but allowing for automatic translation into code that can be executed by autonomous executives. In order to author

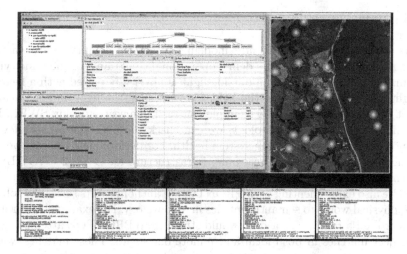

Fig. 10. AMP's tiered control architecture controlling multiple unmanned vehicles using PRONTOE's ontology as the system model.

PRL, we have developed the Procedure Integrated Development Environment (PRIDE) [5]. Procedures authored in PRIDE and output in PRL require knowledge about the available commands and state of the ISS subsystems [6]. PRIDE can read in OWL files and create drag-and-drop interfaces for system commands and to verify system state. PRONTOE makes it easy to edit the OWL files to keep the data consistent with current ISS operational needs. For example, if a new piece of equipment is added to ISS, PRONTOE can assist a flight controller in adding just enough information to the ontology to quickly build procedures for the device. If the new device is faulty, a simple change to the ontology will render all procedures that dependent on it invalid. In this way, PRONTOE can act as a verification of procedures with the systems they interact with.

5 Related Work

Ontological engineering (OE) has been a regular activity in the AI community for many years. In 1999 it was considered in its infancy for lack of use of widely accepted methodologies [7], but as late as 2007, the majority of OE researchers still did not use any methodology [8]. Yet, most OE research accepts as fundamental the need for an efficient, consistent paradigm for knowledge engineering ontologies [9].

Work on meta-theories, e.g., [10], may be considered related in that it attempts to view an ontology from a perspective of common concepts and elements. Myers' work on planning domain meta-theories [11] falls in this vein, where she discusses such things as characterizing air/land/water as "transport media", and that movement concepts involve a source and a destination. Our work on a base ontology as distinct from kernel ontologies is similar and our interactive approach will use abstraction levels to make the authoring of models easier for the user.

6   Future Work

In the short term, we plan to add modeling of system commands, telemetry, and flight rules. The flight rules define operational constraints of the underlying system. For example, on ISS an outside light is physically paired with a heater. A flight rule states to prevent the light from freezing, the light and the heater cannot both be off. This simple example can easily be written using SWRL. By streaming live data in the ontology, we can check the both the consistency of the ontology and the underlying system health. For performance reasons, we are investigating using Tractable reasoning infrastructure for OWL 2 (TrOWL) to perform stream reasoning.

We're also experimenting with PRONTOE to anchor objects sensed by robots in the physical world to semantic entities in an ontology. For example, as a robot moves around a room, sensor data would stream into PRONTOE attempting to find object matches in both local and remote ontologies. These matches would allow that robot (or future robots) to track and reason upon these objects.

7   Conclusion

By leveraging domain specific window widgets and an easy to use development environment, PRONTOE allows operators and other domain experts to develop and maintain ontologies. Importing tools and connections to a wide variety of data sources allows PRONTOE to easily capture system data from a wide variety of sources. Kernels, reasoners, and integration into version controls systems and workflows allows a user of PRONTOE to maintain large and dynamic ontologies. Useful ontologies were built for both human and robotic operations. In the future, we plan to add better modeling of system commands and telemetry. All PRONTOE domains will benefit from procedure executives and planners using the developed ontologies. Based on our case study, the benefits for NASA operations by using PRONTOE are:

1. Make available a consistent domain model that need not be reproduced for each automation application;
2. Unify the often disparate sources of EVA and Core ISS System information;
3. Provide for rapid update of ISS configuration information, thus allowing automated services to provide results based on the most recent data;
4. Provide a consistent view of the domain so as to minimize error in operating ISS;
5. Model a set of core concepts for dynamic system monitoring and control that have been proved out in disparate domains such as robotics.

Acknowledgements. This work is funded by a NASA Small Business Innovation Research (SBIR) grant. The authors grateful to Dr. Jeremy Frank of NASA Ames Research Center for his help with this project. The authors also wish to thank the numerous NASA flight controllers who have worked with us over the last several years to get the correct data into our ontology. Kevin Kusy was instrumental in creating import scripts for external databases. The development of an ontology of robot affordances for the simulated Atlas robot was funded under a DARPA Phase I Small Business Technology Transfer (STTR) contract.

References

1. Norman, D.: The Psychology of Everyday Things. Basic Books, New York (1998)
2. Fox, M., Long, D.: PDDL2.1: an extension to PDDL for expressing temporal planning domains. J. Artif. Intell. Res. **20**, 61–124 (2003)
3. Elsaesser, C., Sanborn, J.: An architecture for adversarial planning. IEEE Trans. Syst. Man Cybern. **20**(1), 186–294 (1990)
4. Kortenkamp, D., Bonasso, R.P., Schreckenghost, D.: A procedure representation language for human spaceflight operations. In: Proceedings of the International Symposium on Artificial Intelligence, Robotics and Automation in Space (i-SAIRAS) (2008)
5. Izygon, M., Kortenkamp, D., Molin, A.: A procedure integrated development environment for future spacecraft and habitats. In: Proceedings of the Space Technology and Applications International Forum (STAIF), vol. 969. Available as American Institute of Physics Conference Proceedings (2008)
6. Bell, S., Kortenkamp, D.: Embedding procedure assistance into mission control tools. In: Proceedings of the IJCAI Workshop on AI in Space (2011)
7. Lopez, F.M.: Overview of methodologies for building ontologies. In: Proceedings of the IJCAI Workshop on Ontologies and Problem-Solving Methods (1999)
8. Cardoso, J.: The semantic web vision: where are we? IEEE Intell. Syst. **22**(5), 84–88 (2007)
9. Soares, A., Fonseca, F.: Building ontologies for information systems: what we have, what we need. In: Proceedings of iConference (2009)
10. Herzig, A., Varzincak, I.: Metatheory of action: beyond consistency. Artif. Intell. **171**, 951–984 (2007)
11. Myers, K.: Domain metatheories: enabling user-centric planning. In: Proceedings of the AAAI Workshop on Representational Issues for Real-World Planning Systems (AAAI Technical Report WS-00-07) (2000)

Methodology to Develop Ontological Building Information Model for Energy Management System in Building Operational Phase

Hendro Wicaksono, Preslava Dobreva, Polina Häfner$^{(\boxtimes)}$,
and Sven Rogalski

Institute for Information Management in Engineering,
Karlsruhe Institute of Technology, Zirkel 2, 76131 Karlsruhe, Germany
{hendro.wicaksono,preslava.dobreva,polina.haefner,
sven.rogalski}@kit.edu

Abstract. Energy consumption in building sector has been taking a significant percentage of the total energy consumption on earth. This is due to the development of more advanced and sophisticated building appliances to fulfil the comfort requirements. The EU has responded this trend by requiring zero CO2 emission in building by 2020 and by supporting innovative research approaches for improving energy efficiency in buildings with still considering inhabitants comfort. This chapter describes an approach to develop an intelligent system for building specific energy management that allows occupants and facility managers to monitor and control the energy consumption and also detects the energy wasting points. In this chapter, we explain the methodology to develop ontology based information model for building energy management offering expressive representation and reasoning capabilities. We also highlight an approach to develop the ontology as the knowledge base providing the intelligence of the system. Furthermore we show the improvement of the energy performance analysis with the help of ontology based approach.

Keywords: Energy efficiency in building · Building information model · Ontology engineering · Ontology population

1 Introduction

A study observing building energy consumption held in 2007 showed that the public and residential buildings represent more than 40 % of the whole energy consumption in European Union, of which residential use represents 63 % of total energy consumption in buildings sector [1]. The energy price has been rising due to high building operational costs, and shortage of fossil energetic resources. These reasons force companies and private persons to organize their behaviour in more energy-efficient way and to look for intelligent and long term solutions.

There are several technical possibilities and products on the market aiming to improve energy usage efficiency designed for business and public buildings. European Union has issued the Directive 2002/91/EC about overall energy efficiency of buildings. The directive aims to improve energy efficiency by taking into account outdoor climatic

© Springer-Verlag Berlin Heidelberg 2015
A. Fred et al. (Eds.): IC3K 2013, CCIS 454, pp. 168–181, 2015.
DOI: 10.1007/978-3-662-46549-3_11

and local conditions, as well as indoor climate requirements and cost-effectiveness. The building energy management systems are acknowledged as a significant source for energy costs reduction up to 30 % [2].

Furthermore, in the future, energy savings in buildings can be increased by intelligence improvement of building automation systems. This kind of method is considered in the literature important as the conventional thermal insulation of walls or insulating glazing to improve energy efficiency in buildings [3, 4]. Recently low cost and low energy consuming building automation technologies have already been developed. Recent technologies offer energy measurement and sensors by using small chips that consume less than 10 mW [5]. These chips can be easily installed in building without modifications.

By using these devices, extra energy consumption can be avoided. However, the energy efficiency could be effectively improved, if they are supported by an intelligent software system. In this paper we propose an intelligent system for energy management in buildings by connecting building automation systems and using intelligent information model. The existing Building Information Model (BIM) standards only defines definitions, dictionary and information structure. In this paper, we extend the BIM standard to have more expressiveness and reasoning capabilities. We incorporate rules and axioms to achieve these.

The information model is used in the knowledge base that allows intelligent analysis on the relations between energy consumption, behaviour model (activities and events in the building), building related information (geometry, boundary conditions, etc.) and surrounding factors, such as temperature, weather condition, occupant habits and behaviour. The knowledge base is represented using ontologies. We also introduce the ontology modelling method that is aligned with existing building information modelling standard called Industry Foundation Class (IFC).

This paper is organized as follows. In Sect. 2 we discuss the state of the art and related work. Next, we introduce the developed system of intelligent energy management in Sect. 3. Section 4 describes our approach in generation of ontology as the center point of our intelligent system. In Sect. 5 we give overview how the energy analysis is performed using ontological query. Finally we make our conclusion in Sect. 6.

2 Related Work

In 2009 Electric Power Research Institute USA conducted research of electricity consumption feedbacks in household. They categorized feedback mechanism based on the information availability into standard billing, enhanced billing, real-time feedback, real-time plus feedback, etc. [6]. The research showed that real-time plus feedback leads to the best improvement of energy conservation comparing to the other feedback mechanism, despite the higher cost of implementation. Real-time feedback allows users to monitor their energy consumption and/or control appliances in their home through building automation system (BAS) and home area network (HAN).

Each building automation technology may offer different functionalities, and has its own strength and weakness. For example, the technology digitalSTROM offers good functionality in energy metering, but it does not support occupancy metering. In order to achieve comprehensive energy management by taking into account as much as related conditions and factors, an integration of different building automation systems is required.

Ontology can be used as generic model to facilitate the integration [7]. The ontology is not only used to describe functionalities of building automation systems, but also to represent states of building, and relations with behaviour model and surrounding factors. In this paper, we introduce also method to generate ontology components semi-automatically based on user events and building specific information.

An ontology based approach was introduced in EU funded project ISES based on description logic ontology containing rules and constraints. The ontology is represented with OWL-DL combined with SWRL and is used as the information model for integrated lifecycle energy management in building. The approach addressed not only interoperability issues with other systems, but also allowed quality control by end user using knowledge-based management methods [8].

The EU funded project HESMOS developed an ontology-equipped framework to address the integration of distributed and heterogeneous data from ICT building energy systems. The framework comprises IFC-BIM as a central integration part and a link model to bind the distributed data together. The core link model is represented with OWL, which includes the capabilities of model management and decision support [9]. Both EU projects do not strongly consider the alignment of existing standards, for instance IFC, with the developed information model. They do not include the modelling of occupant behaviour as one of the factors that affects the energy consumption in the building. This paper introduces an approach that addresses these points.

3   Overview of the Concept

In this paper, we propose an intelligent system, which considers different aspects of a building. The system allows the users to have an integrated view of energy consumption in their apartment, office, as well as in entire building. With the help of building automation system and other metering systems installed on the site the energy consumption can be evaluated in different detail levels and quality, for instance, energy consumption per appliance, per group of appliances, per zone in the building, or per user event [10]. Figure 1 depicts the designed approach of the energy management system developed in our recent research project.

As seen in Fig. 1, the generic ontology is created by a building information modelling expert. The generic ontology represents domain knowledge for building holistic energy management. It contains definitions, terminologies (T-box), and taxonomies that are aligned with IFC. The information model contained in the generic ontology is applicable in any building. The generic ontology is then instantiated and enriched with building specific information resulting building specific ontologies. The development of ontology will be explained further in Sect. 4.

The data collector and aggregator module is developed for collecting energy data and sensor data from different building automation systems installed in the building. It contains an interface to communicate with different building automation logic control

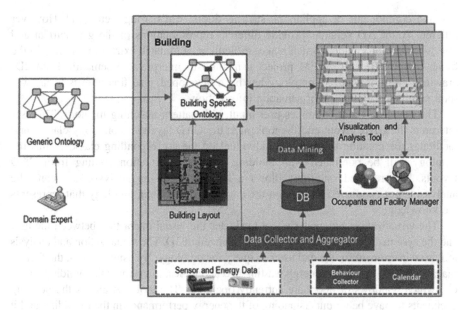

Fig. 1. Overview of the developed energy management framework.

units or gateways via web services. The module is also responsible to collect occupant activities or behaviours in the building. For this, a web-based interface to model occupant activities is developed. Furthermore, an interface to the calendar reflecting the schedules and activities of the occupants is being developed.

The collected data are aggregated and stored in a database. In order to allow visual representation of energy consumption data, we perform necessary data pre-processing such as removing erroneous values, data transformation, data selection and data conversion. The data are prepared to enable an energy consumption analysis in different criteria based on relation between rooms, appliances, time, and user events. Therefore it allows a data-driven analysis that is conducted directly on the collected data by performing SQL-query, simple calculation, or visualization, for instance, energy consumption per time unit and each appliance. The data is provided in such a form to enable the execution of data mining algorithm for finding the energy usage pattern.

The data mining module evaluates energy consumption data that are collected and aggregated and extracts the knowledge in forms of patterns and relationships from the data. Through this module, energy consumptions can be related to device levels, room, and time, which in addition to that, can be combined with relation to occupant behaviours and surrounding conditions. As seen in Fig. 1 the extracted knowledge is incorporated in the building energy management knowledge base represented by building specific ontology. The relationships between data are modelled as rules and represented as SWRL.

A building plan is usually drawn in 2D using CAD software applications, such as AutoCAD. Unfortunately, 2D-drawings only contains geometrical information, for instance, lines, points, curves, circles, etc. the CAD layouts cannot describe any semantics of building components contained in the drawing. We can still understand semantic of the drawing because we already know the symbols representing certain building components or appliances, such as doors, walls, fridge, etc. [11]. However different AutoCAD versions provide different representations of the geometrical and object-related information which makes difficult an automated extraction of data. In the frame of the FP7 KnoHolEM project a method to interpret the semantic from 2D-drawings and populate the ontology has been developed. The final aim is to allow a semi-automated extraction of semantic information.

The developed tool combines user input with pattern matching methods. The user interprets the CAD layout and the tool maps the CAD layouts to ontology classes and facilitates the creation of ontology individual on the corresponding classes. Thus the ontology will be populated with building specific information coming from CAD layouts. The resulted ontology allows a knowledge-driven analysis. It means the analysis is not conducted directly on the data, but by utilizing ontology that represents the knowledge.

The visualization and analysis tool facilitates the visual interaction between the user and the system. The building geometry is visualized in 3D. The visualization and analysis tool is an instrument for the end users to query the ontology. By using the tool the facility manager is able to identify energy wasting and anomalies, to examine the building states, e.g. which windows or doors are currently open, etc. The tool also allows the building occupants to have better understanding of the energy performance in their building and it also empower their engagement in balancing comfort and energy efficiency. The occupant can have an overview of the energy efficiency of the zones, where he is responsible for, thus it increases his awareness to avoid energy wasting and achieve more energy efficiency.

4   Ontology Development

The knowledge base as the centre point of the developed energy management system is represented in OWL (Web Ontology Language), a W3C specified knowledge representation language [12]. Basically there are two types of ontologies that we develop. We develop a generic ontology representing a common information model for building energy management, and then it is populated and extended with building

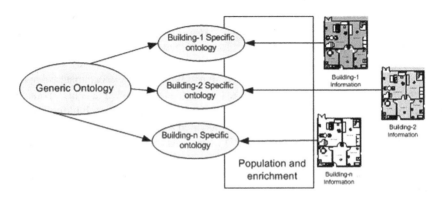

Fig. 2. Generic ontology and building specific ontology.

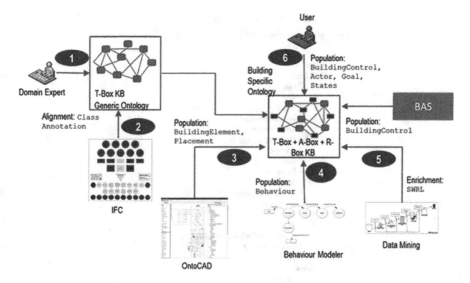

Fig. 3. Ontology development process.

specific information resulting more building specific ontologies corresponding to the specific buildings. It is illustrated in Fig. 2.

In our work, there are six main steps to develop the ontology resulting a building specific ontology. The steps are illustrated in Fig. 3. The following subsections explain each step to develop the ontology.

4.1 Definition of Ontology Main Resources

The ontological classes as well as their attributes and relations representing the resources needed for the energy management in buildings are created manually by experts. It is depicted as step 1 in Fig. 3. The ontology containing these hand-crafted elements is called generic ontology. It only contains the ontological classes or Tbox components that describe the knowledge structure, definitions and terminology. It does not contain any ontological individuals or Abox components and contains no building specific information. Figure 4 depicts the ontology main classes representing the different resources needed for the energy management in building.

The class BuildingElement models the building structures that are observed, examined and analyzed in energy management activities. The building elements are passive entities which have state, but do not have capabilities to measure or to observe their own states. The class BuildingElement and its sub classes represent the fundamental of Building Information Model (BIM). It is aligned with the domain layer in IFC2x4. The class BuildingControl indicates the entities related to building automation system elements in the building. It represents the sensors, actuators, controller, alarm, etc., which are elements of a building automation system. It has capabilities to measure, to observe, and to control the state of BuildingElement. It aligns with entities in the IFC2x4 domain IfcBuildingControlsDomain.

Fig. 4. Ontology main classes.

The class Actor represents the human actors having behavior that can affect the states of BuildingElement. The Actor can be organizations or persons, who have name, postal address, telecom address, etc. It is aligned with the IfcActorResource in IFC2x4. The class Behaviour represents behaviour performed by Actor. The Behaviour can affect the state of BuildingElement. There are two methods to model the behavior in the building, i.e. bottom up and top down. This will be described further in Sect. 4.4. The class State represents the state of BuildingElements. It can be divided as ComplexState and SimpleState. Examples of ComplexState are ComfortState and EnergyEfficiencyState, whereas examples of SimpleState are WindowState, DoorState, etc.

4.2   Explicit IFC-OWL Mapping

The modern building drawing already contains semantic information represented using IFC entities. To support the ontology population from IFC drawing containing semantic information, we develop a method to map the IFC entity to OWL class explicitly. We use class annotation to perform the mapping. As seen in Fig. 5, the class annotation correspondToIfcEntry maps the IFC entity, for instance IfcWall to OWL class Wall, and the class annotation correspondToIfcEnumerationElement maps the IFC enumeration value STANDARD to OWL class StandardWall.

The explicit IFC-OWL class mapping accelerates the ontology population process from IFC drawing containing semantic information. If we have an entity in our IFC drawing, by querying the ontology using SPARQL, we can find the corresponding OWL class. For example, the following SPARQL statement finds the OWL class Wall, if we have the IFC entity IfcWall. (see Fig. 6).

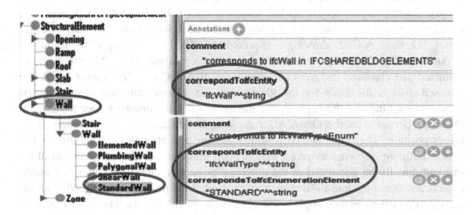

Fig. 5. IFC-OWL explicit mapping.

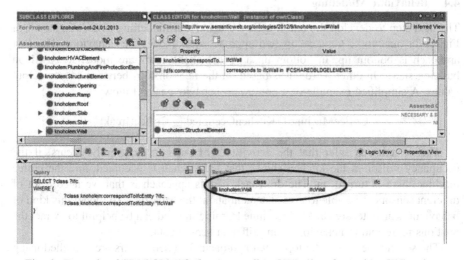

Fig. 6. Example of SPARQL to find corresponding OWL class from given IFC entity.

4.3   Population of BuildingElement Using Developed Tool OntoCAD

A layout of a building is commonly created as a two dimensional drawing or sketch using CAD software applications, such as AutoCAD. Further AutoCAD-based software tools are used to plan and model many domains of a building, such as ventilation, heating, access controls and photovoltaic [13]. The number of elements in a sketch and its complexity may vary [14]. CAD is one of the easiest and oldest technologies used in the industry. At the same time CAD is the least effective technology when it comes to accomplishing building information modelling because it demands a great amount of effort. Recent research shows evidence that it does not ensure high quality, reliable, and coordinated information that the higher level of BIM produces [15].

We develop a tool to extract the semantic information from CAD drawing and populate the ontology using the extracted semantic information. The tool is called OntoCAD. First, A CAD drawing is exported to DXF. Then we import the primitives in our tool OntoCAD from the exported construction layout files. The primitives are extracted and clustered in layers like they were in AutoCAD. This vector based data representation is the basis for the viewer and the pattern matching algorithms. An important user input at the beginning is the mapping of the ontology specific data and object properties with the OntoCAD functions, for instance the computation of the object position. The implemented pattern matching and classification algorithms recognize building elements based on user defined templates. The user selects an object that can directly be populated in the ontology or he can choose to search for similar objects and then populate all of them at once. He has the possibility to directly validate the result and if necessary apply some corrections to the results. The results are continuously and automatically saved to the ontology.

4.4   Behaviour Modelling

The next step to develop the ontology is the behaviour modelling as seen in step 4 of Fig. 3. In our work, we develop two approaches to model the behaviour. The first approach is bottom up. In bottom up approach, behaviours are modelled based on building states. In other words, it is based on the relationships between sensor output values. A simplified example of a bottom up behaviour is as follow:

$$\text{CoffeeMachine} \sqcap \text{KitchenOccupied} \sqsubseteq \text{CoffeeBreak} \tag{1}$$

The system can infer that the activity CoffeeBreak currently occurs, if the occupancy sensor in the kitchen shows that the kitchen is occupied, and the coffee machine there is turned on. The drawback of this approach is that we need a lot of different sensors to be able to give statement about the activities. To model this kind of behaviour is an extensive task. A machine learning method can be helpful to extract the relations representing behaviour from different sensor data.

The second method is the top down approach. The behaviours are modelled using common modelling approaches, such as UML or BPMN diagram. Behaviours are defined as sequences of different sub activities. The drawback of the approach is the impossibility to identify the occurring activities automatically. The system cannot know what kinds of activities are currently occurring in the building. Therefore it needs a manual activity instantiation from the occupants. This kind of manual activity logging causes extensive work to the occupants.

4.5 SWRL Generation Using Data Mining

In our work, data mining algorithms are used to identify energy consumption patterns and their dependencies. Data mining is defined as the entire method-based computer application process with the purpose to extract hidden knowledge from data [16]. In our work, we use different data mining procedures to generate knowledge for recognition of energy usage anomalies, energy wasting and also to predict the energy consumption.

In this work, we relate the energy consumption with the behavioural occupant's pattern in the building. We aim to recognize energy consumptions that do not occur normally in the building. To perform this task, we have to know how energies are consumed normally regarding occupant events and surroundings. For example, normally when an occupant is currently working and the outside temperature is comfortable, let us say greater than 20 degrees Celsius, total energy consumption in the building is low, for instance lower than 10 kWh. If in the same pre condition total energy consumption in building is more than 10 kWh, then it is considered as a usage anomaly.

It is difficult for users if they always have to log their activities. In our work we use simple sensors to recognize user activities automatically. Simple sensor can provide important hint about user activity. For instance, an occupancy sensor in a kitchen can strongly give a clue whether somebody is currently cooking. Of course it should be combined with information of appliance states in the kitchen.

The rules representing normal energy consumptions are obtained through data mining classification rules algorithm. The algorithm is based on a divide-and-conquer approach. The created rule (2) shows the probability of 67 % about how often it could happen if the activity is working while outside temperature is greater than 20 degree with total energy consumption is lower than 10. This value is called confidence. The rules described in (2) represent a condition that normally occurs. The rule is transformed to (3), in order to represent an anomaly condition, by negating the consequent part of the rule [11].

$$\text{Event} = \text{``Working''} \wedge \text{OutsideTemperature}_{-} \geq 20$$
$$\rightarrow \text{TotalEnergyConsumption} < 10 \, (\text{conf} : 0.67) \tag{2}$$

$$\text{Event} = \text{``Working''} \wedge \text{OutsideTemperature}_{-} \geq 20$$
$$\rightarrow \text{TotalEnergyConsumption} \geq 10 \tag{3}$$

The rules created by data mining algorithm are stored in ontology as SWRL. SWRL rule (4) represents the transformed rule (3), which is stored in ontology. The class UsageAnomaly is a sub class of ComplexState.

UserBehaviour(?e) ∧ hasName(?e, "*Working*") ∧ OutsideThermometer (?ot)∧
hasValue(?ot, ?otv) ∧ swrlb : greaterThanOrEqual(?otv, 20) ∧ SmartMeter(?sm) ∧   (4)
hasValue(?sm, ?smv) ∧ swrlb : greaterThanOrEqual(?smv, 10) → UsageAnomaly(?e)

The rules resulted from data mining algorithm do not always have 100 % confidence. Therefore we represent the rules as SWRL in order to enable verification by using SWRL editor such as Protégé. The editor enables users to add, modify and delete the resulted rules.

4.6   Modelling States

In our work we divide the state to SimpleState and ComplexState. To model both kinds of states, we formulate a set of competency questions. Fehler! Verweisquelle konnte nicht gefunden warden. Gives examples of competency questions in order to model a complex state EnergyInefficient.

The SWRL illustrated in (5) implies a complex state of energy inefficiency, if a window is opened, a heater is turned on, and they both are located in a closed zone. The necessary classes are created based on the formulated competency questions. For example, as seen in (5), the classes HeaterSwitchOn and OpeningOpen are created based on competency questions Q1.1 and Q1.2.

Q1.1HeaterSwitchedOn(?h) ∧ Q1.2OpeningOpen(?o) ∧ Inside(?z) ∧

Q1.3isLocatedIn (?o, ?z) ∧ Q.1.3isLocatedIn (?h, ?z) → Q1EnergyInefficient(?h)

$$(5)$$

The simple state class OpeningOpen is represented as axiom (6). It implies that if an opening sensor gives the value true, and it is installed on a certain opening, it can be inferred that the opening is currently open Table 1.

Table 1. Competency questions to model states.

Question-ID	Competency questions	Example of answers
Q1	Which heaters are currently in **energy inefficient** state in the building?	Heater2, Heater3
Q1.1	Which heaters are currently **switched on**?	Heater1, Heater2, Heater3
Q1.2	Which openings are currently **open**?	Window2, Door3, Window4
Q1.3	Which heaters and openings are **currently located in a same closed zone**?	Heater1 and Door1, Heater2 and Window2, Heater3 and Door3, Heater4 and Window4

Opening and (hasSensor some (OpeningSensor and (hasBinaryValue value true)))
⊑ OpeningOpen
$$\quad(6)$$

Analogue to the `OpeningOpen` the simple state `HeaterSwitchedOn` is represented using the axiom (7).

Heater and (hasSensor some (EnergyMeter) and (hasAnalogValue > 0))
⊑ HeaterSwitchedOn
$$\quad(7)$$

5 Energy Analysis Through Ontology Query

In knowledge base represented in ontology, all conditions of energy wasting and anomalies are represented as SWRL. Periodically data acquisition module requests real-time data from building automation gateway. These data contain states given by all installed building automation devices. SWRL rules are used to decide whether these incoming data correspond to complex states, e.g. energy inefficiency and anomaly condition. We develop a rule engine based on SWRLJessBridge to support the execution of SWRL rules combined with Protégé API that provides functionality in managing OWL ontology.

First the attribute values of relevant ontological instance are set to values corresponding to incoming data. For example, if opening sensor attached to window gives a state "Open", then the attribute `hasState` of corresponding ontology instance of concept `OpeningSensor` is set to "Open". After that the rule engine executes the SWRL rules and automatically assigns individuals to the ontology classes defined in the rule's consequent. For example for rule (5) the instance of class Heater is

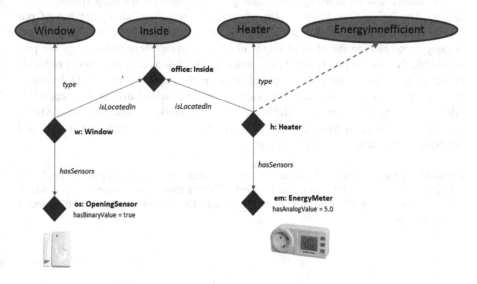

Fig. 7. Example of energy inefficiency identification through ontological reasoning.

additionally assigned to `EnergyInefficient` class (see Fig. 7) and for rule (4) the instance "Working" of class `UserBehaviour` to class `UsageAnomaly`. SPARQL is used to evaluate whether energy inefficient condition or energy usage anomaly occurs. Which appliances cause the energy wasting can be retrieved as well. It is performed by querying all individuals of `EnergyInefficient` or `Usage-Anomaly` class. If individuals of these classes are found, the affected individuals are visualized and marked in the visualization and analysis tool. With this mechanism, user can have more awareness in order to avoid more energy wasting. SPARQL is also used to perform further analysis for example to retrieve all windows or doors that are currently open. The SPARQL interface is part of visualization and analysis tool.

6   Conclusion

In this paper we have presented a system of comprehensive intelligent energy analysis in building. In the developed system, we combined classical data-driven energy analysis with novel knowledge-driven energy analysis that supported by ontology. The analysis is performed on information collected from building automation devices. The ontology supported analysis approach provides intelligent assistance to improve energy efficiency in households or public buildings, by strongly considering individual user behavior and current states in the building. Users do not have to read the whole energy consumption data or energy usage profile curves in order to understand their energy usage pattern. The system will understand the energy usage pattern, and notify user when energy inefficient conditions occur.

We have presented also an approach to develop the ontology as the knowledge base of the intelligent energy management system. There are different methods and steps to generate the ontology. We differentiated between generic ontology as generic information model and building specific ontology containing the building specific information. The generic ontology is aligned with IFC to allow interoperability of our system with existing industry standards. We introduced the main resources of the ontology representing the main elements in energy management in building. We presented briefly a tool called OntoCAD to perform semi-automatic extraction of semantic information and population of building elements in the ontology from CAD drawings. We also introduced our approach to model occupant behaviour and building states that affect the energy performance of the building. In this work, we also integrated SWRL rules that are extracted from different data, i.e. energy consumption, sensor data, and behaviour using data mining algorithms.

Acknowledgements.   Research activities presented in this paper have been partially funded by the German Ministry of Education and Research (BMBF) through the research project KEHL within the program KMU-Innovativ and the European Commission trough the FP7 research project KnoHolEM.

References

1. Balaras, C.A., Gaglia, A.G., Georgopoulou, E., Mirasgedis, S., Sarafidis, Y., Lalas, D.P.: European residential buildings and empirical assessment of the Hellenic building stock, energy consumption emissions and potenctal energy savings. Build. Environ. **42**(3), 1298–1314 (2007)
2. Smithson, D.: Financing a route to reducing energy costs. February 07, 2013. http://www. modbs.co.uk/news/fullstory.php/aid/11483/Financing_a_route_to__reducing_energy_ costs_.html
3. Lonmark, Energieeffizienz durch LON. March 14, 2008. http://www.lonmark.de/technik/ energieeff.asp
4. Spelsberg, J.: Chancen mit dem Gebäudeenergiepass. Energie sparen durch Gebäudeautomation. Technik am Bau **37**, 74–77 (2006)
5. Watteco, WPC Product Description (2009)
6. Neenan, B., Robinson, B., Boisvert, R.N.: Residential Electricity Use Feedback: A research Synthesis and Economic Framework. EPRI, Palo Alto (2009)
7. Reinisch, C., Granzer, W., Praus, F., Kastner, W.: Integration of heterogeneous building automation systems using ontologies. In: Proceedings of 34th Annual Conference of the IEEE Industrial Electronics Society (IECON 2008), pp. 2736–2741 (2008)
8. Scherer, R.J., Katranuschkov, P., Kadolsky, M., Laine, T.: Ontology-based building information model for integrated lifecycle energy management, EEbuilding data models – energy efficiency vocabularies and ontologies. In: Proceedings of the European Conference of Product and Process Modelling (ECPPM) 2012, Reykjavik, Iceland, 25–27 July 2012, pp. 30–41 (2012)
9. Guruz, R., Katranuschkov, P., Schrerer, R.J., Kaiser, J., Grunewald, J., Hensel, B., Kabitzsch, K., Liebich, T.: Ontological specification for the model integration in ICT building energy systems, EEBuilding data models – energy efficiency vocabularies and ontologies. In: Proceedings of the European Conference of Product and Process Modelling (ECPPM) 2012, Reykjavik, Iceland, 25th–27th July 2012, pp. 6–29 (2012)
10. Wicaksono, H., Rogalski, S., Kusnady, E.: Knowledge-based intelligent energy management using building automation system. In: IPEC 2010 Conference Proceedings, pp. 1140–1145 (2010)
11. Wicaksono, H., Aleksandrov, K., Rogalski, S., An intelligent system for improving energy efficiency in building using ontology and building automation systems. In: Kongoli, F. (ed.), Automation, pp. 531–548. InTech, Open Access Publisher (2012)
12. Smith, M., Welly, C., McGuinness, D.: OWL web ontology language guide. In: W3C Recommendation, February 2004
13. Krahtov, K., Rogalski, S., Wacker, D., Gutu, D., Ovtcharova, J.: A generic framework for life-cycle-management of heterogenic building automation systems. In: 19th International Conferemce (FAIM 2009) Proceedings to Flexible Automation and Intteligent Manufacturing (2009)
14. Donath, D.: Bauaufnahme und Planung im Bestand, pp. 35–36. Vieweg + Teubner Verlag, Wiesbaden (2008)
15. Vanlande, R., Nicolle, C., Cruz, C.: IFC and building lifecycle management. Autom. Constr. **18**(1), 70–78 (2008)
16. Kantardzic, M.: Data Mining: Concepts, Methods, and Algorithms. John Wiley, Models (2003)

Combining Ontological and Qualitative Spatial Reasoning: Application to Urban Images Interpretation

François de Bertrand de Beuvron[1]([✉]), Stella Marc-Zwecker[2],
Cecilia Zanni-Merk[1], and Florence Le Ber[3]

[1] Icube Laboratory, BFO Team, INSA de Strasbourg, CNRS, Strasbourg, France
{debeuvron,merk,stella}@unistra.fr
[2] Icube Laboratory, BFO Team, Strasbourg University, CNRS, Strasbourg, France
[3] Icube Laboratory, BFO Team, Strasbourg University/ENGEES, CNRS,
Strasbourg, France
florence.leber@engees.unistra.fr

Abstract. In this paper, we develop a qualitative spatial reasoning (QSR) applied to the interpretation of urban satellite images. This reasoning is integrated within an ontology of urban objects, which has been implemented in cooperation with expert geographers. The spatial concepts mainly include the set of topological relations of the Region Connection Calculus theory (RCC8), and the set of computational primitives (CM8) which have been defined for computing the RCC8 relations on raster images. Our approach relies on a reified representation of the RCC8 relationship and of the CM8 primitives, within a lattice of concepts, implemented in OWL (Ontology Web Language). It provides a straightforward representation of concepts corresponding to conjunctions or disjunctions of spatial relations, and thus offers the advantage to overcome some drawbacks of the existing approaches in OWL, where spatial relations are represented as roles. Indeed, the OWL language does not allow the expression of the disjunction of roles. We can then implement a reasoning on the RCC8 relationship, which in particular allows the computation of the composition table and its transitive closure. As the reification of roles precludes the use of role's properties, such as symmetry and transitivity, we propose to implement RCC8 inferences through SWRL rules (Semantic Web Rule Language).

1 Introduction

The increasing availability of High Spatial Resolution satellite images is an opportunity to characterize and identify urban objects. Image analysis methods using object-based approaches relying on the use of domain knowledge are necessary to classify data. A major issue in these approaches is domain knowledge formalization and exploitation. The use of formal ontologies seems a judicious choice to deal with these issues. OWL [8] is a major language to implement such ontologies.

© Springer-Verlag Berlin Heidelberg 2015
A. Fred et al. (Eds.): IC3K 2013, CCIS 454, pp. 182–198, 2015.
DOI: 10.1007/978-3-662-46549-3_12

We have developed an ontology concerning urban objects (streets, houses, worker or residential housing, etc.) to assist the experts in their interpretation of satellite images and to be included, in the future, in a processing chain whose goal would be the automatic interpretation of the semantics of satellite images [4].

In our ontology, the urban objects are defined by intension, that means they are derived from the conceptualization of a dictionary defined by expert geographers. However, to take into account the actual data from the image, it is necessary, also, to conceptualize spatial relations between the objects. The RCC8 model [17] defines eight basic topological relations between spatial regions. Our proposition is to reify and implement these relations in OWL, and to use the SWRL rules for reasoning on spatial relation between objects and help the interpretation of complex structures on urban satellite images.

The reification of the topological relations is needed because of several drawbacks in OWL, as highlighted by several authors [20]. Reference [12] presented one of the first attempts to represent RCC8 in OWL, proposing to extend the OWL reasoners with the functionality to operate with reflexive roles. Reference [9] claimed that OWL lacked essential features such as role negations, conjunctions, disjunctions and role inclusion axioms, to effectively represent RCC8. They proposed to use a more specific logic to express some of those constructs. Reference [11] formalized 2D spatial concepts and operations into a spatial ontology, implemented as a plug-in for Protégé[1]. Unfortunately, the spatial relations retained by these authors are not the RCC8 ones, although the querying possibilities using SWRL rules seem an interesting approach, close to ours. Reference [1] proposed an ontology for representing and reasoning over spatio-temporal information in OWL. The ontology enabled representation of static as well as of dynamic information, such as objects whose position evolves in time and space. It was built upon well established standards of the semantic web (OWL 2.0, SWRL), as the one we propose here. However, our approach focuses mainly on the definition of the RCC8 relations and gives a complete specification of the TBox and ABox axioms and a minimal set of SWRL rules allowing, in this way, complete reasoning over any composition table.

The paper is organized as follows. After recalling the principles of the RCC8 theory, we present a set of eight primitives, or computational operations (called CM8), that are calculated by image processing routines. In Sect. 3, we present our proposition that is to reify the RCC8 relationships and the CM8 primitives in a lattice of concepts implemented in OWL. We then propose to use the SWRL rules [10] in order to reason about these concepts, and in particular, in order to deduce the composition table of the RCC8 relationships and its transitive closure. Section 4 is a conclusion.

2   The RCC8 Model for Qualitative Spatial Reasoning

2.1   The RCC8 Set of Topological Relations

The set of eight topological relations defined in the RCC8 theory provides a conceptual basis for qualitative spatial reasoning [17]. These relations are binary

---

[1] http://protege.stanford.edu.

and they apply to a couple of regions x and y in a n-dimensional space. The elementary RCC8 relations are exhaustive and mutually exclusive, which means that any configuration of two spatial regions can be described by this set, and that if one of these relations is true, then the others are false. The eight RCC8 relations are:

- $EQ(x, y)$ "x is identical to y"
- $TPP(x, y)$ "x is a tangential proper part of y"
- $TPP^{-1}(x, y)$ "y is a tangential proper part of x"
- $NTPP(x, y)$ "x is a non-tangential proper part of y"
- $NTPP^{-1}(x, y)$ "y is a non-tangential proper part of x"
- $PO(x, y)$ "x partially overlaps y"
- $EC(x, y)$ "x is externally connected with y"
- $DC(x, y)$ "x is disconnected from y"

The following mathematical operations are useful to make inferences on topological relations:

- the *inverse* relation of a relation r is the relation r^{-1} such that $\forall x, \forall y, r(x, y) \Leftrightarrow r^{-1}(y, x)$
- the relations r_1 and r_2 are *disjoint* if $\forall x, \forall y, r_1(x, y) \Rightarrow \neg r_2(x, y)$
- the *complement* of a relation r is the relation rc such that: r and rc are disjoint and $\forall x, \forall y, r(x, y) \lor rc(x, y)$ is true.
- given three spatial regions x, y, z, and a pair of relations r_1 and r_2, the *composition* of $r_1(x, y)$ and $r_2(y, z)$ is the disjunction $r(x, z)$ of all the possible relations holding between x and z.

The composition operation is particularly interesting because it allows the inferences of the possible relations between regions x and z from the known relations holding between the regions x and y on the one hand, and between the regions y and z on the other hand. The rules of composition on the topological RCC8 relations are represented in composition tables [17].

2.2　The CM8 Set of Computational Operations

In many situations, methods are needed to check topological relations on images or spatial databases. Computational operations have been defined in [3,6]. They are based on the interiors and boundaries of spatial regions and are linked to formal models of topological relations. They allow to establish the link between these formal models and the actual data from the image (problem known as the semantic gap).

In [6], a method is defined to deal with vector data, that are mainly used in geographical information systems. A n dimensional x region is characterized by its $x°$ interior set (same dimension) and its ∂x boundary set ($n - 1$ dimension). Intersecting these sets for two regions allows to define four operations: $\partial x \cap \partial y$, $x° \cap y°$, $\partial x \cap y°$ et $x° \cap \partial y$. A topological relation between two regions is then characterized in a unique way by the result values of the four operations.

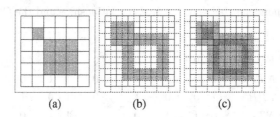

(a)   (b)   (c)

Fig. 1. Defining the boundary of a region: (a) the interior of a region is made of real pixels (hatched); (b) the boundary of the region is made of abstract pixels standing across the real pixels (inverse hatched); (c) the combination of the interior and the boundary [14].

More recently, [5] proposed another set of operations based on the regions themselves, their interior and their boundary sets. The so-called ID-model uses set intersections, $x° \cap y°$, $\partial x \cap \partial y$, and set differences $x - y$, $y - x$.

These approaches are very interesting since they allow to express the RCC8 relations in terms of necessary and sufficient conditions on the regions. However a characterization of the regions in terms of interior and boundary is needed and this is a problem on raster images. When the raster representation is considered on its own, the boundary of a region can be defined by the pixels that are externally connected to the region: if the region is made of only one pixel, then its boundary is made of its eight neighbours. The boundary can also be internal, and composed of the pixels that are externally connected to the complementary set of the region (its exterior). These definitions do not meet the Clarke's definition of connection, *x and y share a point* [2], that underlies RCC8 relationships: if the boundary is internal, two externally connected regions do not share a boundary point; conversely, if the boundary is external, two disconnected regions may share a boundary point.

In [14] the boundary is defined by *abstract pixels*, standing across four real pixels, as shown in Fig. 1. The boundary intersection of two regions is then easily obtained, but relies on two images, the original image – containing the regions – and the boundaries image. Based on this representation four computational operations were introduced: the intersection of the interior sets, $x° \cap y°$; the intersection of the boundary sets, $\partial x \cap \partial y$; the two differences of the interior sets, $x° - y°$ and $y° - x°$.

From these four operations were derived the eight following conditions, called CM8 primitives [14]:

- $x° - y° = \emptyset$, x is a part of y, denoted by $P(x, y)$
- $x° - y° \neq \emptyset$, x is not a part of y, denoted by $NP(x, y)$
- $y° - x° = \emptyset$, x contains y, denoted by $P^{-1}(x, y)$
- $y° - x° \neq \emptyset$, x does not contain y, denoted by $NP^{-1}(x, y)$
- $x° \cap y° = \emptyset$, x is discrete y, denoted by $DR(x, y)$
- $x° \cap y° \neq \emptyset$, x overlaps y, denoted by $O(x, y)$
- $\partial x \cap \partial y = \emptyset$, x does not share a boundary with y, denoted by $NA(x, y)$
- $\partial x \cap \partial y \neq \emptyset$, x shares a boundary with y, denoted by $A(x, y)$

Table 1. Correspondence between the RCC8 relationships and the CM8 primitives.

	P	NP	P^{-1}	NP^{-1}	DR	O	NA	A
EQ	1	0	1	0	0	1	0	1
$NTPP$	1	0	0	1	0	1	1	0
TPP	1	0	0	1	0	1	0	1
$NTPP^{-1}$	0	1	1	0	0	1	1	0
TPP^{-1}	0	1	1	0	0	1	0	1
PO	0	1	0	1	0	1	0	1
EC	0	1	0	1	1	0	0	1
DC	0	1	0	1	1	0	1	0

The CM8 primitives are expressed in terms of the RCC8 relationships, and vice versa, as shown in Table 1.

Indeed, this table can be interpreted in the following way:

- according to the lines: any RCC8 relation is expressed as a conjunction of CM8 primitives. For instance, $EQ \Leftrightarrow P \wedge P^{-1} \wedge O \wedge A$, means that two regions x and y are equal, if and only if, x is a part of y, AND y is a part of x, AND x and y overlap, AND the intersection between the boundaries of x and y is non empty.
- according to the columns: any CM8 primitive is expressed as a disjunction of RCC8 relations. For instance, $P \Leftrightarrow EQ \vee NTPP \vee TPP$, means that x is a part of y, if and only if, x is equal to y, OR x is a tangential proper part of y, OR x is a non-tangential proper part of y.

3 Implementation of a Qualitative Spatial Reasoning Using OWL and SWRL

As shown in Sect. 1, the existing approaches for reasoning on spatial qualitative relations outline the lack of expressiveness of OWL for the negation, conjunction or disjunction of roles. However, these features are essential to implement the composition rules of the RCC8 relations. In this section, we first present the principle of reification and we show how it is used to implement a lattice of RCC8 and CM8 concepts. We then introduce a set of SWRL rules, allowing to compute the relation composition.

3.1 Reification of Spatial Relations

Reification is widely used in conceptual modelling. Reifying a relation means viewing it as an entity, that describes the relation's characteristics.

Although RCC8 relationships are simple binary relations, there are other spatial relations that require additional attributes. A simple example, used in our satellite image analysis application, is the distance between objects (either minimal distance or distance between barycentres) that requires a numeric value. A coherent processing of all the spatial relations will require to reify them all.

Moreover, reasoning on RCC8 relationships causes the creation of new relations, by applying conjunction or disjunction. However, conjunction or disjunction of roles are beyond the expressive power of OWL. This is why many specific extensions have been proposed, as shown in Sect. 1.

In our approach, spatial relations are represented by concepts of the ontology. Let us note SR, the top concept of the hierarchy of spatial relations. It is straightforward to combine the concepts of this hierarchy by using the complete set of logic operators (and, or, not), which are already available in the $\mathcal{ALC}$ description logics. Each instance of the SR concept represents a relation between two geographical objects. As RCC8 relations are generally not symmetric, we associate two distinct functional roles to the SR concept: a spatial relation takes place *from* a first geographical object *to* a second geographical object.

$$SR \sqsubseteq \neg GeoObject$$
(objects and spatial relations are disjoint)
$$SR \sqsubseteq (= 1\,from.GeoObject) \sqcap (= 1\,to.GeoObject)$$
(any spatial relation associates exactly two objects)

We now define the general $RCC8$ concept, included in SR, which subsumes all the concepts resulting from the combination of the RCC8 basic relations.

The eight elementary RCC8 relations form a complete and disjoint partition of the $RCC8$ concept.

The CM8 primitives are defined as disjunctions of the RCC8 basic relations. For example, the primitives denoted by P (inclusion) and O (overlapping) are expressed in the following way:

$$P \equiv EQ \sqcup NTPP \sqcup TPP$$
$$O \equiv EQ \sqcup NTPP \sqcup TPP \sqcup NTPP^{-1} \sqcup TPP^{-1} \sqcup PO$$

The reification of spatial relations among objects provokes no particular problems, although the notation is slightly more complex. For example, we want to state that a workers housing estate consists of a set of adjoining houses within an urban area.

$$uArea \sqsubseteq GeoObject, House \sqsubseteq GeoObject$$

Without reification (Pr and ECr are roles corresponding to the P and EC spatial relations):

$$wHousing \equiv GeoObject \sqcap \exists Pr.uArea \sqcap$$
$$\forall Pr^{-1}.(\neg House \sqcup \exists ECr.House)$$

After reification (P and EC are $RCC8$ sub-concepts corresponding to the P and EC spatial relations):

$$wHousing \equiv GeoObject \sqcap$$
$$\exists from^{-1}.(P \sqcap \exists to.uArea) \sqcap$$
$$\forall to^{-1}.(\neg P \sqcup \forall from.(\neg House \sqcup$$
$$\exists to^{-1}.(EC \sqcap \exists from.House)$$

The reified expressions are cumbersome, but the translation from the non-reified form to the reified form can be easily automated.

At ABox level, additional individuals must be created to represent the reified spatial relation between every couple of objects (see Algorithm 2). Figure 2 gives the ABox corresponding to an urban area containing one workers housing $wh1$, containing two adjacent houses $h1$ and $h2$. sr_i individuals represent the reified spatial relations. Undefined spatial relations (of RCC8 type) are omitted for brevity.

Our representation allows a precise description of spatial relationships between individuals. For example:

- o_1 : $(= 2from^{-1}.\neg DC)$, exactly two objects are connected with o_1. The precise nature of the connection ($EQ, NTTP, P \ldots$) is unknown (for exemple, the image analysis system has not been able to determine it)
- o_2 : $(<= 3from^{-1}.O) \sqcap \exists from^{-1}.P$, at most three objects overlap with o_2, at least one of them contains o_2.

We have presented so far the good properties of our model, deliberately leaving aside its disadvantages, some of which are significant. First, there is no way to express the correspondence between a spatial relation and its inverse: in the previous example, we stated that house $h1$ was adjacent (EC) to house

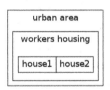

$h1 : House, h2 : House, wh1 : wHousing, ua1 : uArea$

$sr_1 : EC, from(sr_1, h1), to(sr_1, h2)$

$sr_2 : P, from(sr_2, h1), to(sr_2, wh1)$

$sr_3 : P, from(sr_3, h2), to(sr_3, wh1)$

$sr4 : P, from(sr4, wh1), to(sr4, ua1)$

Fig. 2. An ABox for an urban area containing only two houses within a worker's housing.

$h2$, But we cannot deduce the fact, however obvious, that $h2$ is adjacent to $h1$. If spatial relations were represented by roles, this problem could be solved by declaring the role as symmetric in OWL.

There is also no way to handle transitivity: in our example, the house $h1$ is included (P) in the workers' housing $wh1$; $wh1$ itself is included in the urban area $ua1$. No valid inference will deduce the obvious fact that $h1$ is included in $ua1$ (transitivity of inclusion). These disadvantages can be overridden by the use of SWRL rules.

3.2   Using SWRL Rules to Simulate Role Properties

Two types of rules must be defined to reflect respectively inverse relations and the composition table. The inverse relations were not a problem before reification, since OWL allows the definition of inverse, symmetric and transitive roles. Unfortunately, after reification, it is no longer possible to express the correspondences between a relation and its inverse in OWL. We therefore propose to use rules. As we have seen in Sect. 2, it is straightforward to express RCC8 inverse relations in First Order Logic:

$$\forall o_1 \forall o_2 \, EQ(o_1, o_2) \Rightarrow EQ(o_2, o_1)$$
$$\text{(symmetric relation)}$$
$$\forall o_1 \forall o_2 \, TPP(o_1, o_2) \Rightarrow TPP^{-1}(o_2, o_1):$$
$$\text{(explicit inverse relation)}$$

After reification, spatial relations become objects of the logical universe, and the above rules must be expanded as:

$$\forall r_1 \forall r_2 \forall o_1 \forall o_2 \, EQ(r_1) \wedge from(r_1, o_1) \wedge to(r_1, o_2) \wedge$$
$$from(r_2, o_2) \wedge to(r_2, o_1) \Rightarrow EQ(r_2)$$
$$\forall r_1 \forall r_2 \forall o_1 \forall o_2 \, TPP(r_1) \wedge from(r_1, o_1) \wedge to(r_1, o_2) \wedge$$
$$from(r_2, o_2) \wedge to(r_2, o_1) \Rightarrow TPP^{-1}(r_2)$$

The same principle is used to represent the RCC8 composition of relations. For example, the result of composing PO with TPP is either[2] TPP, $NTPP$ or PO. After reification, this should be represented by:

$$\forall r_1 \forall r_2 \forall r_3 \forall o_1 \forall o_2 \forall o_3$$
$$PO(r_1) \wedge from(r_1, o_1) \wedge to(r_1, o_2) \wedge$$
$$TPP(r_2) \wedge from(r_2, o_2) \wedge to(r_2, o_3) \wedge$$
$$from(r_3, o_1) \wedge to(r_3, o_3)$$
$$\Rightarrow TPP(r_3) \vee NTPP(r_3) \vee PO(r_3)$$

---

[2] $PO(x, y) \wedge TPP(y, z) \Rightarrow TPP(x, z) \vee NTPP(x, z) \vee PO(x, z)$.

SWRL is based on Horn clauses. However, the head of a Horn clause can not be a disjunction. So, in order to represent disjunctions, a new concept is created each time a specific disjunction appears in the conclusion of a composition rule. For the rule example above :

$$TPP\_NTPP\_PO \equiv TPP \sqcup NTPP \sqcup PO$$

The SWRL rule simply becomes (in human readable SWRL syntax):

$$PO(?r_1) \wedge from(?r_1, ?o_1) \wedge to(?r_1, ?o_2) \wedge$$
$$TPP(?r_2) \wedge from(?r_2, ?o_2) \wedge to(?r_2, ?o_3) \wedge$$
$$from(?r_3, ?o_1) \wedge to(?r_3, ?o_3)$$
$$\Rightarrow TPP\_NTPP\_PO(?r_3)$$

More formally, we need to determine the set of rules to ensure complete reasoning over the composition table. We will denote by $\mathcal{ER}$ the set of elementary spatial relations and by $\mathcal{R} = 2^{\mathcal{ER}}$ the power set of $\mathcal{ER}$, where each set in $\mathcal{R}$ denotes a disjunction of elementary spatial relations. We will note $er_i \in \mathcal{ER}$ the elementary spatial relations, and $sr_i \in \mathcal{R}$ the disjunctive spatial relations. An elementary composition table $\mathcal{ECT}$ is a function $\mathcal{ECT} : \mathcal{ER} \times \mathcal{ER} \to \mathcal{R}$. A generalized composition table $\mathcal{CT} : \mathcal{R} \times \mathcal{R} \to \mathcal{R}$ can be derived from $\mathcal{ECT}$ by :

$$\mathcal{CT}(sr_1, sr_2) = \bigsqcup_{er_1 \in sr_1} \bigsqcup_{er_2 \in sr_2} \mathcal{ECT}(er_1, er_2)$$

A subset $\mathcal{S} \subseteq \mathcal{R}$ is closed under a composition table $\mathcal{CT}$ if $\forall sr_1, sr_2 \in \mathcal{S}$: $\mathcal{CT}(sr_1, sr_2) \in \mathcal{S}$. From any subset $\mathcal{S} \subseteq \mathcal{R}$, it is straightforward to compute its closure $\mathcal{C}_{\mathcal{CT}}(\mathcal{S})$ by repeatedly applying $\mathcal{CT}$ rules.

The generalized composition table $\mathcal{CT}$ has to be translated into a set of SWRL rules, where each rule is a triple $[sr_1, sr_2, sr_3 = \mathcal{CT}(sr_1, sr_2)]$. Since each sr_i is a disjunction of elementary RCC8 relations, some rules in $\mathcal{CT}$ may be logicaly redundant. More precisely, a rule $r_1 = [sr_1^1, sr_2^1, sr_3^1]$ is more general than a rule $r_2 = [sr_1^2, sr_2^2, sr_3^2]$, noted $r_2 \preceq r_1$, if and only if :

$$r_2 \preceq r_1 \Leftrightarrow \left(sr_1^2 \subseteq sr_1^1 \wedge sr_2^2 \subseteq sr_2^1 \wedge sr_3^1 \subseteq sr_3^2 \right)$$

We will denote by $\mathcal{G}_{\preceq}(\mathcal{CT})$ the set of greatest elements of $\mathcal{CT}$ for the partial order $\preceq$.

Depending on the application, one may want to consider only a subset $\mathcal{UR} \subseteq \mathcal{R}$ of spatial relations. In our approach, $\mathcal{UR}$ corresponds to the set of RCC8 relationships augmented by the set of CM8 primitives.

Therefore, given a qualitative reasoning system defined by a set of elementary relations $\mathcal{ER}$ and an elementary composition table $\mathcal{ECT}$, the set of rules for a user defined subset $\mathcal{UR} \subseteq 2^{\mathcal{ER}}$ is given by:

$$\mathcal{G}_{\preceq} \left(\mathcal{CT} \restriction_{\mathcal{C}_{CT}(\mathcal{UR})} \right)$$

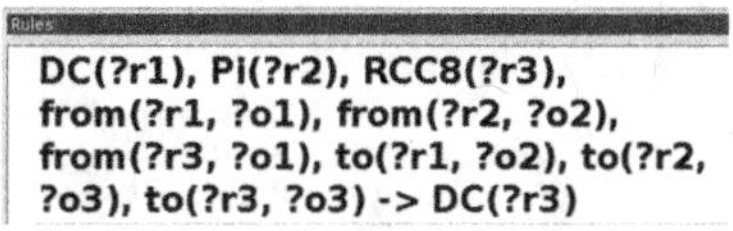

```
Rules
DC(?r1), PI(?r2), RCC8(?r3),
from(?r1, ?o1), from(?r2, ?o2),
from(?r3, ?o1), to(?r1, ?o2), to(?r2,
?o3), to(?r3, ?o3) -> DC(?r3)
```

Fig. 3. A SWRL rule for the composition of DC and Pi.

where $f \restriction_A$ denotes the restriction of f to the domain A. The corresponding algorithm is given in Algorithm 1.

In our application, a set of 94 SWRL rules is generated for the creation of the composition table and its transitive closure. It is interesting to note that the consideration of the CM8 primitives slightly decreases the number of SWRL rules generated (100 rules for a lattice with only the RCC8 relationships). Indeed, the introduction of the CM8 primitives allows the factorisation of some rules (according to the $\preceq$ partial order between the rules).

For example, in the initial RCC8 composition table, there are three rules stating that, for three spatial objects o_1, o_2, o_3, if o_1 is disconnected (DC) from o_2 and o_3 is either identical (EQ), a tangential $(TPP^{-1}$[3]) or non-tangential $(NTPP^{-1})$ proper part of o_2, then o_1 is disconnected from o_3. As $P^{-1} \Leftrightarrow EQ \vee NTPP^{-1} \vee TPP^{-1}$, these three rules will be factored into the single SWRL rule of Fig. 3.

Figure 4 presents the lattice of concepts containing the eight RCC8 relations, the CM8 primitives allowing to compute RCC8 relations from the image, as well as all intermediate concepts generated by the rules that implement the transitive closure of the composition table. We can note that the lattice's hierarchy reflects the correspondence between the RCC8 relationships and the CM8 primitives (see Table 1) : a RCC8 relation is a conjunction of its CM8 ancestors (e.g. $EQ \Leftrightarrow P \wedge PI \wedge O \wedge A$), and a CM8 primitive is a disjunction of its RCC8 descendants (e.g. $P \Leftrightarrow EQ \vee NTPP \vee TPP$).

3.3 Reasoning

Let us continue the example introduced in Sect. 4.1 to show how our model can be effectively used to infer new information linking spatial reasoning to the domain ontology. The definition of a workers housing estate (wHousing) as a set of adjoining houses within an urban area is presented in Fig. 5 using the Manchester OWL Syntax of Protégé. Suppose now that the $wh1$ instance has not been associated with wHousing in the ontology i.e.:

$$h1 : House, h2 : House, ua1 : uArea$$
$$wh1 : GeoObject,$$

Any sound OWL reasoner will not recognize $wh1$ as an instance of wHousing. This negative result would be surprising if working with an object oriented database, but it is consistent with the OWL-DL semantics which is based on

---

[3] In general, for every spatial relation SR, SR^{-1} represents its inverse.

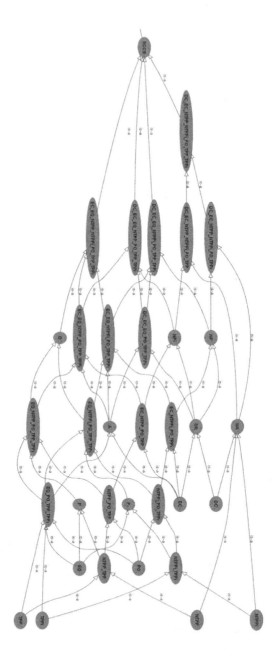

Fig. 4. The lattice of the RCC8 spatial relations, with the CM8 primitives. Links between the atoms (i.e. the eight relations of RCC8) and ⊥ are not represented.

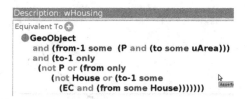

Fig. 5. Workers housing estate definition.

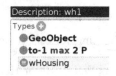

Fig. 6. wh1 is inferred to be a workers housing estate.

the open world assumption. Actually, the OWL-DL reasoner infers that $h1$ and $h2$ are adjoining houses, but with the open world assumption, the existence of another house $h3$ cannot be excluded. Indeed, if we had postulated that $h3$ was within $wh1$ but not adjacent to $h1$ nor $h2$, then $wh1$ would not have been a wHousing.

Therefore in our example, the image recognition software must state that the houses $h1$ and $h2$ are the only objects included in $wh1$. For this purpose, the $wh1$ instance description must be extended with this knowledge by a cardinality constraint (see Fig. 6). The reasoner (Hermit 1.3.6) now correctly infers that $wh1$ actually is a workers housing estate.

Suppose we introduce a third house in our workers housing:

$$isolatedHouse : House$$

$$sr5 : P, from(sr5, isolatedHouse), to(sr5, wh1)$$

Remember that reasoning in OWL is under the open world assumption: our isolated house is not explicitly adjacent to another house in the ABox, but neither is it explicitly defined that it is not adjacent to $h1$ or $h2$, or even to any other house which is not defined in the ABox. Therefore, the ABox is still consistent, even if we have required that all the houses must have at least one adjacent house in the definition of the $wHousing$ concept. Fortunately, cardinality constraints on roles allow to simulate closed world reasoning. If we are sure that the house is isolated, we can for example add in the ABox:

$$isolatedHouse :\leq 0 from^{-1}.EC$$

This time the ABox is inconsistent.

3.4   Preliminary Experiments

To ensure decidability, SWRL rules must follow the DL-safety condition [16]. This condition is ensured if the domain of the variables is restricted to individuals

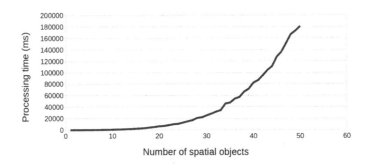

Fig. 7. Processing time for consistency checking.

of the ABox. Current reasoners as Pellet [19] or Hermit [7] manage SWRL rules under the DL-safety assumption. As a consequence, individuals representing the spatial relations between spatial objects must be defined in the ABox in order for the SWRL rules to effectively perform RCC8 inferences. Since RCC8 relations are exhaustive, every pair of spatial objects are related by a RCC8 relationship. Moreover, in our representation, it is always possible to create a relation of type RCC8 (which subsume every elementary relation) if the actual relation between two spatial objects is undetermined. In a domain containing n spatial objects individuals, we must create n^2 RCC8 individuals, representing all the relations between spatial objects.

The purpose of these preliminary experiments is only to test how current OWL reasoners behave with the kind of mixed OWL + SWRL ontology we propose in this article. Ontologies have been created using the OWL-API[4] 3.2.4. For this first experiment, a TBox containing the concepts presented in Fig. 4 and the 94 SWRL rules computed by Algorithm 1 is automatically created. Algorithm 2 creates a Abox for a varying number of individuals. From the basic ABox containing n spatial objects $o_1 \cdots o_n$, we created an inclusion chain $o_1\, TPP\, o_2\, \cdots\, TPP\, o_n$ and added that o_1 is disconnected from o_n. The resulting ABox is obviously inconsistent. The reasoner must repeatedly apply the $TPP \circ TPP$ composition rule to detect this inconsistency.

Figure 7 shows the computing time taken by the Hermit 1.3.6 reasoner [7] for checking the consistency of the Aboxes. This graph calls for some remarks:

- the current version of the Hermit reasoner cannot deal efficiently with about a hundred SWRL rules and a few thousands individuals.
- processing time is quadratic with regard to the number of spatial objects. Since the RCC8 relations created in this example are very simple, processing time almost only depends on the number of rules and on the n^2 spatial relations we created in order to link the n spatial objects. A comprehensive study of reasoning with large set of spatial objects with intricate RCC8 relationships (as in [18]) is outside the scope of this paper, and would be difficult because of the long computation time of current OWL+SWRL reasoners.

---

[4] http://owlapi.sourceforge.net/.

Algorithm 1. RCC8 TBox creation.

Given a set of Elementary Spatial Relation
$ESR = \{NTPP, NTPP^{-1}, TPP, TPP^{-1}, PO, EQ, EC, DC\}$
Given a function creating disjunction names
    $disjunctName : 2^{conceptName} \rightarrow conceptName$
Given a composition table
    $compTable : ESR \times ESR \rightarrow 2^{ESR}$
Given an initial set $\mathcal{UR} \subseteq 2^{ESR}$ $(\mathcal{UR} = ESR \bigsqcup CM8)$

Assert $RCC8$ top concept
Assert $\mathcal{UR}$ concepts as a disjunction of ESR concepts
Assert functional roles to and $from$ and their inverses to^{-1} and $from^{-1}$
for all $SR \in \{PO, EQ, EC, DC\}$ **do**
    assert SWRL rule: */* symmetrical relations*/*
    $SR(?r_1) \wedge from(?r_1, ?o_1) \wedge to(?r_1, ?o_2)$
    $\wedge from(?r_2, ?o_2) \wedge to(?r_2, ?o_1) \Rightarrow SR(?r_2)$
end for
for all $[SR, SRi] \in \{[TPP, TPP^{-1}], [NTPP, NTPP^{-1}]\}$ **do**
    assert SWRL rules: */* inverse relations */*
    $SR(?r_1) \wedge from(?r_1, ?o_1) \wedge to(?r_1, ?o_2)$
    $\wedge from(?r_2, ?o_2) \wedge to(?r_2, ?o_1) \Rightarrow SRi(?r_2)$
    $SRi(?r_1) \wedge from(?r_1, ?o_1) \wedge to(?r_1, ?o_2)$
    $\wedge from(?r_2, ?o_2) \wedge to(?r_2, ?o_1) \Rightarrow SR(?r_2)$
end for
$CSR \leftarrow ESR$ */* CSR set of all created concepts */*
$NewConceptCreated \leftarrow true$
while *(NewConceptCreated)* **do**
    $NewConceptCreated \leftarrow false$
    for all $[SR1, SR2] \in CSR \times CSR$ **do**
        $D \leftarrow \bigsqcup_{er1 \in SR1} \bigsqcup_{er2 \in SR2} compTable(er1, er2)$
        $NSR \leftarrow disjunctName(D)$
        if $NSR \notin CSR$ **then**
            Assert concept $NSR \equiv \bigsqcup_{SR \in D} SR$
            $CSR \leftarrow CSR \bigsqcup \{NSR\}$;
            $NewConceptCreated \leftarrow true$
            let $r = \{SR1, SR2, NSR\}$
            if $\nexists r' = \{SR1', SR2', NSR'\}$ such that
      $SR1 \subseteq SR1' \wedge SR2 \subseteq SR2' \wedge NSR' \subseteq NSR$ **then**
              assert SWRL rule: */* factorization */*
              $SR1(?r_1) \wedge SR2(?r_2)\wedge$
              $from(?r_1, ?o_1) \wedge to(?r_1, ?o_2)\wedge$
              $from(?r_2, ?o_2) \wedge to(?r_2, ?o_3)\wedge$
              $from(?r_3, ?o_1) \wedge to(?r_3, ?o_3) \Rightarrow NSR(?r_3)$
            end if
        end if
    end for
end while

Algorithm 2. RCC8 ABox preprocessing.

Given the set OS of SpatialObject individuals
Given a function creating individual names for spatial relations:
$SR_{Name} : OS \times OS \to individualName$
for all $o_1 \in OS$ **do**
   for all $o_2 \in OS$ **do**
      if $o_1 = o_2$ **then**
         Assert individual: $SR_{Name}(o_1, o_2) : EQ$
      else
         Assert individual: $SR_{Name}(o_1, o_2) : RCC8$
      end if
   end for
end for

Domain knowledge can be used to circumvent this complexity problem, e.g. to select a spatial context where the relations between objects will be computed. For instance the relations between houses should be computed inside a road delimited area. This idea has been used in [15] for querying the geospatial web. The relation computation can also be processed according to a hierarchical domain model as done in [13] for classifying agricultural spatial structures.

4   Conclusions

In this paper we have described the implementation of a qualitative spatial reasoning on topological RCC8 relationships, based on OWL. Our approach copes with the lack of expressiveness of OWL for implementing the negation, conjunction or disjunction of roles. This approach relies on the reification of roles, and is completed by the use of SWRL rules to overcome the loss, due to reification, of some roles properties in OWL, such as symmetry or transitivity.

A disadvantage of this approach is the additional complexity induced by the process of reification, which causes the creation of n^2 instances of spatial relations for n geographic objects. Thus, a reasoner that would be optimized for executing SWRL rules involving many instances would be required.

However, the interest of our proposal is twofold. On the one hand, we have demonstrated the feasibility of the implementation based on OWL and SWRL, of a complete reasoning for calculating the composition table of the RCC8 relationships and its transitive closure. On the other hand, our model integrates the expression of the RCC8 relationships in terms of the CM8 computational primitives: this opens interesting perspectives for the extraction of the topological relations existing among image objects, in the context of satellite images recognition. Other sets of primitives could be included as well. Furthermore, we plan to implement the CM8 computational primitives, and to integrate them in the image classification software that has been developed in our research team (the MUSTIC platform[5]).

---

[5] http://icube-bfo.unistra.fr/fr/index.php/Plateformes.

References

1. Batsakis, S., Petrakis, E.G.M.: SOWL: a framework for handling spatio-temporal information in owl 2.0. In: Bassiliades, N., Governatori, G., Paschke, A. (eds.) RuleML 2011 - Europe. LNCS, vol. 6826, pp. 242–249. Springer, Heidelberg (2011)
2. Clarke, B.L.: A calculus of individuals based on 'connection'. Notre Dame J. Formal Logic **22**(3), 204–218 (1981)
3. Clementini, E., Di Felice, P., van Oosterom, P.: A small set of formal topological relationships suitable for end-user interaction. In: Abel, David J., Ooi, Beng-Chin (eds.) SSD 1993. LNCS, vol. 692, pp. 277–295. Springer, Heidelberg (1993). http://dl.acm.org/citation.cfm?id=647223.718898
4. Cravero, M., Zanni-Merk, C., de Bertrand de Beuvron, F., Marc-Zwecker, S.: A description logics geographical ontology for effective semantic analysis of satellite images. In: 16th International Conference on Knowledge-Based and Intelligent Information and Engineering Systems, KES 2012. vol. 243, pp. 1573–1582. IOS Press, September 2012
5. Deng, M., Cheng, T., Chen, X., Li, Z.: Multi-level topological relations between spatial regions based upon topological invariants. GeoInformatica **11**, 239–267 (2007). http://dx.doi.org/10.1007/s10707-006-0004-x
6. Egenhofer, M.J.: A formal definition of binary topological relationships. In: Litwin, W., Schek, H.-J. (eds.) FODO 1989. LNCS, vol. 367, pp. 457–472. Springer, Heidelberg (1989)
7. Glimm, B., Horrocks, I., Motik, B.: Optimized description logic reasoning via core blocking. In: Giesl, J., Hähnle, R. (eds.) IJCAR 2010. LNCS, vol. 6173, pp. 457–471. Springer, Heidelberg (2010)
8. Grau, B.C., Horrocks, I., Motik, B., Parsia, B., Patel-Schneider, P., Sattler, U.: OWL 2: the next step for OWL. Web Semant. Sci. Serv. Agents World Wide Web **6**(4), 309–322 (2008). http://www.sciencedirect.com/science/article/pii/S1570826808000413, Semantic Web Challenge 2006/2007
9. Hogenboom, F., Borgman, B., Frasincar, F., Kaymak, U.: Spatial knowledge representation on the semantic web. In: Proceedings of the 2010 IEEE Fourth International Conference on Semantic Computing, ICSC 2010, pp. 252–259 IEEE Computer Society, Washington, DC, USA (2010). http://dx.doi.org/10.1109/ICSC.2010.31
10. Horrocks, I., Patel-Schneider, P.F., Boley, H., Tabet, S., Grosof, B., Dean, M.: SWRL: a semantic web rule language combining OWL and RuleML. W3c member submission, World Wide Web Consortium (2004). http://www.w3.org/Submission/SWRL
11. Jitkajornwanich, K., Elmasri, R., Li, C., McEnery, J.: Formalization of 2-D spatial ontology and OWL/Protégé realization. In: Proceedings of the International Workshop on Semantic Web Information Management, SWIM 2011, pp. 9:1–9:7 ACM, New York (2011). http://doi.acm.org/10.1145/1999299.1999308
12. Katz, Y., Grau, B.C.: Representing qualitative spatial information in OWL-DL. In: Proceedings of OWL: Experiences and Directions (2005)
13. Le Ber, F., Napoli, A.: The design of an object-based system for representing and classifying spatial structures and relations. J. Univers. Comput. Sci. **8**(8), 751–773 (2002). Special issue on Spatial and Temporal Reasoning
14. Le Ber, F., Napoli, A.: Design and comparison of lattices of topological relations for spatial representation and reasoning. J. Exp. Theor. Artif. Intell. **15**(3), 331–371 (2003)

15. Miron, A.-D., Gensel, J., Villanova-Oliver, M.: Semantic analysis for the geospatial web – application to OWL-DL ontologies. In: Bertolotto, M., Ray, C., Li, X. (eds.) W2GIS 2008. LNCS, vol. 5373, pp. 37–49. Springer, Heidelberg (2008)
16. Motik, B., Sattler, U., Studer, R.: Query answering for owl-dl with rules. Web Semant. Sci. Serv. Agents World Wide Web 3(1), 41–60 (2005). http://www.sciencedirect.com/science/article/pii/S157082680500003X, Rules Systems
17. Randell, D.A., Cui, Z., Cohn, A.G.: A spatial logic based on regions and connection. In: Proceedings 3rd International Conference on Knowedge Representation and Reasoning (1992)
18. Renz, J., Nebel, B.: Efficient methods for qualitative spatial reasoning. In: Proceedings of the 13th European Conference on Artificial Intelligence, pp. 562–566. Wiley (1998)
19. Sirin, E., Parsia, B., Grau, B., Kalyanpur, A., Katz, Y.: Pellet: a practical OWL-DL reasoner. Web Semant. Sci. Serv. Agents World Wide Web 5(2), 51–53 (2007). http://dx.doi.org/10.1016/j.websem.2007.03.004
20. Wessel, M.: Obstacles on the way to qualitative spatial reasoning with description logics: some undecidability results. In: Proceedings of the International Workshop in Description Logics 2001 (DL 2001), pp. 96–105 (2001)

Design Patterns for Description-Logic Programs

L. Cruz-Filipe[1], G. Gaspar[2,3], and I. Nunes[2,3](✉)

[1] Department of Mathematics and Computer Science, University of Southern
Denmark, Odense, Denmark
lcfilipe@gmail.com
[2] Faculdade de Ciências da Universidade de Lisboa, Lisboa, Portugal
[3] LabMAg, Lisboa, Portugal
{gg,in}@di.fc.ul.pt

Abstract. Originally proposed in the mid-90s, design patterns for software development played a key role in object-oriented programming not only in increasing software quality, but also by giving a better understanding of the power and limitations of this paradigm. Since then, several authors have endorsed a similar task for other programming paradigms, in the hope of achieving similar benefits. In this paper we present a set of design patterns for Mdl-programs, a hybrid formalism combining several description logic knowledge bases via a logic program. These patterns are extensively applied in a natural way in a large-scale example that illustrates how their usage greatly simplifies some programming tasks, at the level of both development and extension.

We also discuss some limitations of this formalism, examining some usual patterns in other programming paradigms that have no parallel in Mdl-programs.

1 Introduction

In the mid-nineties, the Gang of Four's work on software design patterns [1] paved the way for important advances in software quality; presently, many valuable experienced designers' "best practices" are not only published but effectively used by the software development community. From very basic, abstract, patterns that can be used as building blocks of several more complex ones, to business-specific patterns and frameworks, dozens of design patterns have been proposed, e.g. [2–8], establishing a kind of common language between development teams, which substantially enriches their communication, and hence the whole design process.

Despite their widespread usage in the object-oriented paradigm, on which a lot of the work has been focused, effort has also been made in adapting these best practices to other paradigms – service-oriented [3], functional [9–11], logic [12] and others – and in finding new paradigm-specific patterns. As several of these authors observed, studying design patterns in different programming paradigms

Work partially supported by Fundação para a Ciência e Tecnologia under contracts
PEst-OE/MAT/UI0209/2011 and PEst-OE/EEI/UI0434/2011.

© Springer-Verlag Berlin Heidelberg 2015
A. Fred et al. (Eds.): IC3K 2013, CCIS 454, pp. 199–214, 2015.
DOI: 10.1007/978-3-662-46549-3_13

is far from being a trivial task: each paradigm has its specific features, meaning that patterns that are very straightforward in one paradigm can be very complex in another, and vice-versa. In this spirit, we carried the task of identifying several basic and other, more complex, patterns in the paradigm of Mdl-programs [13] – which join description logics with rules (expressed as a Datalog-like logic program) –, a powerful and expressive approach to reasoning over general knowledge bases or ontologies that generalizes the original dl-programs [14]. The goal of this paper is to extend the original presentation in [15] with a more detailed analysis of the limitations that arise in this framework.

This work should be seen as quite distinct from that on ontology design patterns [16]. In the setting of Mdl-programs, ontologies are seen as immutable, being used and not changed, under the coordination of a set of rules. Our patterns focus therefore almost exclusively on these rules; thus, ontology design patterns and design patterns for Mdl-programs should in general be seen as two complementary techniques, and not as alternatives.

1.1 Motivation

The usefulness of combining description logics with rule-based reasoning systems led to the introduction of dl-programs [14,17], which couple a description logic knowledge base with a generalized logic program, interacting by means of special atoms, the *dl-atoms*. These programs were later generalized to include several knowledge bases, yielding Mdl-programs [13].

Looking at Mdl-programs, it is clear that they represent a completely different programming paradigm – not only are they closely related to the logic programming paradigm, but they involve description logic knowledge bases, in the presence of which the study of design patterns attains a different quality: on the one hand, some patterns become trivial (such as FAÇADE) or meaningless (such as DYNAMIC BINDING or SINGLETON), on the other hand some patterns pose totally new problems that have not been addressed in other paradigms where they do not arise (such as PROXY, which we will discuss in Sect. 5).

Mdl-programs, combining description logic knowledge bases and a logic-based rule language, provide the adequate setting for the study of design patterns for the Semantic Web. Indeed, description logics *are* at the core of the Semantic Web, with a huge effort being currently invested in the interchange between OWL – an extension of the description logic SROIQ and a W3C recommendation – and a diversity of rule languages [18]. The components of an Mdl-program are kept independent, giving them nice modularity properties; furthermore, Mdl-programs keep ontologies separate, which is much more convenient than e.g. merging them: not only is it simpler to have independent knowledge bases (which might even be physically separated, or independently managed), but merging ontologies is in itself a mighty task with its own specific problems [19,20].

On the other hand, Mdl-programs limit heterogeneity to two different frameworks: description logics for the knowledge bases part and logic programming for the rule part; the latter somehow represents the "conductor" that "coordinates" the other parts. However, they fully support non-monotonicity (even at the level of the description logic knowledge bases as will be seen later by application of a specific basic pattern). Mdl-programs are therefore a simpler framework than other, more powerful, alternatives (such as HEX-programs [21] or multi-context systems [22]), but expressive enough for their use within the Semantic Web.

The remainder of the paper is structured as follows. Section 2 explains Mdl-programs in detail. Section 3 presents seven different design patterns, and Sect. 4 illustrates their combined use by means of a larger example. Section 5 explores limitations and future directions of research, and Sect. 6 summarizes the contributions presented earlier.

2   Mdl-Programs

Multi-description logic programs, the framework in which we will introduce our design patterns, generalize the original definition of dl-programs in [14] to accommodate for several description logic knowledge bases. This construction, introduced in [15] and detailed in [13], is in line with [23], although it sticks to the original operators $\uplus$ and $\cup$ in dl-atoms.

A *dl-atom* relative to a set of knowledge bases $\{\mathcal{L}_1, \ldots, \mathcal{L}_n\}^1$ is

$$DL_i\,[S_1\,op_1\,p_1, \ldots, S_m\,op_m\,p_m; Q]\,(\bar{t})\,,$$

often abbreviated to $DL_i[\chi; Q](\bar{t})$, where: (1) $1 \leq i \leq n$; (2) each S_k, with $1 \leq k \leq m$, is either a concept or a role from $\mathcal{L}_i$ or a special symbol in $\{=, \neq\}$; (3) $op_k \in \{\uplus, \cup\,\}$; (4) p_k are the *input predicate symbols*, which are unary or binary predicate symbols depending on the corresponding S_k being a concept or a role; and (5) $Q(\bar{t})$ is a *dl-query* in the language of $\mathcal{L}_i$, that is, it is either a concept inclusion axiom F or its negation $\neg F$, or of the form $C(t_1)$, $\neg C(t_1)$, $R(t_1, t_2)$, $\neg R(t_1, t_2)$, $= (t_1, t_2)$, $\neq (t_1, t_2)$, where C is a concept, R is a role, t, t_1 and t_2 are terms (variables or constants).

The operators $\uplus$ and $\cup$ are used to extend the knowledge base $\mathcal{L}_i$ locally, with $S_k \uplus p_k$ (resp., $S_k \cup p_k$) increasing S_k (resp., $\neg S_k$) by the extension of p_k. Intuitively, the dl-atom above adds this information to $\mathcal{L}_i$ and then asks this knowledge base for the set of terms satisfying $Q(t)$.[2]

A *Multi Description Logic program* (Mdl-program) is a pair $\langle\{\mathcal{L}_1, \ldots, \mathcal{L}_n\}, \mathcal{P}\rangle$ where: (1) each $\mathcal{L}_i$ is a description logic knowledge base; (2) $\mathcal{P}$ is a set of (normal) *Mdl-rules*, i.e. rules of the form $a \leftarrow b_1, \ldots, b_k, not\ b_{k+1}, \ldots, not\ b_p$ where a is a logic program atom and each b_j, for $1 \leq j \leq p$, is either a logic program atom or a dl-atom relative to $\{\mathcal{L}_1, \ldots, \mathcal{L}_n\}$. Note that $\mathcal{P}$ is a generalized logic program, so negation is the usual, closed-world, negation-as-failure. This is in contrast with the $\mathcal{L}_i$, which (being description logic knowledge bases) come with an open-world semantics.

---

[1] The description logics underlying the $\mathcal{L}_i$s need not be the same.

[2] The precise semantics can be found in [13]; the third operator in [14] is not included, as it can be defined in terms of $\cup$, and this option simplifies the semantics [23].

The semantics of Mdl-programs [13] is a straightforward generalization of the semantics of dl-programs [14] and will not be discussed here, since it will not be needed explicitly.

A common feature of multi-component systems is the need for entities in one component to "observe" entities in another component. In the setting of Mdl-programs, this is achieved by means of observers. An *Mdl-program with observers* is $\langle \{\mathcal{L}_1, \ldots, \mathcal{L}_n\}, \mathcal{P}, \{\Lambda_1, \ldots, \Lambda_n\}, \{\Psi_1, \ldots, \Psi_n\} \rangle$ where: (1) $\langle \{\mathcal{L}_1, \ldots, \mathcal{L}_n\}, \mathcal{P} \rangle$ is an Mdl-program; (2) for $1 \leq i \leq n$, Λ_i is a finite set of pairs $\langle S, p \rangle$ where S is a concept, a role, or a negation of either, from $\mathcal{L}_i$ and p is a predicate from $\mathcal{P}$; (3) for $1 \leq i \leq n$, Ψ_i is a finite set of pairs $\langle p, S \rangle$ where p is a predicate from $\mathcal{P}$ and S is a concept, a role, or a negation of either, from $\mathcal{L}_i$. For each pair in Ψ_i or Λ_i, the arities of S and p must coincide. The sets $\Lambda_1, \ldots, \Lambda_n, \Psi_1, \ldots, \Psi_n$ will occasionally be referred to as the *observers* of $\langle \{\mathcal{L}_1, \ldots, \mathcal{L}_n\}, \mathcal{P} \rangle$. Intuitively, Λ_i contains concepts and roles in $\mathcal{L}_i$ that $\mathcal{P}$ needs to observe, in the sense that $\mathcal{P}$ should be able to detect whenever new facts about them are derived, whereas Ψ_i contains the predicates in $\mathcal{P}$ that $\mathcal{L}_i$ wants to observe. For simplicity, when we consider Mdl-programs with observers that only have one knowledge base, we will omit the braces and refer to them as dl-programs with observers.

Instead of defining formal semantics for Mdl-programs with observers, we introduced a translation of these into (standard) Mdl-programs that reduces observers to syntactic sugar. The above Mdl-program with observers thus implicitly defines the Mdl-program $\left\langle \{\mathcal{L}_1, \ldots, \mathcal{L}_n\}, \mathcal{P}^{\Psi_1, \ldots, \Psi_n}_{\Lambda_1, \ldots, \Lambda_n} \right\rangle$ where $\mathcal{P}^{\Psi_1, \ldots, \Psi_n}_{\Lambda_1, \ldots, \Lambda_n}$ is obtained from $\mathcal{P}$ by: (1) adding rule $p(X) \leftarrow DL_i[; S](X)$ for each $\langle S, p \rangle \in \Lambda_i$, if S is a concept (and its binary counterpart, if S is a role); and (2) in each dl-atom $DL_i[\chi; Q](t)$ (including those added in the previous step), adding $S \uplus p$ to χ for each $\langle p, S \rangle \in \Psi_i$ and $S \cup\!\!\!-\ p$ to χ for each $\langle p, \neg S \rangle \in \Psi_i$.

We now illustrate these concepts by means of a simple example. Consider two knowledge bases $\mathcal{L}_1$, defining travel-related concepts, including that of (tourist) Destination, and $\mathcal{L}_2$, compiling information about wines, including a concept Region identifying some major wine regions throughout the world. We wish to join these ontologies by means of rules to obtain an Mdl-program with observers that reasons about wine-related destinations. This is achieved by taking $\mathcal{P}$ to be

$$
\begin{array}{ll}
\text{wineDest(Tasmania)} \leftarrow & (r_1) \\
\text{wineDest(TamarValley)} \leftarrow & (r_2) \\
\text{wineDest(Sydney)} \leftarrow & (r_3)
\end{array}
$$

$$
\begin{array}{ll}
\text{overnight}(X) \leftarrow DL_1[; \text{hasAccommodation}](X, Y) & (r_4) \\
\text{oneDayTrip}(X) \leftarrow DL_1[; \text{Destination}](X), not\ \text{overnight}(X) & (r_5)
\end{array}
$$

and observers $\Lambda_2 = \{\langle \text{Region}, \text{wineDest} \rangle\}$, $\Psi_1 = \{\langle \text{wineDest}, \text{Destination} \rangle\}$ and $\Lambda_1 = \Psi_2 = \emptyset$.

This very simple program defines a predicate wineDest with three instances obtained from rules (r_1–r_3), corresponding to three wine regions that are interesting tourist destinations, together with all instances of Region from $\mathcal{L}_2$, which are obtained via the only element in Λ_2. Unfolding this observer yields the rule

$$\mathsf{wineDest}(X) \leftarrow DL_2[; \mathsf{Region}](X) \qquad (r_0),$$

which corresponds to this intuitive semantics.

The set Ψ_1 causes Destination $\uplus$ wineDest to be added to the context of every dl-atom querying $\mathcal{L}_1$, extending $\mathcal{P}$'s view of $\mathcal{L}_1$ – namely in rules (r_4) and (r_5). This causes $\mathcal{L}_1$ to answer taking into account not only those instances of Destination in its knowledge base, but also those instances of wineDest that $\mathcal{P}$ knows about (including the ones derived from $\mathcal{L}_2$).

Rule (r_5) identifies the destinations that are only suitable for one-day trips. The possible destinations are obtained by querying $\mathcal{P}$'s extended view of $\mathcal{L}_1$ for all instances of Destination. The result is then filtered using the auxiliary predicate overnight defined in (r_4) as the set of destinations for which some accommodation is known. This uses the role hasAccommodation of $\mathcal{L}_1$, where hasAccommodation(t_1,t_2) holds whenever t_1 is a Destination and t_2 an accommodation facility located in t_1. The reason for resorting to (r_4) at all is the usual one in logic programming: the operational semantics of negation-as-failure requires all variables in a negated atom to appear in non-negated atoms in the body of the same rule. Note the impact of Ψ_1: if Destination were not being updated with the information from wineDest, the program would not be able to infer e.g. oneDayTrip(Tasmania).

Mdl-programs with observers have been implemented [24] as a plugin for the dlvhex tool [25].

3 Design Patterns for Mdl-programs

In this section, we present a first set of seven design patterns for Mdl-programs, introduced in [15]. These are divided in two categories: the three elementary design patterns are the building blocks for the four more complex ones. Together, these seven patterns form a powerful set from which quite complex programs can be designed in a more structured way, simplifying the programmer's task while at the same time yielding more flexible programs that are easier to maintain.

The presentation of each design pattern follows a similar scheme: each is presented as a pair problem/solution within the context of an Mdl-program with observers $\langle \{\mathcal{L}_1, \ldots, \mathcal{L}_n\}, \mathcal{P}, \{\Lambda_1, \ldots, \Lambda_n\}, \{\Psi_1, \ldots, \Psi_n\} \rangle$. In the next section, we present a large-scale example that illustrates how the seven patterns work together, complementing each other.

Elementary Design Patterns. The three basic design patterns for Mdl-programs deal with three simple tasks: transporting information from the logic program to a knowledge base and reciprocally, and giving closed-world semantics to a concept or role in one of the knowledge bases.

We first consider the case when the logic program component systematically wants to import information from a knowledge base in order to define a predicate, keeping track of changes made to the relevant concept or role.

Pattern OBSERVER DOWN.

Problem. Predicate p from $\mathcal{P}$ needs to be updated every time the extent (set of named individuals) of concept or role S (of the same arity as p) in $\mathcal{L}_i$ is changed.
Solution. Add the pair $\langle S, p \rangle$ to Λ_i.

A second scenario occurs when one of the description logics relies on the observation of a predicate from $\mathcal{P}$.

Pattern OBSERVER UP.

Problem. In $\mathcal{P}$'s view, concept or role S from $\mathcal{L}_i$ needs to be updated every time the extent of predicate p (of the same arity as S) in $\mathcal{P}$ is changed.
Solution. Add the pair $\langle p, S \rangle$ to Ψ_i.

The third building block addresses a very typical situation in ontology usage: a concept or role should be given closed-world semantics.

Pattern CLOSED-WORLD.

Problem. In $\mathcal{P}$'s view, concept (or role) S from $\mathcal{L}_i$ should follow closed-world semantics.
Solution.
Choose predicate symbols s^+, s^- not used in $\mathcal{P}$.
Add $\langle S, s^+ \rangle$ to Λ_i, $\langle s^-, \neg S \rangle$ to Ψ_i, and $s^-(X) \leftarrow not\ s^+(X)$ to $\mathcal{P}$.

Derived Design Patterns. We now present a second set of general-purpose design patterns that can be seen as organized combinations of the previous ones, but are also useful as components of more complex patterns.

A useful variant of the Observer design pattern occurs when a description logic's functionality relies on the observation of a predicate in a *different* description logic; this can be achieved by combining both the OBSERVER UP and OBSERVER DOWN patterns, thus making the logic program $\mathcal{P}$ a mediator. A particular case arises when an ontology designed primarily for reasoning interacts with a knowledge base that is mostly about particular instances. This design pattern appears often in combination with DEFINITIONS WITH HOLES below.

Pattern TRANSVERSAL OBSERVER.

Problem. In $\mathcal{P}$'s view, concept (or role) S from $\mathcal{L}_i$ needs to be updated every time the extent of concept (resp. role) R from $\mathcal{L}_j$ is changed ($i \neq j$).
Solution. Choose a predicate symbol p not used in $\mathcal{P}$.
Add $\langle R, p \rangle$ to Λ_j and $\langle p, S \rangle$ to Ψ_i.

The next design patterns allows one to define a predicate in $\mathcal{P}$ abstracting from how it is represented in the knowledge bases.

Pattern SPLIT DEFINITIONS.

Problem. In $\mathcal{P}$ there is a predicate p whose instances are inherited from concepts or roles $S_1, \ldots, S_k$ where each S_j comes from the knowledge base $\mathcal{L}_{\varphi(j)}$, for $1 \leq j \leq n$.
Solution. For each $1 \leq j \leq n$, add the pair $\langle S_j, p \rangle$ to $\Lambda_{\varphi(j)}$.

Note that SPLIT DEFINITIONS consists of a combined application of several OBSERVER DOWN, all with the same observer predicate in $\mathcal{P}$. This pattern deals with a predicate that is kept as independent as possible from its definition. Instead of defining clauses, the instances are plugged in through the use of Mdl-programs with observers, thus externalizing the definition of the predicate – in the spirit of Dynamic Binding. The possibility of using different concepts or roles (possibly even from different knowledge bases) captures the essence of Polymorphism. For this reason, this pattern was originally named POLYMORPHIC ENTITIES [15].

The converse situation yields a different pattern, due to the way Mdl-programs are typically developed: the logic program is written to connect pre-existing knowledge bases. This pattern is particularly useful when in presence of terminological ontologies where some concepts are not defined, and captures a typical way of working with ontologies.

Pattern DEFINITIONS WITH HOLES.

Problem. Concept or role S is needed for reasoning in $\mathcal{L}_i$, but its definition will be in $\mathcal{L}_j$ (with $i \neq j$) or $\mathcal{P}$.

Solution.
    Use S in $\mathcal{L}_i$ without defining it (so the extent of S is empty).
    Later, connect S to its definition using OBSERVER UP, OBSERVER DOWN or TRANSVERSAL OBSERVER, possibly coupled with SPLIT DEFINITIONS.

This pattern corresponds to the Template Method pattern of object-oriented programs [1], and to the Programming with Holes technique of [7]. In the next section we will show an example where the holes are filled in by resorting to SPLIT DEFINITIONS.

The last design pattern in this section applies when several components of an Mdl-program contribute to the definition of a predicate.

Pattern COMBINED DEFINITIONS.

Problem. There is a predicate being defined in some of the $\mathcal{L}_i$s (in the form of concepts or roles S_i) and $\mathcal{P}$ (in the form of two predicates p^+ and p^-, corresponding to the predicate and its negation).

Solution.
    For each i, add $\langle S_i, p^+ \rangle$ and $\langle \neg S_i, p^- \rangle$ to Λ_i.
    For each i, add $\langle p^+, S_i \rangle$ and $\langle p^-, \neg S_i \rangle$ to Ψ_i.

Note that COMBINED DEFINITIONS is essentially different from OBSERVER: in OBSERVER, a predicate is defined in *one* component and used in others; in COMBINED DEFINITIONS, not only the usage, but also the *definition* of the predicate is split among several components, so that one must look at the whole Mdl-program to understand it. This is also part of the reason to include the negations of the predicates involved in the observers: the distributed predicate must end up with the same semantics, both in $\mathcal{P}$ and in all the involved $\mathcal{L}_i$s – at least regarding named individuals.

It is possible to apply this pattern when $\mathcal{P}$ does not participate in the predicate's definition. In this case, $\mathcal{P}$ is simply a mediator, and p^+ and p^- can be any fresh predicate names.

This pattern was originally introduced in [15] under the name LIFTING.

The application of each of the patterns proposed in this section yields localized changes to the Mdl-program: they consist of either changing dl-atoms (by means of adding pairs to Ψ_i) or adding rules to $\mathcal{P}$ (either directly, as in the case of CLOSED-WORLD, or by adding pairs to Λ_i). In all cases, these changes are only reflected in $\mathcal{P}$, and they can be divided into two or three distinct types. This is in line with the whole philosophy of dl-programs: there is an asymmetry between their components where the logic program is the orchestrator between all components as well as its façade: it is the only entity interacting with the outside world.

4   A Comprehensive Example

We now illustrate the usage of the different design patterns introduced so far by means of a more complex example.

Scenario. The software developers at WISHYOUWERETHERE travel agency decided to develop an Mdl-program to manage several of the agency's day-to-day tasks. Currently, WISHYOUWERETHERE has two active partnerships, one with an aviation company, another with a hotel chain. Thus, the Mdl-program to be developed uses three ontologies:

- $\mathcal{L}_A$ is a generic accounting ontology for travel agencies, which is commercially available, and which contains all sorts of rules relating concepts relevant for the business. This ontology is strictly terminological, containing no specific instances of its concepts and roles.
- $\mathcal{L}_F$ is the aviation partner's knowledge base, containing information not only about available flights between different destinations, but also about clients who have already booked flights with that company.
- $\mathcal{L}_H$ is a similar knowledge base pertaining to the hotels owned by the partner hotel chain.

One of the points to take into consideration is that the resulting Mdl-program with observers $\langle \{\mathcal{L}_A, \mathcal{L}_F, \mathcal{L}_H\}, \mathcal{P}, \{\Lambda_A, \Lambda_F, \Lambda_H\}, \{\Psi_A, \Psi_F, \Psi_H\} \rangle$ should be easily extended so that the travel agency can establish new partnerships, in particular with other aviation companies and hotel chains, as long as those provide their own knowledge bases. At the end of this section, we will show how the systematic use of design patterns and observers helps towards achieving this goal.

By establishing partnerships, WISHYOUWERETHERE's client basis is extended with all the clients who have booked services of its partners. In this way, promotions made available by either partner are automatically offered to every partner's clients, as long as the bookings are made through the travel agency. In return, the partners get publicity and more clients, since a person may be tempted to fly with their company or book their hotel due to these promotions, thereby becoming also their client.

Updating the client database. Ensuring that each partner's clients automatically become WISHYOUWERETHERE's clients can be achieved by noting that this is exactly the problem underlying OBSERVER DOWN. Assuming $\mathcal{L}_F$ and $\mathcal{L}_H$ have concepts Flyer and Guest, respectively, identifying their clients, and that the agency's clients will be stored as a predicate client in $\mathcal{P}$, all that needs to be done is to register client as an observer of Flyer and Guest, which, according to the pattern, is achieved by ensuring that $\langle$Flyer, client$\rangle \in \Lambda_F$ and $\langle$Guest, client$\rangle \in \Lambda_H$.

Identifying pending payments. The designers of $\mathcal{L}_A$ resorted intensively to DEF-INITIONS WITH HOLES, since many of the concepts they use can only be defined in the presence of a concrete client database. In particular, $\mathcal{L}_A$ contains a role toPay, about which it contains no membership axioms. The information about the specific purchases a client has made and not paid so far must be collected from the partners' knowledge bases, $\mathcal{L}_F$ and $\mathcal{L}_H$.

There are two ways of completing this definition. The more direct one stems from noting that toPay should be an observer of adequate roles in $\mathcal{L}_F$ and $\mathcal{L}_H$. We will assume that these roles are payFlight and payHotel. Applying twice TRANSVERSAL OBSERVER (which is the adequate pattern), one needs to ensure that

$$\langle\text{payFlight}, \text{toPayF}\rangle \in \Lambda_F \qquad \langle\text{toPayF}, \text{toPay}\rangle \in \Psi_A$$
$$\langle\text{payHotel}, \text{toPayH}\rangle \in \Lambda_H \qquad \langle\text{toPayH}, \text{toPay}\rangle \in \Psi_A .$$

The major drawback of this solution is that it requires adding two dummy predicates to $\mathcal{P}$ whose only purpose is to serve as go-between from both knowledge bases to $\mathcal{L}_A$. An alternative solution is to create a single auxiliary predicate toPay in $\mathcal{P}$ and make toPay from $\mathcal{L}_A$ an observer of this predicate applying OBSERVER UP. In turn, we use the SPLIT DEFINITIONS pattern to connect toPay to payFlight and payHotel. The resulting Mdl-program with observers is such that:

$$\langle\text{payFlight}, \text{toPay}\rangle \in \Lambda_F \qquad \langle\text{payHotel}, \text{toPay}\rangle \in \Lambda_H \qquad \langle\text{toPay}, \text{toPay}\rangle \in \Psi_A .$$

As we will discuss later, this solution will also simplify the process of adding new partners to the agency.

Offering promotions. WISHYOUWERETHERE offers a number of promotions to its special clients. For example, in February the agency offers them a 20 % discount on all purchases. Because of the partnership, the concept of special client is distributed among all partners: a client is a special client if it fulfills one of the partners' requirements – e.g. having traveled some number of miles with the airline partner, or booked a family holiday in one of the partner's hotels, or bought one of the agency's pricey packages. The partnership protocol requires that each knowledge base provide a concept identifying which clients are eligible for promotions, so that the partners can change these criteria without requiring WISHYOUWERETHERE to change its program.

This is a situation where the COMBINED DEFINITIONS design pattern applies. Assuming that $\mathcal{L}_F$ uses TopClient for its special clients, $\mathcal{L}_H$ uses Gold and $\mathcal{P}$

defines special, these three predicates are given the same semantics through COMBINED DEFINITIONS. Intuitively, this means that, in the Mdl-program's view, all three concepts equally denote *all* special clients, regardless of where they originate. The application of the pattern translates to

$$\langle \mathsf{TopClient}, \mathsf{special} \rangle \in \Lambda_F \qquad \langle \neg \mathsf{TopClient}, \mathsf{notSpecial} \rangle \in \Lambda_F$$

$$\langle \mathsf{special}, \mathsf{TopClient} \rangle \in \Psi_F \qquad \langle \mathsf{notSpecial}, \neg \mathsf{TopClient} \rangle \in \Psi_F$$

and four similar observers in Λ_H and Ψ_H, with Gold in place of TopClient.

Furthermore, in order to determine whether a particular client is entitled to promotions, it is useful to give closed-world semantics to these predicates. Since they are all equivalent, we can do this very simply in $\mathcal{P}$ by adding the rule

$$\mathsf{notSpecial}(X) \leftarrow not\ \mathsf{special}(X).$$

Note that we did not need to apply the CLOSED-WORLD pattern because special is a predicate from $\mathcal{P}$, where the semantics is closed-world: the application of COMBINED DEFINITIONS ensures that Gold and TopClient, being equivalent, also have closed-world semantics.

In order for one of the partner companies to make its clients eligible for special promotions, its ontology just needs to contain inclusion axioms partially characterizing special clients. For example, one could have

$$\exists \mathsf{flies}.10000\mathsf{OrMore} \sqsubseteq \mathsf{TopClient} \qquad \in \mathcal{L}_F$$

$$\mathsf{familyBooking} \sqsubseteq \mathsf{Gold} \qquad \in \mathcal{L}_H$$

$$\mathsf{special}(X) \leftarrow \mathsf{booked}(X,Y), \mathsf{expensive}(Y) \qquad \in \mathcal{P}$$

A subtle issue now appears regarding the consistency problems that may arise from the use of the COMBINED DEFINITIONS pattern. Since this pattern identifies concepts from different knowledge bases, it does not *a priori* guarantee that the resulting knowledge bases are consistent. In particular, if one of the partners grants special status to a client and another denies this status to the same client, an inconsistency will arise. More sophisticated variations of the COMBINED DEFINITIONS pattern can be developed to detect and avoid this kind of situation, but such a discussion is beyond the scope of this presentation.

An example of a promotion offered by WISHYOUWERETHERE to special clients would be

$$20\%\mathsf{Discount}(X) \leftarrow \mathsf{special}(X).$$

All special clients will benefit from this discount, regardless of who (the travel agency, the hotel partner or the aviation company) decided that they should be special clients. However, in some cases partners may want to deny their promotions to particular clients. For example, the aviation company is offering 100 bonus miles to special costumers booking a flight on a Tuesday, but this promotion does not apply to its workers. In order to allow this kind of situation, partners may define a dedicated concept identifying the non-eligible clients. Since all clients external to that partner are automatically eligible, this concept needs to have closed-world semantics so that (in our example) $\mathcal{L}_F$ can include the rules

$$\text{100BonusMilesWinner} \sqsubseteq \text{TopClient} \sqcap \neg\text{Blocked}$$
$$\text{Worker} \sqsubseteq \text{Blocked}$$

still giving the promotion to all clients from the other partners. Although each knowledge base can enforce this semantics in its domain, in order to extend it to other clients the CLOSED-WORLD pattern must be applied, so we will have

$$\langle \text{Blocked}, \text{blockedF} \rangle \in \Lambda_F \qquad \langle \text{nonBlockedF}, \neg\text{Blocked} \rangle \in \Psi_F$$
$$\text{nonBlockedF}(X) \leftarrow not\ \text{blockedF}(X) \in \mathcal{P}$$

Suppose that airline employee Ann qualifies for WISHYOUWERETHERE promotions because she spent three weeks in Jamaica with her husband and their five children, hence Gold(Ann) holds in $\mathcal{L}_H$ and therefore Ann is a special client. She is therefore eligible for WISHYOUWERETHERE's promotions, but she will still not earn the bonus miles because it is $\mathcal{L}_F$ who decides whether someone gets that particular promotion, and even though TopClient(Ann) holds that knowledge base will not return 100BonusMilesWinner(Ann). However, she will earn the 20%Discount, since it is offered directly by WISHYOUWERETHERE.

Adding new partnerships. We now discuss briefly how new partners can be easily added to the system later on, as this illustrates quite well the advantages of working both with design patterns and in the context of Mdl-programs with observers.

Summing up what we have so far relating to the partnerships, the sets $\Lambda_F, \Lambda_H, \Psi_F$ and Ψ_H are:

Λ_F: $\langle \text{Flyer}, \text{client} \rangle$
$\qquad\langle \text{payFlight}, \text{toPay} \rangle$
$\qquad\langle \text{TopClient}, \text{special} \rangle$
$\qquad\langle \neg\text{TopClient}, \text{notSpecial} \rangle$
$\qquad\langle \text{Blocked}, \text{blockedF} \rangle$

Λ_H: $\langle \text{Guest}, \text{client} \rangle$
$\qquad\langle \text{payHotel}, \text{toPay} \rangle$
$\qquad\langle \text{Gold}, \text{special} \rangle$
$\qquad\langle \neg\text{Gold}, \text{notSpecial} \rangle$
$\qquad\langle \text{Blocked}, \text{blockedH} \rangle$

Ψ_F: $\langle \text{special}, \text{TopClient} \rangle$
$\qquad\langle \text{notSpecial}, \neg\text{TopClient} \rangle$
$\qquad\langle \text{nonBlockedF}, \neg\text{Blocked} \rangle$

Ψ_H: $\langle \text{special}, \text{Gold} \rangle$
$\qquad\langle \text{notSpecial}, \neg\text{Gold} \rangle$
$\qquad\langle \text{nonBlockedH}, \neg\text{Blocked} \rangle$

Also, the application of the design patterns added the following rules to $\mathcal{P}$.

$$\text{nonBlockedF}(X) \leftarrow not\ \text{blockedF}(X)$$
$$\text{nonBlockedH}(X) \leftarrow not\ \text{blockedH}(X)$$
$$\text{notSpecial}(X) \leftarrow not\ \text{special}(X)$$

The similarity between Λ_F and Λ_H, and between Ψ_F and Ψ_H, is a clear illustration of the changes required when future partners of WISHYOUWERETHERE are added to the system. Furthermore, the names they use for each concept or role are not relevant – they just need to indicate how they identify their clients, their clients' debts, their special clients, and the clients they wish to exclude from their promotions.

5   Beyond these Patterns

The set of design patterns we presented does not by any means claim to be exhaustive or all-powerful. Design patterns for several programming paradigms have been around for more than two decades, and dozens of different patterns have been proposed and applied, often in very specific contexts. Our goal was to show how some of the most common general-purpose design patterns can be implemented within the framework of Mdl-programs, thereby illustrating the potential of this formalism. Among the more elaborate design patterns, our selection took into account the ones that can be more naturally formalized using Mdl-programs with observers. In this section we explore some limitations and discuss future directions for our work.

In a practical context, it is not uncommon to have a function in $\mathcal{P}$ whose definition is unstable in the sense that it may vary, for example rotating cyclically among several possibilities. Also, there are cases where one foresees possible future variations which are not contemplated in the existing requirements. The following pattern provides a clean way to implement such functions in a way that minimizes undesirable impact on the other elements of the program.

Pattern INDIRECTION.

Problem.   There is a predicate p (in $\mathcal{P}$) whose definition may vary, but $\mathcal{P}$ should be protected from these variations, in the sense that it should suffer minimal and easily identifiable modifications.

Solution.   Create a stable interface through a dedicated knowledge base $\mathcal{L}_I$ in $\mathcal{KB}$.

Define p with a set of rules, each one protected by a query to $\mathcal{L}_I$ on a concept or role S.

Define S in $\mathcal{L}_I$ such that the satisfiable clauses of p are the ones corresponding to its current definition.

Different applications of this pattern may share the same dedicated knowledge base $\mathcal{L}_I$, since each of them only looks at a particular concept or role.

The INDIRECTION design pattern captures some aspects of the principle of Protected Variations in object-oriented programming [5], which is a root principle motivating most of the mechanisms and patterns in programming and design that provide flexibility and protection from variations. This pattern was not originally introduced in [15].

As a particular case, it may happen that a component of a system is not known or available at the time of implementation of others, yet it is necessary to query it. A way to get around this is to use a prototype knowledge base that will later on be connected to the concrete component in a straightforward way.

The same problem may also arise if one wishes to be able to replace a knowledge base with another with a similar purpose, but whose concept and role names may be different.

Both of these situations reflect another aspect of the same principle of Protected Variations mentioned above, but now the point of variation in $\mathcal{KB}$ that $\mathcal{P}$ is being protected from lies in one of the knowledge bases and not in $\mathcal{P}$ itself.

Pattern ADAPTER.

Problem. One wants to work with $\mathcal{L}_k$ independently of its particular syntax.

Solution. Decide the names to use in $\mathcal{P}$ for the concepts and roles involved, and add an empty interface knowledge base $\mathcal{L}_I$ to $\mathcal{KB}$ using these names.

    Later, connect each concept and role in $\mathcal{L}_I$ with its counterpart in $\mathcal{L}_k$ by means of an application of the TRANSVERSAL OBSERVER pattern.

There is one important characteristic of this implementation of the usual Adapter design pattern: the Mdl-program syntax for local extensions to dl-queries only works in the particular case where the query is over a concept or role being directly extended. Because all queries go through the interface knowledge base, where no axioms exist, any other extensions are lost. The following example illustrates this situation.

Consider an interface $\mathcal{L}_I$ specifying two concepts P and Q, which are made concrete in $\mathcal{L}_C$ as A and B. Furthermore, $\mathcal{L}_C$ also contains the inclusion axiom $A \sqsubseteq B$. Finally, $\mathcal{P}$ contains the single fact thisIsTrue(ofMe). In $\mathcal{P}$, the direct query $DL_C[A \uplus \text{thisIsTrue}; B](X)$ returns the answer $X = \text{ofMe}$, since $\mathcal{L}_C$ is extended with $A(\text{ofMe})$ in the context of this query. However, the corresponding indirect query (i.e. the same query, but passing through the adapter)

$$DL_I[P \uplus p^+, P \cup p^-, Q \uplus q^+, Q \cup q^-, P \uplus \text{thisIsTrue}; Q](X)$$

after extending $\mathcal{P}$ with the rules

$$p^+(X) \leftarrow DL_C[; A](X) \qquad\qquad q^+(X) \leftarrow DL_C[; B](X)$$
$$p^-(X) \leftarrow DL_C[; \neg A](X) \qquad\qquad q^-(X) \leftarrow DL_C[; \neg B](X)$$

– introduced by the concretization of the observer sets – returns no answer, since $\mathcal{L}_I$ only knows the facts about B that are directly given by $\mathcal{L}_C$ through q^+.

Note, however, that the dl-atom $DL_C[A \uplus \text{thisIsTrue}; A](X)$ *is* equivalent to

$$DL_I[P \uplus p^+, P \cup p^-, Q \uplus q^+, Q \cup q^-, P \uplus \text{thisIsTrue}; P](X),$$

since the query is directly on the concept whose extent was altered. This is a restriction with respect to the full power of Mdl-programs; but we see it as a *feature* of the ADAPTER design pattern. Should a context arise where such flexibility is essential, then this is not the right design pattern to apply. In practice, situations where a ADAPTER is applicable are common enough to make it a useful pattern.

A more problematic situation arises when one wants to control or restrict access to a resource, for example a database containing sensitive information – a problem typically addressed by means of a proxy. In practice, this is not very different from the ADAPTER design pattern – but ADAPTER is an algorithm-free pattern that just defines interfaces, whereas an entity implementing PROXY is expected to do some processing before passing on the information it receives.

An implementation along the lines we have followed so far would explore the possibility of a proxy knowledge base to serve as a mediator between two components. In the setting of Mdl-programs, however, this is actually not possible to achieve directly, since all queries must go through the logical program. The only other option is to encode the proxy in the logic program itself, forcing every dl-query to the protected resource to be immediately preceded by some atoms implementing the proxy – which from the PROXY design pattern perspective is not completely satisfactory, since the person who develops the logic program can access the implementation of the proxy.

There would be ways to go around these problems, namely by extending Mdl-programs with appropriate syntactic constructions besides observers. Indeed, our motivation for defining Mdl-programs with observers was, primarily, to guarantee that *all* dl-queries were appropriately extended, *even the ones that were written after deciding that a concept or role should be observing a predicate*. As it turned out, this construction is powerful enough to allow for elegant implementations of all the design patterns discussed earlier. There is an aspect that cannot be over-stressed: the sets Λ_i and Ψ_i are syntactic sugar. As such, they do not add to the expressive power of Mdl-programs, but they substantially increase their legibility and internal structure. By working with an Mdl-program with observers, one can more easily understand the core of the program (which is the logic program $\mathcal{P}$) without being disturbed by the presence of myriads of rules that connect $\mathcal{P}$ with the several knowledge bases. In particular, most of the design patterns we presented can be expressed simply as adding specific pairs to carefully chosen observer sets Λ_i and Ψ_i – yielding a clean program that is also very easy to maintain and extend. At the end of the day, though, the Mdl-program with observers simply translates into an Mdl-program.

To deal with a full ADAPTER or PROXY, one would have to extend the syntax of Mdl-programs in a non-conservative way; but doing this would remove their simplicity, which was the main motivation for using them in the first place. Therefore, this section can be summarized as follows: the design patterns here introduced allow one to write Mdl-programs in a clean and elegant way, thereby obtaining programs of a better quality than *ad hoc* designed solutions. If one needs to go beyond the power of Mdl-programs, namely to implement a full-fledged proxy, then one should not use them at all, but rather move to a more powerful formalism.

6   Conclusions

The purpose of the present study, at this stage, is to show that design patterns have a place in the world of the Semantic Web. One can foresee a future where there is a widespread usage of systems combining description logics with rules, and the availability of systematic design methodologies is a key ingredient to making this future a reality.

This paper extends the original presentation in [15] by discussing an initial set of design patterns for Mdl-programs, together with a large-scale example illustrating their application, and showing the inherent limitations of this programming framework.

An aspect that will have to be addressed in the future relates to the practical issues of the usage of design patterns. *Ad hoc* solutions to specific problems may be more efficient than the application of systematic methods, but they tend to yield less generalizable and less extensible software applications. Also, the use of observers (essential to many of the patterns proposed) introduces higher complexity, especially when non-stratified negation is involved. It is important to understand the compromise between efficiency and quality obtained by a systematic use of design patterns by means of a practical evaluation using the prototype implementation of Mdl-programs within `dlvhex` [24].

The mechanisms herein discussed can be applied to multi-context systems, in view of the similarities between these and Mdl-programs. A preliminary study of this connection has been undertaken in [13].

References

1. Gamma, E., Helm, R., Johnson, R., Vlissides, J.: Design Patterns: Elements of Reusable Object-Oriented Software. Addison-Wesley, Reading (1995)
2. Adams, M., Coplien, J., Gamoke, R., Hanmer, R., Keeve, F., Nicodemus, K.: Fault-tolerant telecommunication system patterns. In: Vlissides, J.M., Coplien, J.O., Kerth, K.L. (eds.) Pattern Languages of Program Design, vol. 2, pp. 549–562. Addison-Wesley Longman Publishing Co. Inc., Boston (1996)
3. Erl, T.: SOA Design Patterns. Prentice Hall, New York (2009)
4. Fowler, M.: Patterns of Enterprise Application Architecture. Addison-Wesley, Boston (2002)
5. Larman, C.: Applying UML and Patterns, 3rd edn. Prentice-Hall, Upper Saddle River (2004)
6. Mattson, T., Sanders, B., Massingill, B.: Patterns for Parallel Programming. Addison-Wesley, Reading (2005)
7. Meyer, B.: Object-Oriented Software Construction, 2nd edn. Prentice-Hall, Upper Saddle River (1997)
8. Schmidt, D., Stal, M., Rohnert, H., Buschmann, F.: Pattern-Oriented Software Architecture - Patterns for Concurrent and Networked Objects. Wiley, New York (2000)
9. Antoy, S., Hanus, M.: Functional logic design patterns. In: Hu, Z., Rodríguez-Artalejo, M. (eds.) FLOPS 2002. LNCS, vol. 2441, pp. 67–87. Springer, Heidelberg (2002)
10. Gibbons, J.: Design patterns as higher-order datatype-generic programs. In: Hinze, R. (ed.) Proceedings of WGP 2006, pp. 1–12. ACM (2006)
11. Norvig, P.: Design patterns in dynamic programming. Tutorial slides presented at Object World, Boston, MA, May 1996. http://norvig.com/design-patterns/
12. Sterling, L.: Patterns for prolog programming. In: Kakas, A.C., Sadri, F. (eds.) Computational Logic: Logic Programming and Beyond. LNCS (LNAI), vol. 2407, pp. 374–401. Springer, Heidelberg (2002)

13. Cruz-Filipe, L., Henriques, R., Nunes, I.: Description logics, rules and multi-context systems. In: McMillan, K., Middeldorp, A., Voronkov, A. (eds.) LPAR-19 2013. LNCS, vol. 8312, pp. 243–257. Springer, Heidelberg (2013)
14. Eiter, T., Ianni, G., Lukasiewicz, T., Schindlauer, R., Tompits, H.: Combining answer set programming with description logics for the semantic web. Artif. Intell. **172**(12–13), 1495–1539 (2008)
15. Cruz-Filipe, L., Nunes, I., Gaspar, G.: Patterns for interfacing between logic programs and multiple ontologies. In: Filipe, J., Dietz, J. (eds.) KEOD 2013, pp. 58–69. INSTICC (2013)
16. Gangemi, A., Presutti, V.: Ontology design patterns. In: Staab, S., Studer, R. (eds.) Handbook on Ontologies. International Handbooks on Information Systems, 2nd edn, pp. 221–243. Springer, Heidelberg (2009)
17. Eiter, T., Ianni, G., Lukasiewicz, T., Schindlauer, R.: Well-founded semantics for description logic programs in the semantic Web. ACM Trans. Comput. Logic **12**(2), 1–41 (2011). Article 11
18. Kifer, M., Boley, H. (eds.): RIF overview. W3C Working Group Note, June 2010. http://www.w3.org/TR/2010/NOTE-rif-overview-20100622/
19. Bruijn, J.D., Ehrig, M., Feier, C., Martíns-Recuerda, F., Scharffe, F., Weiten, M.: Ontology mediation, merging, and aligning. In: Davies, J., Studer, R., Warren, P. (eds.) Semantic Web Technologies: Trends and Research in Ontology-based Systems. Wiley, Chichester (2006)
20. Grau, B., Parsia, B., Sirin, E.: Combining OWL ontologies using e-connections. J. Web Semant. **4**(1), 40–59 (2005)
21. Eiter, T., Ianni, G., Schindlauer, R., Tompits, H.: Effective integration of declarative rules with external evaluations for semantic-web reasoning. In: Sure, Y., Domingue, J. (eds.) ESWC 2006. LNCS, vol. 4011, pp. 273–287. Springer, Heidelberg (2006)
22. Brewka, G., Eiter, T.: Equilibria in heterogeneous nonmonotonic multi-context systems. In: Proceedings of AAAI 2007, pp. 385–390. AAAI Press (2007)
23. Wang, K., Antoniou, G., Topor, R., Sattar, A.: Merging and aligning ontologies in dl-programs. In: Adi, A., Stoutenburg, S., Tabet, S. (eds.) RuleML 2005. LNCS, vol. 3791, pp. 160–171. Springer, Heidelberg (2005)
24. Henriques, R.: Integration of ontologies with programs based on rules. Master's thesis, FCUL, November 2013
25. Eiter, T., Ianni, G., Schindlauer, R., Tompits, H.: Towards efficient evaluation of HEX programs. In: Dix, J., Hunter, A. (eds.) Proceedings of NMR 2006, ASP Track, pp. 40–46. Institut für Informatik, TU Clausthal, Germany (2006)

Integration of and a Solution for Proof Problems and Query-Answering Problems

Kiyoshi Akama[1]([⊠]) and Ekawit Nantajeewarawat[2]

[1] Information Initiative Center, Hokkaido University, Sapporo, Hokkaido, Japan
akama@iic.hokudai.ac.jp
[2] Computer Science Program, Sirindhorn International Institute of Technology,
Thammasat University, Pathumthani, Thailand
ekawit@siit.tu.ac.th

Abstract. Proof problems have long been the main target for logical problem solving. A problem in this class is a "yes/no" problem concerning with checking whether one logical formula is a logical consequence of another logical formula. Meanwhile, the importance of anther class of problems, query-answering problems (QA problems), has been increasingly recognized. A QA problem is an "all-answers finding" problem concerning with finding all ground instances of a query atomic formula that are logical consequences of a given logical formula. Several specific subclasses of QA problems have been addressed based on solution techniques for proof problems, without success of finding general solutions. In order to establish solution methods for proof problems and QA problems, we integrate these two classes of problems by embedding proof problems into QA problems. Construction of low-cost embedding mappings from proof problems to QA problems is demonstrated. By such embedding, proof problems can be solved using a procedure for solving QA problems. A procedure for solving QA problems based on equivalent transformation is presented. The presented work provides a new framework for integration of proof problems and QA problems and a solution for them by the general principle of equivalent transformation.

Keywords: Query-answering problems · Proof problems · Equivalent transformation · Solving logical problems

1 Introduction

Given a first-order formula K, representing background knowledge, and an atomic formula (atom) a, representing a query, a *query-answering problem* (*QA problem*) is to find the set of all ground instances of a that are logical consequences of K. Characteristically, it is an "all-answers finding" problem, i.e., all ground instances of the query atom satisfying the requirement must be found. A *proof problem*, by contrast, is a "yes/no" problem; it is concerned with checking whether or not one given logical formula is a logical consequence of another given logical formula.

© Springer-Verlag Berlin Heidelberg 2015
A. Fred et al. (Eds.): IC3K 2013, CCIS 454, pp. 215–229, 2015.
DOI: 10.1007/978-3-662-46549-3_14

Historically, works on logic-based automated reasoning have been centered around proof problems [5–7,10]. Methods for solving proof problems were developed, e.g., tableau-based methods [4] and resolution-based methods [11], and they have been subsequently adapted to address other classes of logical problems, including some specific subclasses of QA problems, e.g., QA problems on definite clauses [8]. As opposed to such a proof-centered approach, we present in this paper a direct approach towards solving QA problems on the basis of the *equivalent transformation* (*ET*) principle. We show that proof problems can naturally be considered as QA problems of a special form; therefore, a method for solving QA problems also lends itself to solve proof problems in a straightforward way.

In order to clearly understand the relation between proof problems and QA problems, we introduce the notion of an embedding mapping from one problem class to another problem class. Using an embedding mapping, we demonstrate that proof problems can be formulated as a subclass of QA problems. We propose a framework for solving QA problems by ET. A given input QA problem on first-order logic is converted into an equivalent QA problem on an extended clause space, called the ECLS$_F$ space, through meaning-preserving Skolemization [1]. The obtained QA problem is then successively transformed on the ECLS$_F$ space by application of ET rules until the answer to the original problem can be readily obtained. With an embedding mapping from proof problems to QA problems, this framework can be used for solving proof problems.

To begin with, Sect. 2 formalizes QA problems and proof problems. Section 3 defines an embedding mapping and shows how to embed proof problems into QA problems. Section 4 introduces extended clauses, the extended space ECLS$_F$ and QA problems on this space. Section 5 presents our ET-based procedure for solving QA problems. Section 6 defines unfolding transformation on the ECLS$_F$ space and provides some other ET rules on this space. Section 7 illustrates application of our framework. Section 8 concludes the paper.

2   QA Problems and Proof Problems

2.1   Interpretations and Models

In this paper, an atom occurring in a first-order formula can be either a usual atom or a constraint atom. The semantics of first-order formulas based on a logical structure given in [2] is used. The set of all ground usual atoms, denoted by $\mathcal{G}$, is taken as the interpretation domain. An *interpretation* is a subset of $\mathcal{G}$. A ground usual atom g is true with respect to an interpretation I iff g belongs to I. Unlike ground usual atoms, the truth values of ground constraint atoms are predetermined independently of interpretations. A *model* of a first-order formula E is an interpretation that satisfies E. The set of all models of a first-order formula E is denoted by *Models*(E). Given first-order formulas E_1 and E_2, E_2 is a *logical consequence* of E_1 iff every model of E_1 is a model of E_2.

2.2   QA Problems

A *query-answering problem* (*QA problem*) is a pair $\langle K, a \rangle$, where K is a first-order formula, representing background knowledge, and a is a usual atom, representing a query. The answer to a QA problem $\langle K, a \rangle$, denoted by $answer_{qa}(\langle K, a \rangle)$, is defined as the set of all ground instances of a that are logical consequences of K. Using $Models(K)$, the answer to a QA problem $\langle K, a \rangle$ can be equivalently defined as

$$answer_{qa}(\langle K, a \rangle) = (\bigcap Models(K)) \cap rep(a),$$

where $rep(a)$ denotes the set of all ground instances of a. Accordingly, a QA problem can also be seen as a model-intersection problem. When no confusion is caused, $answer_{qa}(\langle K, a \rangle)$ is often written as $answer_{qa}(K, a)$.

2.3   Proof Problems

A *proof problem* is a pair $\langle E_1, E_2 \rangle$, where E_1 and E_2 are first-order formulas, and the answer to this problem, denoted by $answer_{pr}(\langle E_1, E_2 \rangle)$, is defined by

$$answer_{pr}(\langle E_1, E_2 \rangle) = \begin{cases} \text{"yes" if } E_2 \text{ is a logical} \\ \qquad \text{consequence of } E_1, \\ \text{"no" otherwise.} \end{cases} \tag{1}$$

It is well known that a proof problem $\langle E_1, E_2 \rangle$ can be converted into the problem of determining whether $E_1 \wedge \neg E_2$ is unsatisfiable [5], i.e., whether $E_1 \wedge \neg E_2$ has no model. As a result, $answer_{pr}(\langle E_1, E_2 \rangle)$ can be equivalently defined by

$$answer_{pr}(\langle E_1, E_2 \rangle) = \begin{cases} \text{"yes" if } Models(E_1 \wedge \neg E_2) \\ \qquad \text{is the empty set,} \\ \text{"no" otherwise.} \end{cases} \tag{2}$$

When no confusion is caused, $answer_{pr}(\langle E_1, E_2 \rangle)$ is often written as $answer_{pr}(E_1, E_2)$.

3   Embedding Proof Problems into QA Problems

3.1   Embedding Mappings

The notion of a class of problems and that of an embedding mapping are formalized below.

Definition 1. A *class* **C** *of problems* is a triple $\langle \text{PROB}, \text{ANS}, answer \rangle$, where

1. PROB and ANS are sets,
2. *answer* is a mapping from PROB to ANS.

The sets PROB and ANS are called the *problem space* and the *answer space*, respectively, of **C**. Their elements are called *problems* and *(possible) answers*, respectively, in **C**. Given a problem $prb \in \text{PROB}$, $answer(prb)$ is the answer to prb in **C**.   □

Definition 2. Let $\mathbf{C}_1 = \langle \text{PROB}_1, \text{ANS}_1, \textit{answer}_1 \rangle$ and $\mathbf{C}_2 = \langle \text{PROB}_2, \text{ANS}_2, \textit{answer}_2 \rangle$ be classes of problems. An *embedding mapping* from $\mathbf{C}_1$ to $\mathbf{C}_2$ is a pair $\langle \pi, \alpha \rangle$, where π is an injective mapping from PROB_1 to PROB_2 and α is a partial mapping from ANS_2 to ANS_1 such that for any $prb \in \text{PROB}_1$, $\textit{answer}_1(prb) = \alpha(\textit{answer}_2(\pi(prb)))$. □

Let $\mathbf{C}_1$ and $\mathbf{C}_2$ be classes of problems. Suppose that (i) there exists an embedding mapping $\langle \pi, \alpha \rangle$ from $\mathbf{C}_1$ to $\mathbf{C}_2$, (ii) there exists a procedure P for solving problems in $\mathbf{C}_2$, and (iii) there also exist a procedure P_π for realizing π and a procedure P_α for realizing α. Then a procedure for solving problems in $\mathbf{C}_1$ can be obtained by making the composition of the procedures P_π, P and P_α. $\mathbf{C}_1$ is regarded as a *subclass* of $\mathbf{C}_2$ iff there exists an embedding mapping $\langle \pi, \alpha \rangle$ from $\mathbf{C}_1$ to $\mathbf{C}_2$ such that π and α can be realized at low computational cost.

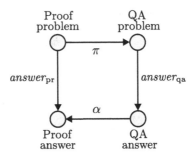

Fig. 1. Embedding proof problems into QA problems.

3.2   Embedding Proof Problems into QA Problems

Next, we show how to embed proof problems into QA problems. Assume that:

- $\mathbf{C}_{\text{qa}} = \langle \text{PROB}_{\text{qa}}, \text{ANS}_{\text{qa}}, \textit{answer}_{\text{qa}} \rangle$ is the class of QA problems defined by Sect. 2.2, i.e., PROB_{qa} is the set of all QA problems, ANS_{qa} is the power set of $\mathcal{G}$, and $\textit{answer}_{\text{qa}} : \text{PROB}_{\text{qa}} \to \text{ANS}_{\text{qa}}$ is given by Sect. 2.2.
- $\mathbf{C}_{\text{pr}} = \langle \text{PROB}_{\text{pr}}, \text{ANS}_{\text{pr}}, \textit{answer}_{\text{pr}} \rangle$ is the class of proof problems defined by Sect. 2.3, i.e., PROB_{pr} is the set of all proof problems, $\text{ANS}_{\text{pr}} = \{\,\text{"yes", "no"}\,\}$, and $\textit{answer}_{\text{pr}} : \text{PROB}_{\text{pr}} \to \text{ANS}_{\text{pr}}$ is given by Sect. 2.3.

Figure 1 gives a pictorial view of an embedding mapping from $\mathbf{C}_{\text{pr}}$ to $\mathbf{C}_{\text{qa}}$. In order to construct such an embedding mapping, we want to construct from any arbitrary given proof problem $\langle E_1, E_2 \rangle$ a QA problem $\langle K, yes \rangle$ such that $\textit{answer}_{\text{pr}}(E_1, E_2) = $ "yes" iff $\textit{answer}_{\text{qa}}(K, yes) = \{yes\}$, where yes is a 0-ary predicate symbol and the atom yes occurs in neither E_1 nor E_2. The following approaches can be taken for constructing such a formula K:

- Construct K such that every model of K contains yes iff $\textit{answer}_{\text{pr}}(E_1, E_2) = $ "yes".
- Construct K such that K has no model iff $\textit{answer}_{\text{pr}}(E_1, E_2) = $ "yes".

We refer to the first approach as *positive* construction, and the second one as *negative* construction. They are given below.

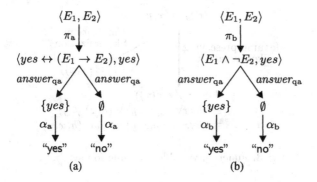

Fig. 2. Embedding proof problems into QA problems: (a) Using positive construction; (b) Using negative construction.

Embedding Using Positive Construction. Positive construction of an embedding mapping from $\mathbf{C}_{pr}$ to $\mathbf{C}_{qa}$ can be obtained by Proposition 1.

Proposition 1. Let E_1 and E_2 be first-order formulas. Assume that:

1. *yes* is a 0-ary predicate symbol and *yes* occurs in neither E_1 nor E_2.
2. prb_1 is the proof problem $\langle E_1, E_2 \rangle$.
3. prb_2 is the QA problem $\langle yes \leftrightarrow (E_1 \rightarrow E_2), yes \rangle$.

Then $answer_{pr}(prb_1) = $ "yes" iff $answer_{qa}(prb_2) = \{yes\}$. □

As depicted by Fig. 2(a), Proposition 1 determines an embedding mapping $\langle \pi_a, \alpha_a \rangle$ from $\mathbf{C}_{pr}$ to $\mathbf{C}_{qa}$ as follows:

- For any proof problem $\langle E_1, E_2 \rangle$, $\pi_a(\langle E_1, E_2 \rangle) = \langle yes \leftrightarrow (E_1 \rightarrow E_2), yes \rangle$.
- $\alpha_a(\{yes\}) = $ "yes" and $\alpha_a(\emptyset) = $ "no".

Embedding Using Negative Construction. The next proposition illuminates negative construction of an embedding mapping from $\mathbf{C}_{pr}$ to $\mathbf{C}_{qa}$.

Proposition 2. Let E_1 and E_2 be first-order formulas. Assume that:

1. *yes* is a 0-ary predicate symbol and *yes* occurs in neither E_1 nor E_2.
2. prb_1 is the proof problem $\langle E_1, E_2 \rangle$.
3. prb_2 is the QA problem $\langle E_1 \wedge \neg E_2, yes \rangle$.

Then $answer_{pr}(prb_1) = $ "yes" iff $answer_{qa}(prb_2) = \{yes\}$. □

As shown in Fig. 2(b), Proposition 2 determines an embedding mapping $\langle \pi_b, \alpha_b \rangle$ from $\mathbf{C}_{pr}$ to $\mathbf{C}_{qa}$ as follows:

- For any proof problem $\langle E_1, E_2 \rangle$, $\pi_b(\langle E_1, E_2 \rangle) = \langle E_1 \wedge \neg E_2, yes \rangle$.
- $\alpha_b(\{yes\}) = $ "yes" and $\alpha_b(\emptyset) = $ "no".

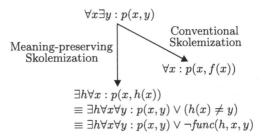

$$\forall x \exists y : p(x, y)$$

Meaning-preserving
Skolemization

Conventional
Skolemization

$$\forall x : p(x, f(x))$$

$$\exists h \forall x : p(x, h(x))$$
$$\equiv \exists h \forall x \forall y : p(x, y) \vee (h(x) \neq y)$$
$$\equiv \exists h \forall x \forall y : p(x, y) \vee \neg func(h, x, y)$$

Fig. 3. Function variables vs. function symbols.

4   QA Problems on an Extended Space

To solve a QA problem $\langle K, a \rangle$ on first-order logic, the first-order formula K is usually converted into a conjunctive normal form. The conversion involves removal of existential quantifications by Skolemization, i.e., by replacement of an existentially quantified variable with a Skolem term determined by a relevant part of a formula prenex. The classical Skolemization, however, does not preserve the logical meaning of a formula—the formula resulting from Skolemization is not necessarily equivalent to the original one [5]. In [1], a theory for extending the space of first-order formulas was developed and how meaning-preserving Skolemization can be achieved in the obtained extended space was shown. A procedure for converting first-order formulas into extended conjunctive normal forms in an extended clause space, called the ECLS$_F$ space, was also presented.

The basic idea of meaning-preserving Skolemization [1] is to use existentially quantified function variables instead of usual Skolem functions. Figure 3 illustrates the basic difference between meaning-preserving Skolemization and the conventional Skolemization, where h is a unary function variable, *func* is a built-in predicate symbol, and f is a usual unary Skolem function. Function variables, *func*-atoms, extended clauses, extended conjunctive normal forms, and QA problems on ECLS$_F$ are introduced below.

4.1   Function Constants, Function Variables and *func*-Atoms

A usual function symbol, say f, in first-order logic denotes an unevaluated function; it is used for constructing from existing terms, say $t_1, \ldots, t_n$, a syntactically new term, e.g., $f(t_1, \ldots, t_n)$, possibly recursively, without evaluating the new term $f(t_1, \ldots, t_n)$. A different class of functions is used in the extended space. A function in this class is an actual mathematical function, say h, on ground terms; when it takes ground terms, say $t_1, \ldots, t_n$, as input, $h(t_1, \ldots, t_n)$ is evaluated for determining an output ground term. We called a function in this class a *function constant*. Variables of a new type, called *function variables*, are introduced; each of them can be instantiated into a function constant or a function variable, but not into a usual term.

In order to clearly separate function constants and function variables from usual function symbols and usual terms, a new built-in predicate symbol *func* is introduced. Given any n-ary function constant or n-ary function variable $\bar{f}$, an expression $func(\bar{f}, t_1, \ldots, t_n, t_{n+1})$, where the t_i are usual terms, is considered as an atom of a new type, called a *func -atom*. When $\bar{f}$ is a function constant and the t_i are all ground, the truth value of this atom is evaluated as follows: it is true iff $\bar{f}(t_1, \ldots, t_n) = t_{n+1}$.

4.2   Extended Clauses

An *extended clause* C is a closed formula of the form

$$\forall v_1, \ldots, \forall v_m : (a_1 \vee \cdots \vee a_n \vee \neg b_1 \vee \cdots \vee \neg b_p \vee \neg \mathbf{f}_1 \vee \cdots \vee \neg \mathbf{f}_q),$$

where $v_1, \ldots, v_m$ are usual variables, each of $a_1, \ldots, a_n, b_1, \ldots, b_p$ is a usual atom or a constraint atom, and $\mathbf{f}_1, \ldots, \mathbf{f}_q$ are *func*-atoms. It is often written simply as $(a_1, \ldots, a_n \leftarrow b_1, \ldots, b_p, \mathbf{f}_1, \ldots, \mathbf{f}_q)$. The sets $\{a_1, \ldots, a_n\}$ and $\{b_1, \ldots, b_p, \mathbf{f}_1, \ldots, \mathbf{f}_q\}$ are called the *left-hand side* and the *right-hand side*, respectively, of the extended clause C, denoted by $lhs(C)$ and $rhs(C)$, respectively. When $n = 0$, C is called a *negative extended clause*. When $n = 1$, C is called an *extended definite clause*, the only atom in $lhs(C)$ is called the *head* of C, denoted by $head(C)$, and the set $rhs(C)$ is also called the *body* of C, denoted by $body(C)$. When $n > 1$, C is called a *multi-head extended clause*. All usual variables in an extended clause are universally quantified and their scope is restricted to the clause itself. When no confusion is caused, an extended clause, a negative extended clause, an extended definite clause and a multi-head extended clause will also be called a *clause*, a *negative clause*, a *definite clause* and a *multi-head clause*, respectively.

An extended normal form called *existentially quantified conjunctive normal form* (ECNF) is a formula of the form $\exists v_{h1}, \ldots, \exists v_{hm} : (C_1 \wedge \cdots \wedge C_n)$, where $v_{h1}, \ldots, v_{hm}$ are function variables and $C_1, \ldots, C_n$ are extended clauses. It is often identified with the set $\{C_1, \ldots, C_n\}$, with implicit existential quantifications of function variables and implicit clause conjunction. Function variables in such a clause set are all existentially quantified and their scope covers entirely all clauses in the set.

4.3   QA Problems on ECLS$_F$

The set of all ECNFs is referred to as the *extended clause space* (ECLS$_F$). By the above identification of an ECNF with a clause set, we often regard an element of ECLS$_F$ as a set of (extended) clauses. With occurrences of function variables, clauses contained in a clause set in the ECLS$_F$ space are connected through shared function variables. By instantiating all function variables in such a clause set into function constants, clauses in the obtained set are totally separated.

A QA problem $\langle Cs, a \rangle$ such that Cs is a clause set in ECLS$_F$ and a is a usual atom is called a *QA problem on* ECLS$_F$. Given a QA problem $\langle K, a \rangle$ on first-order logic, the first-order formula K can be converted equivalently by meaning-preserving Skolemization, using the conversion procedure given in [1], into a clause set Cs in the ECLS$_F$ space. The obtained clause set Cs may be further transformed equivalently in this space for problem simplification, by using unfolding and other transformation rules.

5   Solving QA Problems

Using the notation introduced in Sects. 5.1 and 5.2, our ET-based procedure is presented in Sect. 5.3.

5.1   Inclusion of Query Information

The following notation is used. A set A of usual atoms is said to be *closed* iff for any $a \in A$ and any substitution θ for usual variables, $a\theta$ belongs to A. Assume that (i) $\mathcal{A}$ is the set of all usual atoms, (ii) $\mathcal{A}_1$ and $\mathcal{A}_2$ are disjoint closed subsets of $\mathcal{A}$, and (iii) ϕ is a bijection from $\mathcal{A}_1$ to $\mathcal{A}_2$ such that for any $a \in \mathcal{A}_1$ and any substitution θ for usual variables, $\phi(a\theta) = \phi(a)\theta$. For any $i, j \in \{1, 2\}$, an extended clause C is said to be from $\mathcal{A}_i$ to $\mathcal{A}_j$ iff all usual atoms in $rhs(C)$ belong to $\mathcal{A}_i$ and all those in $lhs(C)$ belong to $\mathcal{A}_j$.

Let $\langle K, a \rangle$ be a QA problem such that K is a first-order formula in which all usual atoms belong to $\mathcal{A}_1$ and $a \in \mathcal{A}_1$. As will be detailed in Sect. 5.3, to solve this problem using ET, K is transformed by meaning-preserving transformation into a set Cs of extended clauses from $\mathcal{A}_1$ to $\mathcal{A}_1$ and a singleton set Q consisting only of the clause $(\phi(a) \leftarrow a)$ from $\mathcal{A}_1$ to $\mathcal{A}_2$ is constructed from the query atom a. The resulting QA problem $\langle Cs \cup Q, \phi(a) \rangle$ is then successively transformed using ET rules.

5.2   Triples for Transformation

In order to make a clear separation between a set of extended clauses from $\mathcal{A}_1$ to $\mathcal{A}_1$ and a set of those from $\mathcal{A}_1$ to $\mathcal{A}_2$ in a transformation process of QA problems, the following notation is introduced: Given a set Cs of extended clauses from $\mathcal{A}_1$ to $\mathcal{A}_1$, a set Q of extended clauses from $\mathcal{A}_1$ to $\mathcal{A}_2$ and an atom b in $\mathcal{A}_2$, let the triple $\langle Cs, Q, b \rangle$ denote the QA problem $\langle Cs \cup Q, b \rangle$. A QA problem $\langle Cs, Q, b \rangle$ can be transformed by changing Cs, by changing Q, or by changing both Cs and Q.

Definition 3. A transformation of a QA problem $\langle Cs, Q, b \rangle$ into a QA problem $\langle Cs', Q', b \rangle$ is *equivalent transformation* (*ET*) iff $answer_{qa}(Cs \cup Q, b)$ and $answer_{qa}(Cs' \cup Q', b)$ are equal.   □

5.3   A Procedure for Solving QA Problems by ET

Let $\mathcal{A}_1$ be a closed set of usual atoms. Assume that a QA problem $\langle K, a \rangle$ is given, where K is a first-order formula in which all usual atoms belong to $\mathcal{A}_1$ and $a \in \mathcal{A}_1$. To solve the QA problem $\langle K, a \rangle$ using ET, perform the following steps:

1. Transform K by meaning-preserving Skolemization into a clause set Cs in the ECLS$_F$ space.
2. Determine (i) a closed set $\mathcal{A}_2$ of usual atoms such that $\mathcal{A}_1$ and $\mathcal{A}_2$ are disjoint and (ii) a bijection ϕ from $\mathcal{A}_1$ to $\mathcal{A}_2$ such that for any $a \in \mathcal{A}_1$ and any substitution θ for usual variables, $\phi(a\theta) = \phi(a)\theta$.

3. Successively transform the QA problem $\langle Cs, \{(\phi(a) \leftarrow a)\}, \phi(a)\rangle$ in the ECLS$_F$ space using unfolding and other ET rules (see Sect. 6).
4. Assume that the transformation yields a QA problem $\langle Cs', Q, \phi(a)\rangle$. Then:
    (a) If $Models(Cs') = \emptyset$, then output $rep(a)$ as the answer.
    (b) If $Models(Cs') \neq \emptyset$ and Q is a set of unit clauses such that the head of each clause in Q is an instance of $\phi(a)$, then output as the answer the set

$$\phi^{-1}(\bigcup_{C \in Q} rep(head(C))).$$

    (c) Otherwise stop with failure.

It is shown in [3] that the obtained answer is always correct.

The set $\mathcal{A}_2$ and the bijection ϕ satisfying the requirement of Step 2 can be determined as follows: First, introduce a new predicate symbol for each predicate symbol occurring in $\mathcal{A}_1$. Next, let $\mathcal{A}_2$ be the atom set obtained from $\mathcal{A}_1$ by replacing the predicate of each atom in $\mathcal{A}_1$ with the new predicate introduced for it. Finally, for each atom $a \in \mathcal{A}_1$, let $\phi(a)$ be the atom obtained from a by such predicate replacement.

6   ET Rules on ECLS$_F$

Next, ET rules for unfolding and definite-clause removal are presented, along with some other ET rules.

6.1   Unfolding Operation on ECLS$_F$

Assume that (i) Cs is a set of extended clauses, (ii) D is a set of extended definite clauses, and (iii) occ is an occurrence of an atom b in the right-hand side of a clause C in Cs. By unfolding Cs using D at occ, Cs is transformed into

$$(Cs - \{C\}) \cup (\bigcup\{resolvent(C, C', b) \mid C' \in D\}),$$

where for each $C' \in D$, $resolvent(C, C', b)$ is defined as follows, assuming that ρ is a renaming substitution for usual variables such that C and $C'\rho$ have no usual variable in common:

- If b and $head(C'\rho)$ are not unifiable, then $resolvent(C, C', b) = \emptyset$.
- If they are unifiable, then $resolvent(C, C', b) = \{C''\}$, where C'' is the clause obtained from C and $C'\rho$ as follows, assuming that θ is the most general unifier of b and $head(C'\rho)$:
    • $lhs(C'') = lhs(C\theta)$
    • $rhs(C'') = (rhs(C\theta) - \{b\theta\}) \cup body(C'\rho\theta)$.

The resulting clause set is denoted by $\text{UNFOLD}(Cs, D, occ)$.

6.2   ET by Unfolding and Definite-Clause Removal

Let $Atoms(p)$ denote the set of all atoms having a predicate p. ET rules on $ECLS_F$ for unfolding and for definite-clause removal are described below.

ET by Unfolding. Let $\langle Cs, a \rangle$ be a QA problem on $ECLS_F$. Assume that:

1. q is the predicate of the query atom a.
2. p is a predicate such that $p \neq q$.
3. D is a set of extended definite clauses in Cs that satisfies the following conditions:
   (a) For any $C \in D$, $head(C) \in Atoms(p)$.
   (b) For any $C' \in Cs - D$, $lhs(C') \cap Atoms(p) = \emptyset$.
4. occ is an occurrence of an atom in $Atoms(p)$ in the right-hand side of an extended clause in $Cs - D$.

Then $\langle Cs, a \rangle$ can be equivalently transformed into the QA problem $\langle \text{UNFOLD} (Cs, D, occ), a \rangle$.

ET by Definite-Clause Removal. Let $\langle Cs, a \rangle$ be a QA problem on $ECLS_F$. Assume that:

1. q is the predicate of the query atom a.
2. p is a predicate such that $p \neq q$.
3. D is a set of extended definite clauses in Cs that satisfies the following conditions:
   (a) For any $C \in D$, $head(C) \in Atoms(p)$.
   (b) For any $C' \in Cs - D$, $lhs(C') \cap Atoms(p) = \emptyset$.
4. For any $C' \in Cs - D$, $rhs(C') \cap Atoms(p) = \emptyset$.

Then $\langle Cs, a \rangle$ can be equivalently transformed into the QA problem $\langle Cs - D, a \rangle$.

6.3   Some Other ET Rules on $ECLS_F$

Next, ET rules for merging *func*-atoms having the same call pattern, for removing isolated *func*-atoms, and for removing subsumed clauses are presented. They are used in examples in Sect. 7.

Merging*func*-Atoms with the Same Invocation Pattern. Let $\langle Cs, a \rangle$ be a QA problem on $ECLS_F$. Suppose that $C \in Cs$ and $rhs(C)$ contains *func*-atoms $\mathbf{f}_1$ and $\mathbf{f}_2$ that differ only in their last arguments. Then:

1. If the last arguments of $\mathbf{f}_1$ and $\mathbf{f}_2$ are unifiable, with their most general unifier being θ, and C' is an extended clause such that
   – $lhs(C') = lhs(C\theta)$, and
   – $rhs(C') = (rhs(C) - \{\mathbf{f}_2\})\theta$,

then $\langle Cs, a \rangle$ can be equivalently transformed into the QA problem $\langle ((Cs - \{C\}) \cup \{C'\}, a \rangle$.
2. If their last arguments are not unifiable, then $\langle Cs, a \rangle$ can be equivalently transformed into the QA problem $\langle Cs - \{C\}, a \rangle$.

Elimination of Isolated *func* -Atoms. A *func*-atom $func(h, t_1, \ldots, t_n, v)$, where v is a usual variable, is said to be *isolated* in an extended clause C iff there is only one occurrence of v in C.

Now let $\langle Cs, a \rangle$ be a QA problem on $ECLS_F$. Suppose that:

1. $C \in Cs$ such that C contains a *func*-atom that is isolated in C.
2. C' is the extended clause obtained from C by removing all *func*-atoms that are isolated in C.

Then $\langle Cs, a \rangle$ can be equivalently transformed into the QA problem $\langle ((Cs - \{C\}) \cup \{C'\}, a \rangle$.

Elimination of Subsumed Clauses. An extended clause C_1 is said to *subsume* an extended clause C_2 iff there exists a substitution θ for usual variables such that $lhs(C_1)\theta \subseteq lhs(C_2)$ and $rhs(C_1)\theta \subseteq rhs(C_2)$.

A subsumed clause can be removed as follows: Let $\langle Cs, a \rangle$ be a QA problem on $ECLS_F$. If Cs contains extended clauses C_1 and C_2 such that C_1 subsumes C_2, then $\langle Cs, a \rangle$ can be equivalently transformed into the QA problem $\langle Cs - \{C_2\}, a \rangle$.

7   Examples

Example 1 demonstrates how the procedure in Sect. 5.3 solves a QA problem using the ET rules in Sect. 6. Example 2 shows how to apply the procedure to solve a proof problem based on the embedding mapping in Sect. 3.2.

Example 1. Consider the "Tax-cut" problem discussed in [9]. This problem is to find all persons who can have discounted tax, with the knowledge that (i) any person who has two children or more can get discounted tax, (ii) men and women are not the same, (iii) a person's mother is always a woman, (iv) Peter has a child named Paul, (v) Paul is a man, and (vi) Peter has a child, who is someone's mother. This background knowledge is represented in first-order logic as the formulas F_1–F_6 below, assuming that hc, ns, tc, mn, wm and mo stand, respectively, for *hasChild*, *notSame*, *TaxCut*, *Man*, *Woman* and *motherOf*:

F_1: $\forall x: ((\exists y_1 \exists y_2: (hc(x, y_1) \wedge hc(x, y_2) \wedge ns(y_1, y_2))) \rightarrow tc(x))$
F_2: $\forall x \forall y: ((mn(x) \wedge wm(y)) \rightarrow ns(x, y))$
F_3: $\forall x: ((\exists y: mo(x, y)) \rightarrow wm(x))$
F_4: $hc(Peter, Paul)$
F_5: $mn(Paul)$
F_6: $\exists x: (hc(Peter, x) \wedge (\exists y: mo(x, y)))$

Accordingly, the "Tax-cut" problem is formulated as the QA problem $\langle K, tc(x) \rangle$, where K is the conjunction of F_1–F_6. Using the meaning-preserving Skolemization procedure given in [1], the first-order formula K is transformed into a clause set Cs consisting of the following extended clauses:

C_1: $tc(x) \leftarrow hc(x, y_1), hc(x, y_2), ns(y_1, y_2)$
C_2: $ns(x, y) \leftarrow mn(x), wm(y)$
C_3: $wm(x) \leftarrow mo(x, y)$
C_4: $hc(Peter, Paul) \leftarrow$
C_5: $mn(Paul) \leftarrow$
C_6: $hc(Peter, x) \leftarrow func(h_1, x)$
C_7: $mo(x, y) \leftarrow func(h_1, x), func(h_2, y)$

The clauses C_6 and C_7 together represent the first-order formula F_6, where h_1 and h_2 are 0-ary function variables.

Assume that all usual atoms occurring in Cs belong to $\mathcal{A}_1$, ans is a newly introduced unary predicate symbol, all ans-atoms belong to $\mathcal{A}_2$, and for any term t, $\phi(tc(t)) = ans(t)$. Let

$$C_0 = (ans(x) \leftarrow tc(x)).$$

To solve the QA problem $\langle K, tc(x) \rangle$, the QA problem $\langle Cs, \{C_0\}, ans(x) \rangle$ is successively transformed by applying the ET rules in Sect. 6 as follows:

1. By unfolding C_0 at $tc(x)$ using $\{C_1\}$, C_0 is replaced with:

   C_8: $ans(x) \leftarrow hc(x, y_1), hc(x, y_2), ns(y_1, y_2)$

2. By unfolding C_8 at the last body atom using $\{C_2\}$, C_8 is replaced with:

   C_9: $ans(x) \leftarrow hc(x, y_1), hc(x, y_2), mn(y_1), wm(y_2)$

3. By unfolding C_9 at the third body atom using $\{C_5\}$, C_9 is replaced with:

   C_{10}: $ans(x) \leftarrow hc(x, Paul), hc(x, y_2), wm(y_2)$

4. By unfolding C_{10} at the last body atom using $\{C_3\}$, C_{10} is replaced with:

   C_{11}: $ans(x) \leftarrow hc(x, Paul), hc(x, y_2), mo(y_2, z)$

5. By unfolding C_{11} at the last body atom using $\{C_7\}$, C_{11} is replaced with:

   C_{12}: $ans(x) \leftarrow hc(x, Paul), hc(x, y_2), func(h_1, y_2), func(h_2, z)$

6. By removing an isolated $func$-atom, C_{12} is replaced with:

   C_{13}: $ans(x) \leftarrow hc(x, Paul), hc(x, y_2), func(h_1, y_2)$

7. By unfolding C_{13} at the first body atom using $\{C_4, C_6\}$, C_{13} is replaced with:

   C_{14}: $ans(Peter) \leftarrow hc(Peter, y_2), func(h_1, y_2)$
   C_{15}: $ans(Peter) \leftarrow func(h_1, Paul), hc(Peter, y_2), func(h_1, y_2)$

8. By merging *func*-atoms with the same invocation pattern, C_{15} is replaced with:

C_{16}: $ans(Peter) \leftarrow func(h_1, Paul), hc(Peter, Paul)$

9. Since C_{16} is subsumed by C_{14}, C_{16} is removed.
10. By unfolding C_{14} at the first body atom using $\{C_4, C_6\}$, C_{14} is replaced with:

C_{17}: $ans(Peter) \leftarrow func(h_1, Paul)$
C_{18}: $ans(Peter) \leftarrow func(h_1, y_2), func(h_1, y_2)$

11. By definite-clause removal, C_1–C_7 are removed.
12. By merging *func*-atoms with the same invocation pattern, C_{18} is replaced with:

C_{19}: $ans(Peter) \leftarrow func(h_1, y_2)$

13. By removing an isolated *func*-atom, C_{19} is replaced with:

C_{20}: $ans(Peter) \leftarrow$

14. Since C_{17} is subsumed by C_{20}, C_{17} is removed.

The resulting QA problem is $\langle \emptyset, \{C_{20}\}, ans(x) \rangle$. Since $Models(\emptyset) \neq \emptyset$ and C_{20} is a unit clause whose head is an instance of $\phi(tc(x))$, the answer to the "Tax-cut" problem $\langle K, tc(x) \rangle$ is determined by

$$\phi^{-1}(\bigcap\{rep(head(C_{20}))\}) = \phi^{-1}(\{ans(Peter)\}) = \{tc(Peter)\},$$

i.e., Peter is the only one who gets discounted tax.   □

Example 2. Refer to the description of the "Tax-cut" problem, the first-order formulas F_1–F_6, the clauses C_0–C_{20} and the clause set $Cs = \{C_1, \ldots, C_7\}$ in Example 1. From the background knowledge of the "Tax-cut" problem, suppose that we want to prove the existence of someone who gets discounted tax. This problem is formulated as the proof problem $\langle E_1, E_2 \rangle$, where E_1 is the conjunction of F_1–F_6 and E_2 is the first-order formula $\exists x : tc(x)$.

Using Proposition 2, this proof problem is converted into the QA problem $\langle E_1 \wedge \neg E_2, yes \rangle$. Using the procedure in Sect. 5.3, this QA problem is solved as follows:

– Convert $E_1 \wedge \neg E_2$ by meaning-preserving Skolemization, resulting in the clause set $Cs \cup \{C'_0\}$, where C'_0 is the negative clause $(\leftarrow tc(x))$.
– Transform the QA problem

$$\langle Cs \cup \{C'_0\}, \{(\phi(yes) \leftarrow yes)\}, \phi(yes) \rangle$$

using ET rules. By following the transformation Steps 1–14 in Example 1 except that the initial target clause is C'_0 instead of C_0, the clauses C'_8–C'_{20} are successively produced, where for each $i \in \{8, \ldots, 20\}$,
- $lhs(C'_i) = \emptyset$, and
- $rhs(C'_i) = rhs(C_i)$,

and C_1–C_7 are removed. As a result, $Cs \cup \{C_0'\}$ is transformed into $\{C_{20}'\}$, where $C_{20}' = (\leftarrow)$, and the QA problem

$$\langle \{C_{20}'\}, \{(\phi(yes) \leftarrow yes)\}, \phi(yes) \rangle$$

is obtained.

– Since C_{20}' is the empty clause, the clause set $\{C_{20}'\}$ has no model, i.e., $Models(\{C_{20}'\}) = \emptyset$. So the procedure outputs $rep(yes) = \{yes\}$ as the answer to the QA problem $\langle E_1 \wedge \neg E_2, yes \rangle$.

It follows from Proposition 2 that the answer to the proof problem $\langle E_1, E_2 \rangle$ is "yes", i.e., there exists someone who gets discounted tax.   □

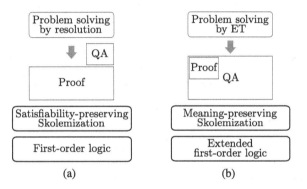

Fig. 4. (a) Conventional proof-centered approaches; (b) The proposed QA-problem-centered approach.

8   Conclusions

As shown in Fig. 4(a), previous approaches to solving QA problems are proof-centered. The classical first-order logic and the conventional Skolemization provide a foundation for solving proof problems. Based on them, several solution methods for specific subclasses of QA problems have been developed; for example, answering queries in logic programming and deductive databases can be regarded as solving QA problems on definite clauses and those on a restricted form of definite clauses, respectively. There has been no general solution method for QA problems on full first-order formulas. The conventional first-order logic and the conventional Skolemization are not enough for developing a general solution for QA problems.

QA problems on full first-order logic are considered in this paper. We introduced the concept of embedding and proposed how to embed proof problems into QA problems. This embedding leads to a unified approach to dealing with proof problems and QA problems, allowing one to use a method for solving QA problems to solve proof problems. It enables a QA-problem-centered approach to solving logical problems, which is shown in Fig. 4(b).

Equivalent transformation (ET) is one of the most fundamental principles of computation, and it provides a simple and general basis for verification of computation correctness. We proposed a framework for solving QA problems by ET. All computation steps in this framework are ET steps, including transformation of a first-order formula into an equivalent formula in the extended clause space $ECLS_F$ and transformation of extended clauses on $ECLS_F$. To the best of our knowledge, this is the only framework for dealing with the full class of QA problems on first-order formulas.

Since many kinds of ET rules can be employed, the proposed ET-based framework opens up a wide range of possibilities for computation paths to be taken. As a result, the framework enables development of a large variety of methods for solving logical problems. The range of possible computation methods can also be further extended by using computation spaces other than $ECLS_F$. Proof by resolution can be seen as one specific example of these possible methods. As demonstrated in [2], it can be realized by using two kinds of ET rules, i.e., resolution and factoring ET rules, on a computation space that differs slightly from $ECLS_F$.

References

1. Akama, K., Nantajeewarawat, E.: Meaning-preserving skolemization. In: 2011 International Conference on Knowledge Engineering and Ontology Development, Paris, France, pp. 322–327 (2011)
2. Akama, K., Nantajeewarawat, E.: Proving theorems based on equivalent transformation using resolution and factoring. In: 2nd World Congress on Information and Communication Technologies, Trivandrum, India, pp. 7–12 (2012)
3. Akama, K., Nantajeewarawat, E.: Embedding proof problems into query-answering problems and problem solving by equivalent transformation. Technical report, Hokkaido University, Sapporo, Japan (2013)
4. Beth, E.W.: Semantic Entailment and Formal Derivability. Noord-Hollandsche Uitg. Mij, Amsterdam (1955)
5. Chang, C.-L., Lee, R.C.-T.: Symbolic Logic and Mechanical Theorem Proving. Academic Press, New York (1973)
6. Fitting, M.: First-Order Logic and Automated Theorem Proving, 2nd edn. Springer, New York (1996)
7. Gallier, J.H.: Logic for Computer Science: Foundations of Automatic Theorem Proving. Wiley, New York (1986)
8. Lloyd, J.W.: Foundations of Logic Programming, 2nd edn. Springer, Heidelberg (1987)
9. Motik, B., Sattler, U., Studer, R.: Query answering for OWL-DL with rules. J. Web Semant. **3**, 41–60 (2005)
10. Newborn, M.: Automated Theorem Proving: Theory and Practice. Springer, New York (2000)
11. Robinson, J.A.: A machine-oriented logic based on the resolution principle. J. ACM **12**, 23–41 (1965)

The Web Knowledge Management:
A Taxonomy-Based Approach

Filippo Eros Pani, Maria Ilaria Lunesu,
Giulio Concas$^{(\boxtimes)}$, and Gavina Baralla

Department of Electrics and Electronics Engineering,
University of Cagliari, Piazza d'Armi, Cagliari, Italy
{filippo.pani,ilaria.lunesu,concas,
gavina.baralla}@diee.unica.it

Abstract. The structure of the Web grows and changes, giving the user the chance to actively participate in its development. This study tries to link the worlds of Social Media and Semantic Web, with the aim of proposing a semantic classification of the information coming from the Web. Our approach consists in a mixed-iterative process, where top-down and bottom-up analyses of the knowledge domain which has to be represented are applied. We start from the concept of the domain knowledge base. The fundamental body of knowledge available on a domain is the knowledge valuable for knowledge users. We need to represent and manage this knowledge, to define a formalization and codification of knowledge in the domain. This kind of representation was created according to the criteria by which information, in a given domain, is structured within the net, by taking the hierarchy proposed in different sites as a reference.

Keywords: Knowledge management · Multimedia content · Semantic web · Knowledge base · Taxonomy · Ontology

1 Introduction and Related Work

Over the last decade a broader knowledge of the Web has strengthened and fostered the developing of new applications: the Web has turned into a multifunctional platform where users no longer get the information passively; in fact, they become authors and makers. This has been mainly possible thanks to the developing of new applications which allow users to add contents without knowing any programming code. The social value which the Web has acquired recently is therefore unquestionable; the Web's structure grows and changes depending on the user's needs, becoming every day more complex.

The new frontier for the Internet is represented by the Web 3.0 [1]: with the evolution of the Web into its semantic version, a transition to a more efficient representation of knowledge is a necessary step [2]. Particularly, data are no longer represented just by the description of their structure (syntax) but also by the definition of their meaning (semantics). In fact, a datum can have a different meaning depending on the contexts; the use of tools like ontologies - explicit formal specifications of terms in

© Springer-Verlag Berlin Heidelberg 2015
A. Fred et al. (Eds.): IC3K 2013, CCIS 454, pp. 230–244, 2015.
DOI: 10.1007/978-3-662-46549-3_15

the domain and relations among them [3–5] - and taxonomies helps the classification of information, as shown also in [6–13].

In recent years, the development of models to formalize knowledge was studied and analyzed. Ontologies take an important part in those formalization approaches. In the Web range, there are many Web site directory services: the most famous is Yahoo. These directory services are large taxonomies which organize websites in categories. Other systems categorize products for e-commerce purposes: the most famous is Amazon. They use an implicit taxonomy to organize the products for sale by type and features. The World Wide Web Consortium (W3C)[1] has developed the Resource Description Framework (RDF) [14, 15], a language for encoding knowledge on Web pages to make it understandable to electronic agents searching for information, as main foreground concept of the Semantic Web. The Defense Advanced Research Projects Agency (DARPA),[2] in cooperation with the W3C, has developed DARPA Agent-Markup Language (DAML) by extending RDF with more expressive constructs aimed at facilitating agent interaction on the Web.

The Web becomes clever and is conceived as a big database in which data are orderly classified. "Information", therefore, is one of the keywords at the base of the success of both search engines (Google, Yahoo, Bing, …), which become more refined in data retrieval and presentation, and Social Networks (YouTube, Facebook, Twitter, Flickr, …), which allow exchange and sharing, creating an interconnection among users and content makers. However, such data, despite being formally available, are often unreachable as for their semantic meaning and cannot be used as real knowledge.

Various proposals to solve these problems can be found in literature, also to overcome the semantic heterogeneity problem [16] and to facilitate knowledge sharing and reuse [17, 18]. In [19] an approach based on the use of an ontology to make annotating photos and searching for specific images more intelligent is described; and in [20] a data-driven approach to investigate semi-automatic construction of multi-media ontologies is used. With the emergence of the Semantic Web, a shared vocabulary is necessary to annotate the vast collection of heterogeneous media: in [11] an ontology is proposed to provide a meaningful set of relationships which may enable this process.

Particularly, in [21, 22] the problem of representing and managing the knowledge which can be found on the Internet is discussed as for the User Generated Content (UGC), classifying this knowledge through a top-down (TD) and bottom-up (BU) combined approach. To reach such target, an ontology was built as a base to define a repository of multimedia contents, putting a special focus on the georeferencing of multimedia objects. As for the TD approach, the standards used for the multimedia objects (XMP, EXIF, etc.) have been defined by selecting data of interest to represent them on the ontology, which in turn was defined through rules of correspondence. As for the BU approach, UGCs of two particularly exemplifying platforms (Flickr and Youtube) have been analysed, in order to extrapolate some structured tags, folksono-mies and attributes of multimedia objects (characteristically EXIF, as for Flickr, and

---

[1] World Wide Web Consortium (W3C), http://www.w3.org.

[2] Defense Advanced Research Projects Agency (DARPA), http://www.darpa.mil.

proprietary tags, as for Youtube). In conclusion, this ontology allowed for the construction of a repository to store the information extracted from UGC systems, where all the information related to multimedia objects are shown (compatibly with the XMP standard) as well as other tags of general interest, apart from representing also the information which can be found on the as-is folksonomies.

This study aims at defining a new approach for the problem of the contents on the Internet, especially semi-structured contents coming from heterogeneous sources referring to a common knowledge domain. Through a combined TD and BU approach, knowledge of a specific domain was extracted defining a common structure through a taxonomy, in order to classify and make the majority of such knowledge available [23].

With the TD approach the knowledge of interest on the domain was defined, following the specifications and the analysis of the ontologies and other classifications, in order to define a reference taxonomy.

On the other hand, the BU approach started from the selection of some websites concerning the domain of interest, to pinpoint the knowledge in them. Then, these contents were classified with the taxonomy previously defined and the mapping rules between contents and taxonomy.

This taxonomy allowed for the definition of a reference knowledge which may later be managed in terms of really usable and interesting knowledge, fostered by the whole knowledge of all the selected websites.

We chose to test this approach on the knowledge domain of Italian wines reviews. As for the validation, we verified how this KMS allowed such knowledge to become available on systems that were compliant with the Wines ontology[3] as defined as an example of Semantic Web by W3C; then we checked other websites of Italian wines reviews, verifying how their contents of interest could be represented and managed on the KMS through some simple mapping rules.

The paper is structured as follows: in the second section of this paper we present our proposed approach for knowledge management and in the fourth we explain the case study. The next section includes the analysis of results and verification. Finally, the fifth section includes the conclusion and reasoning about the future evolution of the work.

2   The Approach to Manage Knowledge

All real-world information belongs to a determined structure invented and defined over the millennia; our brain associates one or more meanings to every single word, that may vary depending on the context and the sentence where the word is placed and used. People are hard-wired to need to categorize and define any type of information, even with a single word, to assign an order and speed up the search. This need has led to the development of many tools that try to simulate human reasoning and learning in an automatic way. The concept of Knowledge Management (KM) is born according to these requirements. It relates to those organizational structures that combine the

---

[3] http://www.w3.org/TR/owl-guide/wine.rdf.

potential to combine data and process information, with the creativity and ability to innovate that belongs to human beings [1].

In the same way, sharing knowledge improves information quality because knowledge is shared and unified, and can provide and receive responses in real time if this is present on the network and a collaborative strategy is in use.

Consequently, such strategy can be achieved by considering the interaction that exist between the user, the used tool and the environment in which the information is present.

If "Knowledge" is defined as a unique resource of interest which should be developed, managed and shared with others, we must consider not only the tools to be used, but also the aspects of maximum usability and accessibility for the user.

Making information shareable, and ensure that it is carried among multiple users, means, therefore, to provide a simple access to knowledge, in real time and without time limits, in order to offer a further possibility to the user: integrate, rectify and interact actively inside of content and information to make "knowledge" as complete as possible.

The proposed approach aims at defining a taxonomy able to represent knowledge through a mixed-iterative approach, where TD and BU analyses of the knowledge domain which has to be represented are applied: these are typical approaches for this kind of problems. In this case, they are applied following an iterative approach which allows, through further refinements, for the efficient definition of the taxonomy able to represent the domain's knowledge of interest.

The knowledge to be represented is the most popular among users of a certain domain. To determine which is the users' knowledge of real interest we chose to select the most used websites by users, the most important and looked up ones. For this definition, websites with a higher ranking on Google among the domain of interest are typically chosen.

2.1 Top-Down Phase

When our knowledge or our expectations are influenced by perception, we refer to schema-driven or TD elaboration. A schema is a model formerly created by our experience.

More general or abstract contents are indicated as higher level, while concrete details (senses input) are indicated as lower level.

The TD elaboration happens whenever a higher level concept influences the interpretation of lower level sensory data. Generally, the TD process is an information process based on former knowledge or acquired mental schemes; it allows us to make inferences: to "perceive" or "know" more than what can be found in data. TD methodology starts, therefore, by identifying a target to reach, and then pinpoints the strategy to use in order to reach the established goal.

Our aim is therefore to begin by a formalization of the reference knowledge (ontology, taxonomy or others) to start classifying the information on the reference domain.

The model could be, for instance, a formalization of one or more classifications of the same domain, formerly made in a logic of metadata. Therefore, the output of this phase will be a table with all the elements of knowledge formalized through the definition of the reference metadata.

2.2   Bottom-Up Phase

With this phase the knowledge to be represented is analysed by pinpointing, among the present information, the ones which are to be represented together with a reference terminology for data description.

When an interpretation emerges from data, it is called data-driven or BU elaboration. Perception is mainly data-driven, as it must precisely reflect what happens in the external world. Generally, it is better if the interpretation coming from a system of information is determined by what is effectively transmitted at sensory level rather than what is perceived as an expectation. Applying this concept, we analysed a set of websites containing the information of the domain of interest; from these websites, both information whose structure needed to be extrapolated and the information in them were pinpointed. Typically, reference websites for that information domain are selected, namely the ones which users mainly use to find information of their interest over the domain itself.

Primary information, important ones, already emerge during the phase of websites analysis and gathering: during a first skimming phase, the minimum, basic information necessary to well describe our domain can be noticed. Then, important information are extrapolated by choosing fields or keywords which best represent the knowledge, in order to create a knowledge base (KB). In this phase, one of the limits could be the creation of the KB itself, because each website is likely to show a different structure and a different way of presenting the same information. Therefore, it will be necessary to pinpoint the present information of interest, defining and outlining them.

After this analysis of gathering of information, a classification is made and it has to reflect, in the most faithful way, the structure of the knowledge proposed by every single website, respecting both its contents and hierarchy.

To analyse data, we chose to build a tabular system for each website coming from a precise identification of each information area existing in every website taken as a knowledge base.

For each website we created a table which accurately gathers and describes the information that can be found in it, with a detailed field of descriptions. With this stage we obtained a complete representation of the knowledge which can be found on the chosen websites, but not a usable one because it had not been classified yet.

2.3   Integration Phase

In these phases we will try to reconcile these two representations of knowledge of the domain, as represented in the former phases.

Thus, we want to pinpoint, for each single TD's metadata, where the information can be found in the table's fields representing the knowledge of each website (which, for us, represents the knowledge we want to represent, considering the semantic concept and not the way to represent it, absolutely subjective for every website).

At this point we check if, in every table, the information of our representation of knowledge coming from the TD can be found in the tables coming from the BU, verifying if it exists as a field or can be found in a field or is missing.

Then, we will create a mapping macro-table of knowledge containing, for each item of the taxonomy, the correspondence if and where that information exists in the various websites and also the information of the websites which are not represented by the taxonomy.

From the macrosystem a KB originates, which is able to represent both the formalization of knowledge and the present knowledge.

2.4 Formalization of Knowledge

Starting from this KB, further iterative refining can be made by re-analysing the information in different phases: (1) with a TD approach, checking if the information which are not represented by the chosen formalization can be formalized; (2) with a BU approach, analysing if some information of the websites can be connected to formalized items; (3) with the mixed phase by which these concepts are reconciled. This is obviously made only for the information to be represented.

The knowledge we want to represent is the one considered of interest by the users for the domain: for this reason, the most important and looked up websites are chosen.

At the end of this analysis we will define a taxonomy able to represent the knowledge of interest for this domain, which may also not have items from the taxonomy (or ontology from which we started in the TD analysis), but may have items which did not exist in it, emerged from the BU analysis.

The final result of this phase will be a reference taxonomy, where, for each item, there is a linked information about where the knowledge of interest can be found on each website.

3   Case Study

In this study we chose as a case study the domain of wines and, particularly, the one belonging to the technical files and/or descriptions of "Italian wines": the choice was not made randomly as the world of wines is rich in contents and complete enough to give a good starting point for our study. In fact, there are thousands of contents which can be found on the Internet; also, there are different studies on the classification of wines from which we can draw on.

3.1   Knowledge Base of Interest

Contents on wine available on the Web are thousands, offering a significant KB.

Our study takes into consideration a subdomain of wine, represented by all the most important reviews which can be found on the Internet. From the analyses of the domain on the web and the Google Ranking of these websites, we chose a list of suitable and representative websites, having considered the popularity and the reliability given by the Web.

The websites we took into consideration are the following:

– Decanter.com
– DiWineTaste.com
– Lavinium.com
– GamberoRosso.it
– Vintrospective.com
– Snooth.com
– Vinix.com.

These websites are considered as representative for our study also because of their own information structures, particularly various and differentiated. Each website has its own structure and a different representation of the information. To correctly define our domain it was therefore necessary to precisely analyse the contents in each of them and the layouts.

The structure of the page showing the review is useful to understand if the same website always uses the same structure and the same items for every review. Unfortunately we saw that some of them show the same information differently depending on the review, using, for instance, different tags for the same datum. This, obviously, is a limit in the process of classification of contents. It is thus necessary to align the different items for the same website, used to represent the same datum.

Below is a brief summary of the main aspects of each site's layout.

– diversity of information and representation, heterogeneous nature of the contents;
– the data extrapolated from some sites are in English;
– one site, for example, uses icons to indicate the fields, and not textual tags;
– the same information is given in different sites with different names;
– information existing on some sites is missing in others;
– information that is embedded in the text and has to be searched manually.

To address all the problems listed above, we built a macrosystem, in the shape of a table. It shows all the fields that appear in different sites, so that every single entry can be analyzed and categorized.

The macrosystem includes all 7 tables, one for each site. Each table is populated with the list of information on the site, including embedded information. For each piece of information a textual entry description was added, which specifies the content type in natural language. Subsequently, after considering all sites, we tried to enrich the macrosystem with the addition of fields that are present on other sites, but not in the ones we considered.

3.2 Top-Down Phase

In this phase we analysed the existing formalizations for the representation of knowledge of this domain.

A very interesting formalization which we pinpointed was the one by the Associazione Italiana Sommelier (AIS), The Italian Sommelier Association, providing a detailed description of all the terms associated with wine. Another important formalization was the one by the European law defining the reference features of a certain wine, such as type, colour, grape variety, etc. From these two, a reference taxonomy for those features was created.

As an additional formalization, we chose a reference scheme, represented by an ontology already existing on the Web and made by An ontology is surely more complex than a taxonomy. It has, apart from class hierarchies, property hierarchies with cardinality ties for the assignable values. It offers a general view of the world of wines, with a less detailed description for certain fields as stated on the reviews found on the Web.

Moreover, from this ontology we took into consideration only the areas of interest existing in our classification, omitting those ones representing elements not of interest (such as, for instance, each winemaker's property).

Starting from these reference formalizations, a first taxonomy was built in which we pinpointed the items to create the reference table. After choosing the items of interest in the reference ontology, we analysed the direct correspondence among tags of the two representations, directly extracting the ontology ones from the OWL code. To standardise our taxonomy we decided to take into consideration the RDF standard indicating, just for the items with a correspondence, its URI. The RDF standard allows to associate a URI also to the properties. website, used to represent the same datum.

3.3   Bottom-Up Phase

The BU analysis required a detailed analysis of the contents of these websites, trying to pinpoint the information we considered as important; then we studied the structure of each single source, useful to see the existing data and their position in the layout of the page.

Once the KB for the domain of interest composed by the websites was defined, the next step was classifying all the chosen information. Such classification is made by creating a classification of the BU contents because it was built from the bottom: information on the websites are thus accurately analysed.

We start from the analysis of the specific to reach a general classification of data. One of the initial steps of our project contemplated the study of the structure of each source, useful to see the existing data and their position on the page's layout.

This procedure happens to be important also at this point of the study, because allows for the evaluation of the classification of information. Both the item "maturazione", but also the organoleptic analysis (visual, olfactory and gustatory test) if existing, are systematically shown on the websites taken into consideration, into the area which we identified as "tasting notes". For this reason, to build the hierarchy we tried to respect the original, already existing one.

The type of classification was also revealed during the data analysis phase, during the study of semantics and uniformation.

With the creation of the tables we tried to represent the knowledge in the shape of fields as faithful as possible to those ones already existing in the samples taken into consideration.

The evaluation of this phase is subjective and left to the intuition of the analyst, which freely interprets the information at their disposal, intuitively obtaining the taxonomic tree. This step happens to be very tricky, because is susceptible to accidental mistakes. However, we could say that the various structures found in the domain which we considered, apart from the caption used to define each field, are not so different, thus the classification did not raise any big doubt, as for the representation.

Thus, the macrosystem made 7 tables, one for each website. Every table has the list of information of the website it represents.

3.4   Mixed Phase

During this phase, the items of the fields existing in the taxonomy defined in the TD phase were compared to the fields of the tables created in the BU phase. To do this, we built a mapping macro-table of knowledge containing, for each item of the taxonomy, the correspondence if and where that information exists on the various websites and also the information existing on the websites which were not represented by the taxonomy.

To each item we thus assigned a numerical value to represent this mapping: (1) existing and extractable information; (2) existing but not extractable information; (3) sometimes existing and extractable information; (4) sometimes existing but not extractable information; (5) always missing information.

For the fields with values 1 and 3, the corresponding field and the mapping rule to extrapolate the information are also indicated.

The information with value 2 and 4 is embedded (hidden in the text) and, therefore, should be specifically looked for with tools of semantic analysis. Anyway, the field in which it exists is indicated.

With this analysis and classification of every single datum we managed to solve the inhomogeneity of the information existing in the Web, as for the domain of interest. This allowed to study both its structure and the type of information existing, giving us the chance to examine how data are presented and the classification given for each website.

When creating the taxonomy, which wants to be a semantic classification, we also tried to represent the structure of data and the existing hierarchies of the sample websites.

3.5   Formalization of Knowledge

This activity was iteratively repeated to best represent the knowledge and its connections described in the macro-system mentioned above. As expected, not all the fields were taken into consideration, neither among those existing in the initial taxonomy nor among the extrapolated ones, and those ones which appear just once in the whole macro-system were rejected (evaluation made considering the field value = 5), such as, for instance, "Bicchiere consigliato" or "Temperatura di servizio consigliata".

The inhomogeneity among the information existing in the different websites was analysed by looking for the semantic correspondences represented in the macrosystem with the column 'field details'. The same principle was used to uniform fields with numerical values. The final range takes into account the classification used by the majority of websites.

A simplifying table summarizing the procedure of classification described above is shown below.

The result of these phases was the knowledge base formalized through the taxonomy. Table 1 shows some items of it, with a field of value 1 or 3 and expressed in textual form (for instance, those ones directly extractable through tags or metadata). Other fields, represented by an icon, were rejected, though their presence was considered.

Table 1. Classification.

Macrosystem items	Final tags
Wine's identification name	Wine
\<Produttore\> \<Winery\> \<Producer\>	Winery: address, telephone, fax, e-mail, web, map, other wine, other info winery
\<classification\> \<denominazione\> \<tipologia\>	Classification: Vino da tavola, IGT, DOC, DOCG
\<tipologia\> \<type\>	Colour: white, rose, red
\<type\> \<tipologia\>	specification
Qualification: embedded	Qualification: classic, reserve, superior
\<typical grape composition\> \<Varietal\> \<vitigni\> \<uve\>	grape
\<titolo alcolometrico\> \<alcohol\> \<alcol\>	Alcohol
Label	Label
\<origin\> \<region\> \<zona\>	State/Region
\<tasting notes\> \<reviews\> \<overview\>	Tasting notes

(Continued)

Table 1. (*Continued*)

Macrosystem items	Final tags
<prezzo enoteca> <prezzo> <starting at> <$> <average bottle price>	Price
<abbinamento> <suggested recipe pairings> <food pairing suggestions>	Food pairing suggestion
<posted by> <source>	Author
<posted on> <inserito> <degustazione in data>	Date
<decanter rating>: max 5 stelle <rated>: max 5 bicchieri <valutazione>: max 5 chiocciole <punteggio>: max 5 diamanti <voto>: max 5 chiocciole Punteggio: max 3 bicchieri	Rate: 60–70; 71–75; 76–80; 81–85; 86–90; 91–100

4   Analysis of Results and Verification

During the validation phase we verified how our KMS made the acquired knowledge usable for the systems compliant with than other ontology of wines and for other websites on Italian wines reviews. We went on verifying how the contents of interest of these websites could be represented and managed on the KMS through some simple mapping rules.

Then, we tried to solve the clear inhomogeneity by paying more attention to the semantic meaning and not to the notation used to represent those contents. In fact, the purpose of the study was not to describe the whole world of wines, but just the part of it represented by the information which can be found on the Web.

After matching the two systems, Ontology and Taxonomy, the information were generalized and made coherent. This allowed us to verify that our system is able to represent and combine specific information, and at the same time understands the main variances between the two systems, namely the difference of some considered information.

This kind of study can also be used to enrich an already existing ontology with fields coming from a general classification, evaluating a possible integration of such information without damaging the existing hierarchy, so that we can have a broader and more accurate view over the analysed domain.

4.1   Choice of Samples

To continue with the phase of verification of the created taxonomy, we decided to take into consideration another set of samples - again, wine reviews which can be found on the Web.

The choice of the websites for the testing phase followed the same criteria used during the analysis of the domain. The main obstacle we found was due to the popularity of the product and the large amount of followers who have a very subjective way of representing the information about wine and the acquired knowledge. Here comes the need of pinpointing sources with clear, easily extractable and objective information.

One of the main features which these sources needed to have was the presence of differentiated fields with a single notation rather than a broad textual field. So, also in this case, all the websites gathering a large quantity of information in just a macro-textual area were rejected. In fact, these kind of websites, though full of contents, were not suitable for the testing phase. The embedded information, though fostering the acquisition of a general knowledge, do not facilitate its own structured classification. Similarly, some apparently suitable sources happened to have very few contents, with a database so poor that it did not mention the most appreciated wines.

After these considerations, the websites we decided to take into consideration for the tests were the following:

- guida-vino.com
- vinogusto.com
- kenswineguide.com
- buyingguide.winemag.com.

4.2 Testing Phase

For each sample website, in this testing phase we verified whether the information in them could be found in the classification proposed by us, and whether our taxonomy could be able to represent them.

For each website, therefore, Table 2 was built, representing the specific fields of information which was the same for every review that we analysed.

In the light of the results obtained in this testing phase, we are satisfied with the taxonomy which we created. In fact, with this testing phase, we saw that the classification defined in our study reflects the type of contents needed. Such classification, therefore, is usable, re-usable and possibly extendible to the domain of interest of wine. It is also possible to find the following information in most of the selected sites (Table 3).

Table 2. Testing phase (1).

Existing information	Field details	Taxonomy item
Label	Label's image	Wine.label
Producer	About the producer	Wine.winery
Classification	IGT, DOC, DOCG	Wine.classification
Grape variety	Grape variety	Wine.grape
Range of prices	Price	Wine.prices
Others years	Others years	Wine.winery.infoWinery.otherWines
Presentation/comments	Wine tasting	Wine.tastingNotes
Rate: max 5 stars	Rate	Wine.rate

Table 3. Testing phase (2).

Vineyard of origin	Information about the land	Wine.tasting Notes. vineyardNotes
Grape harvest	Information about grape harvest	Wine.otherInfoWinery
Wine making	Follow steps for wine making	Wine.otherInfoWinery
Maturation	Maturation	Wine.tastingNotes.aged
Alchol degree	Alcoholic strength	Wine.alchool
Organoleptic test	Information on visual, gustatory, and olfactory examination	Wine.tastingNotes. visualExamination
		Wine.tastingNotes. olfactoryAnalysis
		Wine.tastingNotes.tasteAnalysis
Gastronomy	Wine-pairing	Wine.foodPairingSuggestions
Servizio	Wine Glass	Absent
Price	Price	Wine.price

5   Conclusions

The spread of the Social Web is significantly influencing the evolution of Semantic Web: users themselves are creating rules for the representation of information. The structure of the Web grows and changes giving the user the chance to actively participate in the developing of the Web. For this reason, our study took into consideration this feature with the uniformation of UGCs, trying to link the two worlds: Social Media and Semantic Web. Also the main search engines (Google, Yahoo, Bing, …) and the main Social Network (Youtube, Facebook, Twitter, Flickr, …) are evolving, specializing and interconnecting themselves on data retrieval, presentation, exchange and sharing.

That being so, the basic idea of our study was to propose a solution to the problem of the different contents of the Web, coming from different sources but belonging to the same domain of knowledge.

Our proposal is to define a taxonomy able to represent knowledge through a mixed iterative approach, articulated in a top-down analysis and a bottom-up one of the domain of knowledge which is to be represented. Thus, first we tried to define the knowledge of interest on the domain, depending on the specifications, and through the analysis of the existing ontologies, in order to define a reference taxonomy. Then, the knowledge we considered as important (and as an element of common interest) was extracted from a selection of websites belonging to the domain of interest. These contents are to be classified in the taxonomy mentioned before, also using mapping rules made ad hoc.

The taxonomy created allowed for a definition of the reference knowledge which could then be managed as an actual usable knowledge, fostered by all the information existing on the selected websites. Due to the large amount of the information available, we chose as domain of knowledge a sub-domain of wine, represented by the reviews which can be found on the Web.

From the analysis of the domain on the Web and the Google Ranking of many websites, we chose a list of some suitable and representative ones after considering popularity and reliability given from the Web.

We chose to validate the proposed approach by verifying how the KMS allowed to make the acquired knowledge usable and accessible to the systems compliant with the Ontology of Wines taken into consideration along with other websites of Italian wine reviews, underlining how, also in this case, the collected information could be represented and managed on the KMS through some simple mapping rules.

Such a system could be enriched by deducing an ontology of information existing on the Web to be compared with another ontology representing the same domain. A similar comparison has the advantage to be simple, less disorganized and surely less susceptible to mistakes than the one proposed in our project.

A further, interesting development could be the creation of repositories able to collect the information previously classified and, through a system made ad hoc, they would be presented to the final user in a structured and customized way, depending on the requests, and possibly developing a graphic interface which could be able to draw the curiosity and the interest of the user.

References

1. Berners-Lee, T., Hendler, J., Lassila, O.: The Semantic Web. Scientific American **284**, 28–37 (2001). http://www.scientificamerican.com/2001/0501issue/0501berners-lee.html
2. Horrocks, I.: Ontologies and the Semantic Web. Commun. ACM **51**(12), 58–67 (2008)
3. Noy, N.F., McGuinness, D.L.: Ontology development 101: a guide to creating your first ontology. Stanford Knowledge Systems, Laboratory Technical Report KSL-01-05 (2001)
4. Gruber, T.: A translation approach to portable ontology specification. Knowl. Acquis. **5**, 199–220 (1993)
5. Gruber, T.: Ontology. In: Liu, L., Özsu, M.T. (eds.) Encyclopedia of Database Systems. Springer, Heidelberg (2008)
6. Decker, S., Melnik, S., Van Harmelen, F., Fensel, D., Klein, M., Broekstra, J., Erdmann, M., Horrocks, I.: The Semantic Web: the roles of XML and RDF. IEEE Internet Comput. **4**(5), 63–73 (2000)
7. Maedche, A., Staab, S.: Ontology learning for the Semantic Web. IEEE Intell. Syst. **16**(2), 72–79 (2001)
8. Jacob, E.K.: Ontologies and the Semantic Web. Bull. Am. Soc. Inf. Sci. Technol. **29**(4), 19–22 (2003). Wiley Periodicals
9. Davies, J., Fensel, D., van Harmelen, F.: Towards the Semantic Web: Ontology-Driven Knowledge Management. Wiley, Chichester (2003). ISBN 9780470848678
10. Strintzis, J., Bloehdom, S., Handschuh, S., Staab, S., Simou, N., Tzouvatras, V., Petridis, K., Kompatsiaris, I., Avrithis, Y.: Knowledge representation for semantic multimedia content analysis and reasoning. In: Proceedings of the European Workshop on the Integration of Knowledge, Semantics and Digital Media Technology (2004)
11. Jewell, M.O., Lawrence, K.F., Tuffield, M.M., Prugel-Bennett, A., Millard, D.E., Nixon, M.S., Schraefel, M.C., Shadbolt, N.R.: OntoMedia: an ontology for the representation of heterogeneous media. In: Multimedia Information Retrieval Workshop, ACM SIGIR (2005)

12. Hepp, M.: Ontologies: state of the art, business potential, and grand challenges. In: Hepp, M., De Leenheer, P., de Moor, A., Sure, Y. (eds.) Ontology Management: Semantic Web, Semantic Web Services, and Business Applications, 3–22. Springer, New York (2007). ISBN 978-0-387-69899-1

13. Simperi, E.: Reusing ontologies on the Semantic Web: a feasibility study. Data Knowl. Eng. **68**(10), 905–925 (2009)

14. Brickley, D., Guha, R.V.: Resource Description Framework (RDF) Schema Specification. World Wide Web Consortium (1999). http://www.w3.org/TR/PR-rdf-schema

15. Lassila, O., Swick, R.: Resource Description Framework (RDF): Model and Syntax Specification, Recommendation W3C (1999). http://www.w3.org/TR/REC-rdf-syntax

16. Euzenat, J., Shvaiko, P.: Ontology Matching. Springer, Heidelberg (2007). ISBN 978-3-540-49612-0

17. Fensel, D., Van Harmelen, F., Horrocks, I., McGuinness, D.L., Patel-Schneider, P.F.: OIL: an ontology infrastructure for the Semantic Web. IEEE Intell. Syst. **16**(2), 38–45 (2001)

18. Gómez-Pérez, A., Corcho, O.: Ontology languages for the Semantic Web. Intell. Syst. **17**(1), 54–60 (2002)

19. Schreiber, ATh, Dubbeldam, B., Wielemaker, J., Wielinga, B.: Ontology-based photo annotation. IEEE Intell. Syst. **16**, 66–74 (2001)

20. Jaimes, A., Smith, J.: Semi-automatic, data-driven construction of multimedia ontologies. In: Proceedings of IEEE International Conference on Multimedia and Expo (ICME), vol. 2 (2003)

21. Lunesu, M.I., Pani, F.E., Concas, G.: An approach to manage semantic informations from UGC. In: Proceedings of International Conference on Knowledge Engineering and Ontology Development (KEOD), Paris, France (2011)

22. Lunesu, M.I., Pani, F.E., Concas, G.: Using a standards-based approach for a multimedia knowledge-base. In: Proceedings of International Conference on Knowledge Management and Information Sharing (KMIS), Paris, France (2011)

23. Pani, F.E., Concas, G., Lunesu, M.I., Baralla, G.: An approach to manage the web knowledge. In: Proceedings of International Conference on Knowledge Engineering and Ontology Development (KEOD), Algarve, Portugal (2013)

Chunking Complexity Measurement for Requirements Quality Knowledge Representation

David C. Rine[1] and Anabel Fraga[2(✉)]

[1] George Mason University, Fairfax, VA, USA
davidcrine@yahoo.com
[2] Carlos III of Madrid University, Madrid, Spain
afraga@inf.uc3m.es

Abstract. In order to obtain a most effective return on a software project investment, then at least one requirements inspection shall be completed. A formal requirement inspection identifies low quality knowledge representation content in the requirements document. In software development projects where natural language requirements are produced, a requirements document summarizes the results of requirements knowledge analysis and becomes the basis for subsequent software development. In many cases, the knowledge content quality of the requirements documents dictates the success of the software development. The need for determining knowledge quality of requirements documents is particularly acute when the target applications are large, complicated, and mission critical. The goal of this research is to develop knowledge content quality indicators of requirements statements in a requirements document prior to informal inspections. To achieve the goal, knowledge quality properties of the requirements statements are adopted to represent the quality of requirements statements. A suite of complexity metrics for requirements statements is used as knowledge quality indicators and is developed based upon natural language knowledge research of noun phrase (NP) chunks. A formal requirements inspection identifies low quality knowledge representation content in the requirements document. The knowledge quality of requirements statements of requirements documents is one of the most important assets a project must inspect. An application of the metrics to improve requirements understandability and readability during requirements inspections can be built upon the metrics shown and suggested to be taken into account.

Keywords: Requirements inspections · Chunking and cognition · Complexity metrics · Cohesion · Coupling · NP chunk · Requirements · Software quality · Information retrieval · Natural language understanding and processing

1 Introduction

Steven R. Rakitin [30] states "If you can only afford to do one inspection on a project, you will get the biggest return on investment from a requirements inspection. A requirements inspection should be the one inspection that is never skipped." The formal inspection makes significant knowledge quality improvements to the requirements

© Springer-Verlag Berlin Heidelberg 2015
A. Fred et al. (Eds.): IC3K 2013, CCIS 454, pp. 245–259, 2015.
DOI: 10.1007/978-3-662-46549-3_16

document, or formally a software requirements specification (SRS), which is the single artifact produced through the requirements engineering process. Kinds of knowledge in an SRS include, functional requirements, non-functional requirements, system requirements, user requirements, and so on [34]. The knowledge quality of the SRS document unavoidably is the core of requirements management of which the formal inspection is an important part. And the SRS, which is comprised of requirements, or requirements statements, is a basis for developing or building the rest of the software, including verification and validation phases. Despite abundant suggestions and guidelines on how to write knowledge quality requirements statements, knowledge quality SRS's are difficult to find.

The goal of this research is to improve the knowledge quality of the SRS by the identification of natural language knowledge defects derived from that of prior research [12–15] and applies a set of metrics as quality indicators of requirements statements in an SRS. Many research studies on software quality [6, 11, 16, 21, 23, 26], and various quality factors have been proposed to represent the quality of software. The quality factors adopted in this research are developed by Schneider (2002, 2000a, 2000b) and are named as goodness properties.

Din and Rine [12–15] evaluated knowledge quality by means of Noun Phrase Chunking complexity metrics. That research compared the NPC-Cohesion and NPC-Coupling metrics with the cohesion and coupling metrics proposed earlier [3, 7, 8, 10, 18, 31, 36].

The evidence provided by Din and Rine [15] concludes that the "NPC complexity metrics indicate the content goodness properties of requirements statements." The contribution of the research from [12–15] is "a suite of NP chunk based complexity metrics and the evaluation of the proposed suite of metrics."

The paper is organized as follows. Section 2 presents the research problem statement and the importance of the research problem. The background of the research is summarized in Sect. 3. Section 4 illustrates the detailed process of obtaining the elements of measurement, Noun Phrase (NP) chunks, and then presents the proposed suite of metrics. Section 5 summarizes the contributions of the research and conclusions.

2 Research Problem and Importance

2.1 Research Problem

The research was designed to answer the following question: How can low natural language knowledge quality requirements statements be identified in an SRS? Although the research focuses on SRS's, the conclusions of the research can be applied to other requirements documents such as system requirements documents.

Certain requirements defects are hard to identify. The Fagan's requirements inspection [17] can be used to identify requirements defects, requirements inspections can in the present practice "be effective when sections of an SRS are limited to 8-15 pages so that a requirements quality inspector can perform an inspection of a given section within two hours' time frame" [26, 34].

Defects such as inconsistent or missing requirements statements can easily be missed due to the spatial distance. The current requirements inspection practice does not consider these kinds of requirements defects.

2.2   The Importance of the Research

The suite of NPC complexity metrics is supported by a software tool researched and developed as part of this research [12–15] to identify high knowledge complexity and hence low knowledge quality requirements.

Low quality requirements are not only the source of software product risks but also the source of software development resource risks, which includes cost overrun and schedule delay [12–15].

Quality software "depends on a software manager's awareness of such low quality requirements, their ability to expediently assess the impacts of those low quality requirements, and the capability to develop a plan to rectify the problem" [12–15]. The proposed suite of complexity metrics expedites the process of identifying low quality requirements statements. The subsequent risk analysis of those requirements can be performed earlier. The rectification plan can hence be developed and carried out in a timely manner. This process, from quickly identifying low quality requirements to developing and carrying out the corresponding rectification plan, provides the foundation for the development of high quality software.

3   Background

3.1   Quality and Content Goodness Properties

Schneider, in his Ph.D. Dissertation directed by Rine [32, 33], proposed eleven goodness properties as a better coverage of quality factors 36]: Understandable, Unambiguous, Organized, Testable, Correct, Traceable, Complete, Consistent, Design independence, Feasible, and Relative necessity. Representing quality with a set of properties that are each relatively easier to measure is an important step towards measuring quality. However, the context of the current research focuses upon Understandable, Unambiguous, Organized, and Testable, keys to natural language knowledge representation quality.

3.2   Complexity Metrics and Measurement

Complexity is a Major Software Characteristic That Controls or Influences Natural Language Knowledge Representation Quality. It Has Been Widely Accepted as an Indirect Indicator of Quality [19, 22, 25, 27] and Hence the Content Goodness Properties.

3.3   Readability Index

Difficult words are necessary to introduce new concepts and ideas, especially in education and research.

Coh-Metrix has developed readability indexes based on cohesion relations, inter-action between a reader's skill level, world knowledge, and language and discourse characteristics. "The Coh-Metrix project uses lexicons, part-of-speech classifiers, syntactic parsers, templates, corpora, latent semantic analysis, and other components that are widely used in computational linguistics" [23].

The Coh-Metrix readability index is used to address quality of an entire written document, such as an essay, rather than individual sections of technical documents, such as software requirements documents.

Unfortunately, readability indexes, including Coh-Metrix, are not comparable with this research [12–15] for the following reasons:

(1) The readability metrics are designed for the whole documents, instead of sections of documents.
(2) The readability scores are not reliable indicators when the document under evaluation has less than 200 words (McNamara, 2001). However, many of the requirements statements have less than 50 words.
(3) Although Coh-Metrix attempts to measure the cohesion of texts, the definition of cohesion used by Coh-Metrix is different from the definition of cohesion used in Computer Science, and there are no coupling metrics in Coh-Metrix.
(4) Coh-Metrix does not have a single metric to represent the size, cohesion, or coupling complexity. Coh-Metrix includes more than 50 metrics to measure very specific aspects of texts. No composite metric that combines those specific aspects of a document has been proposed.
(5) Coh-Metrix attempts to measure the cohesion of texts. Future work of Coh-Metrix may address comprehension, or understandability. However, Coh-Metrix will never address the issue of testability and many other goodness properties.

4   NP Chunk Based Complexity Metrics

4.1   Chunking, Cognition and Natural Language Quality

Humans tend to read and speak texts by chunks. Abney [1] proposed chunks as the basic language parsing unit. Several categories of chunks include but are not limited to Noun Phrase (NP) chunks, Verb Phrase (VP) chunks, Prepositional Phrase (PP) chunks, etc. [1]. This research NP chunks and ignores other types of chunks.

4.2   Three Core Metrics

It has been recognized that it is not likely that a single metric can capture software complexity [20, 24]. To deal with the inherent difficulty in software complexity, a myriad of indirect metrics of software complexity have been proposed [29].

Multiple empirical studies indicate that Line of Code (LOC) is better or at least as good as any other metric [2, 16, 35]. All these evidence and findings indicate that counting should be one of the core software metrics. Zuse [37] also identified simple counts in his measurement theory as one of the metrics that possesses all the desired properties of an ideal metric.

Many of the published metrics, either for procedural languages or for object-oriented languages, include some variation of the cohesion and coupling metrics [4, 5]. Furthermore, cohesion and coupling metrics are ubiquitous across a wide variety of measurement situations, including 4GLs, software design, coding and rework. Darcy and Kemerer believe that cohesion and coupling are effective metrics and they can represent the essential complexity measures for the general software design tasks [9]. Hence, NPC-Cohesion and NPC-Coupling are chosen to represent the complexity of requirements. To assist the identification of low quality requirements, a composite metric (NPC-Composite) that combine cohesion and coupling measures is also proposed and studied in the research.

4.3   Requirements Documents Used

Two requirements documents are used as cases of study in this research:

(1) A public domain requirements document for Federal Aviation Agency (FAA). The requirements document is available in Ricker's dissertation [31], and [3, 7, 8, 10, 18, 36].
(2) Versions of the Interactive Matching and Geocoding System II (IMAGS II) requirements documents for U. S. Bureau of Census. The IMAGS II, or IMAGS, project has gone through several iterations of requirements analysis. Four versions of the requirements documents are available for the research.

4.4   Sentence/Requirements Statement Level Complexity

The calculation can be expressed as follows.

$$\text{NPC} - \text{Sentence}(\text{sentence}_j) = \sum_{1 \le i \le N} \frac{\text{Entry}(i,j)}{\sum_{1 \le j \le C} \text{Entry}(i,j)}, \tag{1}$$

where $\text{Entry}(i, j)$ is the number of occurrence of NP chunk NPi in sentencej, $1 \le i \le N, 1 \le j \le C, N$ is the total number of NP chunks, and C is the total number of sentences. Intuitively, NPC-Sentence is a metric that measures the normalized count of NP chunks in a sentence of a document.

The requirements statement level complexity metric, or NPC-Req(reqj), is the aggregation of NPC-Sentence of the component sentences and can be expressed as follows.

$$\text{NPC} - \text{Req}(\text{req}_j) = \sum_{\text{sentence}_i \in \text{req}_j} \text{NPC} - \text{Sentence}(\text{sentence}_i), \tag{2}$$

where $1 \le i \le L_j, 1 \le j \le M$, L_j is the total number of sentences of requirement j, and M is the total number of requirements.

Example - From Partial Parsing to Sentence Level Complexity. The following three requirements (four sentences) are extracted from the IMAGS Version 4 requirements document. The requirements in the IMAGS requirements document are marked with a label (e.g., "IM-WKASSIGN-4") and a short description (e.g., "Assign Users to WAAs"). Note that WAA stands for Work Assignment Area and is a collection of counties. WAA is defined to facilitate the assignment of workload to individual users.

IM-WKASSIGN-4: Assign Users to WAAs. IMAGS II shall track and maintain users' assignment to WAAs. A user can be assigned to one or more WAAs, and a WAA can have more than one user assigned.

IM-WKASSIGN-7: Assign Incoming Addresses on WAAs. IMAGS II shall assign incoming addresses to users based upon their WAAs.

IM-WKASSIGN-8: Assign Multiple WAAs to Multiple Users. IMAGS II shall provide a way to assign a list of WAAs to multiple users at once.

The results of the chunk parsing are as follows.

```
(S:
        (0: <imags/NN> <ii/NN>)
        <shall/MD>
        <track/VB>
        <and/CC>
        <maintain/VB>
        (1: <users/NNS> <'/POS> <assignment/NN>)
        <to/TO>
        <waas/VB>
        <./.>
)
(S:
        (2: <a/DT> <user/NN>)
        <can/MD>
        <be/VB>
        <assigned/VBN>
        <to/TO>
        (3: <one/CD>)
        <or/CC>
        (4: <more/JJR><waas/NNS>)
        <,/,>
        <and/CC>
        (5: <a/DT> <waa/NN>)
        <can/MD>
        <have/VB>
        <more/JJR>
        <than/IN>
        (6: <one/CD> <user/NN>)
        <assigned/VBN>
        <./.>
)
```

```
(S:
        (7: <imags/NN> <ii/NN>)
        <shall/MD>
        <assign/VB>
        <incoming/VBG>
        (8: <addresses/NNS>)
        <to/TO>
        (9: <users/NNS>)
        <based/VBN>
        <upon/IN>
        (10: <their/PRP$><waas/NNS>)
        <./.>
)

(S:
        (11: <imags/NN> <ii/NN>)
        <shall/MD>
        <provide/VB>
        (<12: <a/DT> <way/NN>)
        <to/TO>
        <assign/VB>
        (<13: <a/DT> <list/NN>)
        <of/IN>
        (<14: <waas/NNS>)
        <to/TO>
        (<15: <multiple/NN> <users/NNS>)
        <at/IN>
        <once/RB>
        <./.>
)
```

The chunk parsing results (1-2) in 16 NP chunks (see Table 1), where sentence is abbreviated as "sent." and requirement is abbreviated as "req.". Note that the word "WAAs" in the first sentence is tagged as "VB", a verb. This is an error due to the nature of statistical NLP process, which considers words after "to" as verbs. The stop list indicated in the table is used to filter NP chunks that have little meanings.

Table 1. Example – parsing results.

Stop NP chunks: (a user), (one), (users), (a way), (a list)		req. 1		req. 2	req. 3
		sent. 1	sent. 2	sent. 3	sent. 4
<imags/NN> <ii/NN>	0,7,11	1	0	1	1
<users/NNS> </POS> <assignment/NN>	1	1	0	0	0
<more/JJR> <waas/NNS>	4	0	1	0	0
<a/DT> <waa/NN>	5	0	1	0	0
<one/CD> <user/NN>	6	0	1	0	0
<addresses/NNS>	8	0	0	1	0
<their/PRP$> <waas/NNS>	10	0	0	1	0
<waas/NNS>	14	0	0	0	1
<multiple/NN><users/NNS>	15	0	0	0	1

By applying the stemming and text normalization techniques, the result is depicted in Table 2.

Table 2. Example – Stemming and Text Normalization.

Stop words: (a user), (one), (users), (a way), (a list)		req. 1		req. 2	req. 3
		sent. 1	sent. 2	sent. 3	sent. 4
<imags/NN> <ii/NN>	0,7,11	1	0	1	1
<users/NNS> <'/POS> <assign/NN>	1	1	0	0	0
<waa/NN>	4,5,10,14	0	2	1	1
<user/NN>	6	0	1	0	0
<address/NN>	8	0	0	1	0
<multiple/NN><user/NN>	15	0	0	0	1

For the sake of example, it is assumed that the four sentences constitute the complete requirements document. The resulting NPC-Sentence and NPC-Req are shown in Table 3.

Table 3. Example – NPC-Sentence and NPC-Req.

Stop words: (a user), (one), (users), (a way), (a list)	req. 1		req. 2	req. 3
	sent. 1	sent. 2	sent. 3	sent. 4
<imags/NN> <ii/NN>	1	0	1	1
<users/NNS> <'/POS> <assign/NN>	1	0	0	0
<waa/NN>	0	2	1	1
<user/NN>	0	1	0	0
<address/NN>	0	0	1	0
<multiple/NN><user/NN>	0	0	0	1
NPC-Sentence	1/3+1 =1.3	2/4+1 =1.5	1/3+1/4+1= 1.58	1/3+1/4+1= 1.58
NPC-Req	1.3 + 1.5 = 2.8		1.58	1.58

4.5 Intra-section Level Complexity

The formula (3) for NPC-Cohesion is as follows.

$$\text{NPC} - \text{Cohesion}(S_j) = \begin{cases} \dfrac{\sum\limits_{1 \leq i \leq M_j} \text{ClusterSize}(i,j)}{L_j - 1}, & L_j > 1 \\ 1, & L_j = 1 \end{cases} \tag{3}$$

where M_j is the total number of clusters in section S_j, and L_j is the total number of sentences in section S_j.

If a requirements section consists of a single sentence ($L_j = 1$), the NPC-Cohesion is 1. If all the adjacent sentences have common NP chunks, then the NPC-Cohesion is also 1.

For example, Fig. 1 shows three sections of the requirements. The first section contains three sentences, the second section contains two sentences, and the third section has one sentence. Section 1 has a cluster that covers the whole section, and the size of the cluster is two. Section 2 has a cluster that covers the whole section, and the size of the cluster is one. Section 3 does not have any cluster. Based upon the above formula (3), the values of the NPC-Cohesion metric are as follows (4).

$$NPC - Cohesion(S1) = 2/(3 - 1) = 1,$$
$$NPC - Cohesion(S2) = 1/(2 - 1) = 1, \quad (4)$$
$$NPC - Cohesion(S3) = 1$$

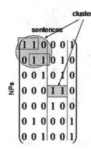

Fig. 1. Clusters of NP chunks.

4.6   Inter-section Level Complexity

The proposed NPC-Coupling metric value is the sum of the spatial distances among NP chunks and clusters that share the same NP chunks. Once a cluster is formed, the cluster represents all the components NP chunks inside the cluster. One possible algorithm to calculate the NPC-Coupling metric is as follows.

The NPC-Coupling metric of section S_j:
x = 0,
remarks: handle clusters coupling
for   every cluster i in section S_j,
   for   every NP_k in cluster i,
      for   every sentence l outside section S_j,
         if   sentence l contains NP_k,
            calculate the distance between sentence l and the centroid of cluster i,
            add the distance to x,
remarks: handle non-clusters coupling
for   every NP_k of section S_j that does not belong to any cluster of S_j,
   for   every sentence l in section S_j,
      if   sentence l contains NP_k,
         for   every sentence m outside section S_j,
            if   sentence m contains NP_k,
               calculate the distance between sentence l and sentence m,
               add the distance to x,
return x

Figure 2 shows the calculation of NPC-Coupling using the same requirements sections in the previous example. The centroid of the cluster of the first section resides in the second sentence. The centroid of the cluster of the second section resides between sentence 4 and 5. Based upon the above formula (4-5), the values of the NPC-Coupling metrics are as follows.

$$\text{NPC - Coupling}(S1) = 4 + 3 + 2 + 4 + 3 = 16,$$
$$\text{NPC - Coupling}(S2) = 3 + 2 = 5, \tag{5}$$
$$\text{NPC - Coupling}(S3) = 4 + 4 + 3 = 11$$

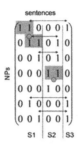

Fig. 2. Calculation of coupling metrics.

4.7 A Composite Metric

It has been recognized that a single metric for a software product does not work. Multiple metrics provide multiple views of the subject, and each view serves its own purpose. Without carefully following engineering disciplines, operation, comparison, combination, or manipulation of different metrics can result in meaningless measures [22].

To identify the low quality requirements sections, this research combines the cohesion and coupling metrics into a single indicator. The formula used in the research is as follows (6).

$$\text{NPC} - \text{Composite}(S_i) = \text{NPC} - \text{Cohesion}_i - a * (\text{NPC} - \text{Coupling}_i - b), \tag{6}$$

for all i, $1 \leq i \leq S$, S is the total number of requirements sections, where $b = \min(\text{NPC - Coupling}_i)$ and $a = 1 / (\max(\text{NPC - Coupling}_i) - b)$

The coefficients (6), a and b, are used to adjust the coupling metric so that (1) the measure falls in the range between −1 and 1, and (2) both the cohesion and coupling metrics use the same unit.

Cohesion Metrics: Based upon the proposed NPC-Cohesion metric defined previously, the NPC-Cohesion measures are depicted in Fig. 3, together with the cohesion measures published in [28]. It is clear that the two metrics are consistent with each other except in one section– Sect. 11 of the FAA requirements document.

Although NPC-Cohesion is able to identify low cohesion requirements in the above example using syntactic categories of words, syntactic categories can sometimes mislead

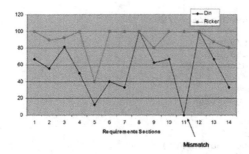

Fig. 3. Cohesion values between two methods.

the analysis. For example, the following two sentences are low cohesion sentences according to NPC-Cohesion. "The computer program shall discretize the continuous function f(t). Then, it shall approximate the integral using the discretization."

One remedy to the weakness of NPC-Cohesion is to parse verb phrase (VP) chunks. Part of the text normalization process is to transform words into its root form. Then "discretize" and "discretization" can be normalized as the same chunk. However, the incorporation of a second type of chunks, i.e., verb phrase (VP) chunks, to the proposed metrics substantially increase the complexity of the parser and hence the processing effort and time. The addition of the parsing process for VP chunks to cope with the weakness that rarely occurs does not seem to be cost effective. Hence, it was decided to focus on NP chunks for the research.

The mismatch between the two cohesion metrics can be explained as follows. Section 11 of the FAA requirements document consists of two sentences. Here are the sentences and the corresponding chunk parsing results.

Sentence 1: "The system shall provide flight plan outputs to a variety of operational positions, collocated processors, and remote facilities."

```
(S:
          (0: <the/DT> <system/NN>)
          <shall/MD>
          <provide/VB>
          (1: <flight/NN> <plan/NN> <outputs/NN>)
          <to/TO>
          (2: <a/DT> <variety/NN>)
          <of/IN>
          (3: <operational/JJ> <positions/NNS>)
          <,/,>
          <collocated/VBN>
          (4: <processors/NNS>)
          <,/,>
          <and/CC>
          (5: <remote/JJ> <facilities/NN>)
          <./.>
)
```

Sentence 2: "The ACCC shall output data periodically, on request, or in accordance with specified criteria (NAS-MD-311 and NAS-MD-314)."

```
(S:
        (6: <the/DT> <accc/NN>)
        <shall/MD>
        <output/VB>
        (7: <data/NNS>)
        <periodically/RB>
        <,/,>
        <on/IN>
        (8: <request/NN>)
        <,/,>
        <or/CC>
        <in/IN>
        (9: <accordance/NN>)
        <with/IN>
        <specified/VBN>
        (10: <criteria/NN>)
        <(/(>
        (11: <nas-md-311/NN>)
        <and/CC>
        (12: <nas-md-314/NN>)
        <)/)>
        <./.>

)
```

There are 13 NP chunks. It is clear that there are no common NP chunks between the two sentences. This is why the NPC-Cohesion metric gives a low cohesion measure for the above requirements section. On the other hand, Ricker uses terms to measure the cohesion of the section, and the word "output" appears in the first sentence as a noun, while the word "output" appears in the second sentence as a verb. Ricker's algorithm does not consider syntactic categories and hence links the two sentences. It is believed that a word in different forms, i.e., verbs and nouns, in different sentences should not always be considered as cohesive, since the two words in the two forms can refer to two totally different objects. By closely examining the two sentences, it can be found that the word "output" in the two sentences indeed refers to two different things or two different concepts. Hence, the proposed cohesion metrics is more effective.

Coupling Metrics: The coupling measures based on the NPC-Coupling metric and the coupling metric in (Pleeger, 1993) are depicted in Fig. 4. The two metrics display consistent results except in one section - Sect. 4 of the requirements document.

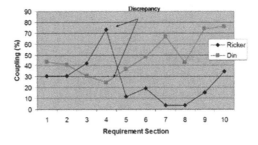

Fig. 4. Coupling values between two methods.

The evaluation criterion for cohesion is whether the two sets of metrics are strongly consistent with each other. The derived data from this case study supports this consistency.

Sensitivity/Accuracy: The NPC-Cohesion metrics are relative measures. They are normalized and fall in the range of 0 to 1. Comparing such relative measures derived from different requirements documents is not logical. In other words, it is not appropriate to compare the sensitivity and accuracy of the NPC-Cohesion metrics with Ricker's metrics.

Although the NPC-Coupling metrics are based upon spatial distance between NP chunks and they are not normalized, comparing it with Ricker's metric which uses different units of measurement does not seem to be logical either.

Summary: Based upon the above analysis, it can be concluded that the derived data from the case study met the evaluation criteria for the consistency hypothesis.

5 Summary of Contributions and Conclusions

This research derived from [12–15] made two contributions: (1) the invention of a suite of complexity metrics to measure the content goodness properties of requirements documents and (2) the empirical case study to evaluate the invented suite of complexity metrics.

The invented complexity metrics are researched and developed to identify low quality requirements statements in SRS's. These metrics are based on the NP chunks in SRS's. In the empirical two phased case study, it is concluded that the proposed metrics can measure the content goodness properties of requirements statements.

This research provides evidence for the feasibility of using NP chunks as the elements of measurement for complexity metrics. In addition the invented suite of complexity metrics provides requirements engineers and managers with a tool to measure the quality of the requirements statements. These metrics can be use to identify low quality requirements statements. They can also be used to identify requirements statements and requirements sections that may require more rigorous testing. Potential flaws and risks can be reduced and dealt with earlier in the software development cycle.

At a minimum, these metrics should lay the groundwork for automated measures of the quality of the requirements statements in SRS's. Because those metrics are constructed by a software tool, their measures are easy to collect, a vital characteristics for a quality measurement program [28].

During the research, and as stated by Genova et al. [22] and Fanmuy et al. [18], a set of metrics to complement the chunk based metrics can be included in the set of requirements metrics indicators in order to create a complete set of metrics of interest for any systems requirement engineer. It includes a set of semantic metrics based on natural language processing (NLP) and semantic notions assisted by a knowledge-based system and requirements patterns.

References

1. Abney, S.: Parsing by chunks. In: Berwick, R., Abney, S., Tenny, C. (eds.) Principle-Based Parsing. Kluwer Academic Publishers, Dordrecht (1991)
2. Basili, V.R.: Qualitative Software Complexity Models: A Summary, Tutorial on Models and Methods for Software Management and Engineering. IEEE Computer Society Press, Los Alamitors (1980)
3. Bøegh, J.: A new standard for quality requirements. IEEE Softw. 25(2), 57–63 (2008)
4. Briand, L.C., Daly, J.W., Wust, J.K.: A unified framework for cohesion measurement in object-oriented systems. IEEE Trans. Softw. Eng. 3(1), 65–117 (1998)
5. Briand, L.C., Daly, J.W., Wust, J.K.: A unified framework for coupling measurement in object-oriented systems. IEEE Trans. Softw. Eng. 25, 91–121 (1999)
6. Cant, S., Jeffery, D.R., Henderson-Sellers, B.: A conceptual model of cognitive complexity of elements of the programming process. Inf. Softw. Technol. 37(7), 351–362 (1995)
7. Chung, L., do Prado Leite, J.C.S.: On non-functional requirements in software engineering. In: Borgida, A.T., Chaudhri, V.K., Giorgini, P., Yu, E.S. (eds.) Conceptual Modeling: Foundations and Applications. LNCS, vol. 5600, pp. 363–379. Springer, Heidelberg (2009)
8. Costello, R.J., Liu, D.-B.: Metrics for requirements engineering. J. Syst. Softw. 29(1), 39–63 (1995)
9. Darcy, D.P., Kemerer, C.F., Software Complexity: Toward a Unified Theory of Coupling and Cohesion, 8 February 2002
10. Davis, A., Overmyer, S., Caruso, J., Dandashi, F., Dinh, A.: Identifying and measuring quality in a software requirements specification. In: Proceedings of the First International Software Metrics Symposium, 21–22 May, pp. 141–152 (1993)
11. Demarco, T.: Controlling Software Projects. Yourdon Press, Englewood Cliffs (1982)
12. Din, C.Y.: Requirements content goodness and complexity measurement based on NP chunks. Ph.D. thesis, George Mason University, Fairfax, VA, 2007, Reprinted by VDM Verlag Dr. Muller (2008)
13. Din, C.Y., Rine, D.C.: Requirements content goodness and complexity measurement based on NP chunks. In: Proceedings, Complexity and Intelligence of the Artificial Systems: Bio-inspired Computational Methods and Computational Methods Applied in Medicine, WMSCI 2008 Conference (2008)
14. Din, C.Y., Rine, D.C.: Requirements metrics for requirements statements stored in a database. In: Proceedings of the 2012 International Conference on Software Engineering Research and Practice, SERP 2012, July 16–19, pp. 1–7 (2012)
15. Din, C.Y., Rine, D.C.: Requirements Statements Content Goodness and Complexity Measurement. International Journal of Next-Generation Computing. 4(1) (2013)
16. Evangelist, W.: Software complexity metric sensitivity to program structuring rules. J. Syst. Softw. 3(3), 231–243 (1983)
17. Fagan, M.: Advances in Software Inspections. IEEE Trans. Softw. Eng. 12(7), 744–751 (1986)
18. Fanmuy, G., Fraga, A., Llorens, J.: Requirements Verification in the Industry. CSDM, Paris, France (2011)
19. Farbey, B.: Software quality metrics: considerations about requirements and requirement specifications. Inf. Softw. Technol. 32(1), 60–64 (1990)
20. Fenton, N.E., Neil, M.: Software metrics: roadmap. In: Proceedings of the International Conference on Software Engineering (ICSE), pp. 357–370 (2000)
21. Fenton, N.E., Pleeger, S.L.: Software Metrics: A Rigorous and Practical Approach, 2nd edn. International Thomson Computer Press, Boston (1997)

22. Genova, G., et al.: A framework to measure and improve the quality of textual requirements. Requirements Eng. **18**(1), 25–41 (2013). doi:10.1007/s00766-011-0134-z. Url: http://dx.doi.org/10.1007/s00766-011-0134-z

23. Graesser, A.C., Mcnamara, D.S., Louwerse, M.M., Cai, Z.: Coh-Metrix: analysis of text on cohesion and language. Behav. Res. Methods Instrum. Comput. **36**(2), 193–202 (2004)

24. Henderson-Sellers, B.: Object-Oriented Metrics textendash Measures of Complexity. Prentice Hall PTR, New Jersey (1996)

25. Kemerer, C.F.: Progress, obstacles, and opportunities in software engineering economics. Commun. ACM **41**, 63–66 (1998)

26. Kitchenham, B.A., Pleeger, S.L., Fenton, N.E.: Towards a framework for software measurement validation. IEEE Trans. Softw. Eng. **21**, 929–943 (1995)

27. Klemola, T.: A cognitive model for complexity metrics, vol. 13 (2000)

28. Mcnamara, D.S.: Reading both high and low coherence texts: effects of text sequence and prior knowledge. Can. J. Exp. Psychol. **55**, 51–62 (2001)

29. Mcnamara, D.S., Kintsch, E., Songer, N.B., Kintsch, W.: Are good texts always better? Text coherence, background knowledge, and levels of understanding in learning from text. Cogn. Instr. **14**, 1–43 (1996)

30. Pleeger, S.L.: Lessons learned in building a corporate metrics program. IEEE Softw. **10**(3), 67–74 (1993)

31. Purao, S., Vaishnavi, V.: Product Metrics for Object-Oriented Systems. ACM Comput. Surv. **35**(2), 191–221 (2003)

32. Rakitin, S.: Software verification and validation: a practitioner's guide (Artech House Computer Library). Artech House Publishers, Norwood (1997). ISBN-10: 0890068895 ISBN-13: 978-0890068892

33. Ricker, M.: Requirements specification understandability evaluation with cohesion, context, and coupling. Ph.D. thesis, George Mason University, Fairfax, VA (1995)

34. Schneider, R.E., Buede D.,: Criteria for selecting properties of a high quality informal requirements document. In: Proceedings of the International Conference on Systems Engineering, Mid-Atlantic Regional Conference, INCOSE-MARC, 5–8 April 2000a, pp. 7.2-1–7.2-5 (2000)

35. Schneider, R.E., Buede D.: Properties of a high quality informal requirements document. In: Proceedings of the Tenth Annual International Conference on Systems Engineering, INCOSE, 16–20 July, 2000b, pp. 377–384 (2000)

36. Weyuker, E.: Evaluating software complexity measures. IEEE Trans. Softw. Eng. **14**(9), 1357–1365 (1988)

37. Wnuk, K., Regnell, B., Berenbach, B.: Scaling up requirements engineering – exploring the challenges of increasing size and complexity in market-driven software development. In: Berry, D. (ed.) REFSQ 2011. LNCS, vol. 6606, pp. 54–59. Springer, Heidelberg (2011)

KODEGEN: A Code Generation and Testing Tool Using Runnable Knowledge

Iaakov Exman$^{(\boxtimes)}$, Anton Litovka, and Reuven Yagel

Software Engineering Department,
The Jerusalem College of Engineering – Azrieli,
POB 3566, 91035 Jerusalem, Israel
{iaakov,robi}@jce.ac.il, antonli@post.jce.ac.il

Abstract. KDE – *Knowledge Driven Engineering* – is a generalization of MDE – Model Driven Engineering – to a higher level of abstraction than the standard UML software models, aiming to be closer to the system designer concepts. But in order to reach an effective technology applicable in industry, one needs to implement it in a tool using *Runnable Knowledge*, i.e. which can be run and tested. This work describes KODEGEN – a KDE tool for testing while generating code – whose input consists of system ontologies, ontology states and scenario files. Incidental concepts not part of the ontologies are replaced by mock objects. The implementation uses a modified Gherkin syntax. The tool is demonstrated in practice by generating the actual code for a few case-studies.

Keywords: KDE · Runnable knowledge · Ontology · Ontology states · Model testing · Mock objects

1 Introduction

Software system development starting from the natural system concepts facilitates the system designer work and its understanding. KDE – *Knowledge Driven Engineering* – is a generalization of MDE (or MDA [6]) to a higher abstraction level than UML models, exactly to support a conceptually neat approach to system development.

From a slightly different point of view, "earlier bug discovery reduces costs", is a widely accepted wisdom [5]. Within KDE earlier means higher abstraction levels. Thus, also from this aspect, KDE offers novel development perspectives.

Exman et al. [9] have recently proposed *Runnable Knowledge* – bare concepts and their states – as the highest system abstraction level. Exman and Yagel [10] made a further step by proposing their Runnable Ontology Model, starting from ontologies and ontology states, and incorporating concrete scenario files for code generation and testing.

This paper embodies the latter proposal in the KODEGEN tool. One assumes for a certain domain the a priori given relevant ontology and its states. KODEGEN generates, from the ontology and its states, classes of the system under development (SUD), while submitting them to tests to be applied according to given scenario specifications.

KODEGEN is being built to gradually develop software systems in a semi-automatic approach, at times with human intervention. The interactions refine the SUD and KODEGEN itself. The ultimate goal in our vision is to automatically generate the running code from the abstract model and its tests.

© Springer-Verlag Berlin Heidelberg 2015
A. Fred et al. (Eds.): IC3K 2013, CCIS 454, pp. 260–275, 2015.
DOI: 10.1007/978-3-662-46549-3_17

1.1   Related Work: From Executable Specifications to Code Generation

A concise literature review is presented here. The Agile software movement has stressed in recent years early testing methods, e.g. Freeman and Pryce [11]. Its main purposes are faster understanding of the software under development obtained by short feedback loops, and guiding the software system development in rapidly changing environments.

Early testing methods stemmed from Test Driven Development (TDD), the unit-testing practice by Beck [4]. In such methods, scripts demonstrate the various system behaviors, instead of just specifying the interface and a few additional modules. Since the referred scripts' execution can be automated, the referred methods are also known as automated functional testing.

Among TDD extensions one finds Acceptance Test Driven Development (ATDD) also known as Agile Acceptance Testing, see e.g. Adzic [2]. Another such extension is Behavior Driven Development (BDD) North [14], emphasizing readability and understanding by stakeholders. Recent representatives are Story Testing, Specification with examples Adzic [3] or Living/Executable Documentation, e.g. Smart [19].

There exist common tools to implement TDD practices. FitNesse by Martin [1] is a wiki-based web tool for non-developers to write formatted acceptance tests, e.g. tabular example/test data. The Cucumber (Wynne and Hellesoy [21], see also [8]) and Spec-Flow [20] tools directly support BDD. They accept stories in plain natural language (English and a few dozen others). They are easily integrated with unit testing and user/web automation tools. Yagel [22] reviews extensively these practices and tools.

An introductory overview of ontologies in the software development context is found in Calero et al. [7]. Ontology-driven software development papers are found in Pan et al. [16]. The combination of ontology technologies with Model Driven Engineering is discussed at length in Pan et al. [16] and in Parreiras [17].

In the remaining of the paper we introduce the Ontology abstraction level (Sect. 2), describe testing with the modified Gherkin syntax of the Cucumber tool (Sect. 3), study code generation implemented in the KODEGEN tool (Sect. 4), describe three case studies (Sect. 5) and conclude with a discussion (Sect. 6).

2   Runnable Knowledge: The Ontology Abstraction Level

Runnable Knowledge (Exman et al. [9]) is an abstraction level above standard UML models. Since UML models separate modeling structure and behavior into different diagrams – typically class diagrams and statecharts – Runnable Knowledge, the highest abstraction level, is also designed to separate structure from behavior.

Ontologies – mathematical graphs with concepts as vertices and relationships as edges – represent the static semantics of software systems. From ontologies one can, by means of appropriate tools, to naturally generate UML structures, viz. classes.

Ontology states – mathematical graphs with concepts' states as vertices and labeled transitions as edges – are our representation of the dynamic semantics of software systems. From ontology states one can, by means of appropriate tools, to naturally generate UML behaviors, viz. statecharts. Ontology states are a higher abstraction of statecharts, abstracting detailed attributes, functions and parameters. Ontology states are not the only alternative to represent dynamic semantics (see e.g. Pan et al. [16]).

For illustration, Fig. 1 displays a graphical representation of a simplified version of an ATM (Automatic Teller Machine) ontology. An ATM appears later on in the case studies – in Sect. 5.

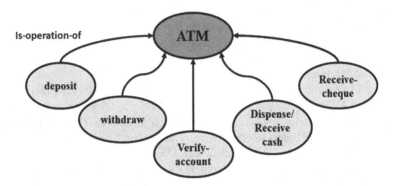

Fig. 1. An ontology for an ATM – Five concepts standing for five possible ATM operations are displayed, besides the ATM concept itself.

3   Modified Gherkin Syntax for Testing

To test ontologies and ontology states, we use a modified Gherkin Syntax specification as in Fig. 2. This file is usually developed by the system's stakeholders.

The keywords shown in blue in Fig. 2 are:

(a) *Feature* – provides a general title to the specification;
(b) *Scenario* – provides a title for a specific walk through;
(c) *Given* – pre-conditions before some action is taken;
(d) *When* – an action that triggers the scenario;
(e) *Then* – the expected outcome.

For further details see [21] and our previous work [10].

```
Feature:   Account Withdrawal

Scenario: Successful withdrawal from an account
    Given an account has a balance of <amount>$100
    When <amount>$20 are withdrawn from an ATM
    Then the account <balance>balance should be $80
```

Fig. 2. ATM withdrawal operation specification – It specifies successful cash withdrawal from an ATM. It is expressed in the modified Gherkin style. Tags are added by the developer – marked in bold red within angular brackets – to facilitate test script generation (see Sect. 5).

Running this specification alone fails as it lacks supporting code. A domain model is needed. A tool like Cucumber can suggest steps to satisfy the given specification. Mock objects could also stand for the concepts missing in the ontologies.

Cucumber's mode of usage is iteration and refinement until the specification is complete. This is checked by test scripts. These may catch software regressions caused by new system features.

KODEGEN goes a step further and fills the generated steps with actual code that exercises the interactions between the ontology classes. The ontology may not be complete, or the specifications, sometimes written by non-technical persons, may contain yet more gaps. KODEGEN is designed to maximize automation with the known ontologies. Thus, KODEGEN hints to the developer to slightly modify the specification with tags to be used to generate the code.

4   KODEGEN Software Modules: Generation of Running Code

KODEGEN, whose software modules are seen in Fig. 3, has three *inputs*:

- Initial Specification – a scenario obtained by elicitation of system requirements;
- Ontology – obtained by specialization of generic domain ontologies;
- Ontology States – obtained by setting transitions between concept states.

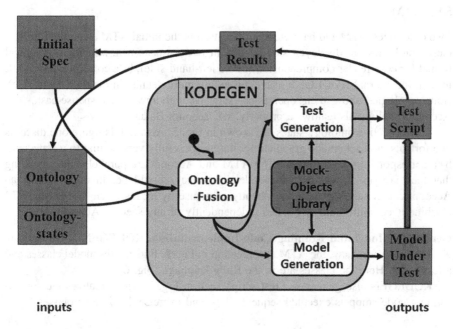

Fig. 3. KODEGEN software modules – modules are round (white) rectangles, while inputs and outputs are regular (yellow) rectangles. Mock-Objects may be needed to complement generated code. The wide arrow upwards means that test-script is used to the test the MUT (Model Under Test) (Color figure online).

The fusion module coordinates two sub-modules: a model generator and a test generator. These use ontology concepts and their states to generate the *outputs*:

- *MUT* – code skeletons of the *Model Under Test*;
- *Test Scripts* – unit tests to test the MUT (using e.g. NUnit [15]).

If there are concepts in the scenario that do not appear in the ontologies, KODEGEN inserts them into the generated code by means of a Mock Objects library (see e.g. Moq [13], RSpec [18]). Mock objects are a fast and efficient addendum to Runnable Knowledge to obtain an *actually* running model.

If the tests results are negative, one should modify the specifications and/or the ontology and then repeat the loop. Otherwise the system model is approved.

The *Runnable Knowledge* model – i.e. the ontologies and their states – is the utmost abstract level in the software layers hierarchy. It is runnable in the sense that, a suitable tool can make transitions between states.

5   Case Studies

Here we describe two case studies from the given input, to the generated code. The first is an ATM, Automatic Teller Machine, with cash withdrawal transactions. We further elaborate this example in Subsect. 5.3.

5.1   ATM

Two ontologies, *ATM* and bank *Account*, are used in the initial ATM example. In Fig. 4 these ontologies are displayed side-by-side in two formats: a- a schematic graphical format for easy reader comprehension (in the left-hand-side); b- a corresponding XML format, for internal KODEGEN usage, providing more details (in the right-hand-side). Both ontologies show their operations, such as withdraw and dispense-cash. The Account ontology also shows a property, viz. account Balance.

The respective ontology states are shown in Fig. 5. Also in this figure one discerns two formats: a- a schematic graphical format, purposefully very similar to a statechart; b- a corresponding XML format. The ATM and Account are parallel states, meaning that they are orthogonal or "independent", as they should be. In other words, an Account can certainly exist independently of its use by means of an ATM. An ATM machine is certainly built and tested independently of any specific Account.

Generated Model and Running Code Implementation. KODEGEN is fed with an XML ontology and say, the ATM specification in Fig. 2. It generates model classes and a test script. Here the classes are in the Ruby language (Fig. 6).

KODEGEN also generates a test script, seen in Fig. 7, which realizes the specification – code snippets executed sequentially – and exercise the various classes.

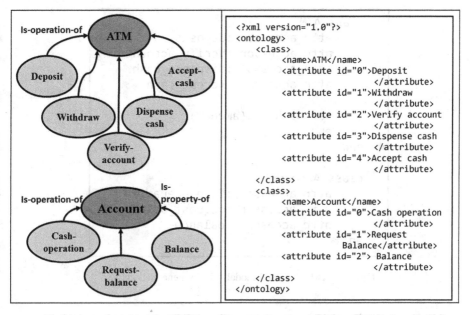

Fig. 4. ATM and bank account ontologies – a graphical representation is in the left hand side. The concepts in the ATM ontology (upper) are operations performed by the ATM. The concepts in the account ontology (lower) are operations (cash-operation and request-balance) and a property (balance) of the account. The XML representation is in the right hand side.

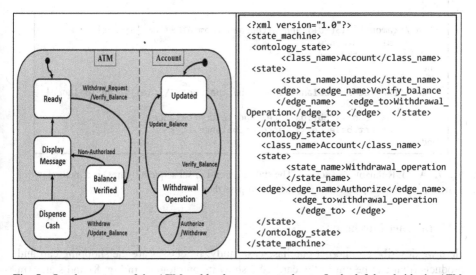

Fig. 5. Ontology states of the ATM and bank account ontology – In the left hand side the *ATM* and *account* parallel states display the states for a cash withdrawal operation. In the right hand side an XML representation of the partial account ontology states, for internal manipulation within KODEGEN.

```
class ATM
        attr_accessor :deposit
        attr_accessor :Verify_account
        attr_accessor :dispense_cash
        attr_accessor :Accept_cash

        def withdraw(amount)
        end
    end

    class Account
        attr_accessor :Cash_operation
        attr_accessor :request_balance
        attr_accessor :balance
    end
```

Fig. 6. ATM: Extracted model – Ruby generated classes.

```
require "test/unit/assertions"
require "/usr/lib/ruby/vendor_ruby/cucumber/rspec/doubles"
World(Test::Unit::Assertions)

Given /^account has a balance of <balance> \$(\d+)$/ do |balance|
        @account = Account.new
        @account.balance = balance
end

When /^<amount> \$(\d+) are withdrawn from ATM$/ do |amount|
        atm = ATM.new
        atm.withdraw( amount )
end

Then /^the account balance should be <balance> \$(\d+)$/ do |balance|

        @account.stub(:balance).and_return(balance)
        assert balance == @account.balance.to_s
end
```

Fig. 7. ATM: Runnable test script – The stub (here and in subsequent figures) is a method with a signature, but not implemented yet, needed to pass the test.

5.2   Internet Purchase

Here we describe an internet purchase case study. Its classes are the shopping cart and products (that can be put in the cart). We show its ontologies (in Fig. 8) and states (in Fig. 9), directly in the internal XML representation. Testing of these classes is shown by a transaction in which two product types are purchased.

A Gherkin specification file is given in Fig. 10.

```
<?xml version="1.0"?>
<ontology>
    <class>
            <name>shopping cart</name>
            <attribute id="0">products</attribute>
            <attribute id="1">items per product</attribute>
            <attribute id="2">tax</attribute>
            <attribute id="3">current price</attribute>
            <attribute id="4">total price</attribute>
    </class>
    <class>
            <name>product</name>
            <attribute id="0">name</attribute>
            <attribute id="1">price</attribute>
            <attribute id="2">serial number</attribute>
            <attribute id="3">part number</attribute>
    </class>
</ontology>
```

Fig. 8. XML representation of shopping cart and product ontologies – The shopping cart ontology shows objects contained by the cart (product and items-per-product) and purchase properties (total-price, current-price and tax). The product concepts are just its properties.

```
<?xml version="1.0"?>
<state_machine>
        <ontology_state>
                <class_name>shopping cart</class_name>
                <state>
                        <state_name>wait</state_name>
                        <edge>
                                <edge_name>add</edge_name>
                                <edge_to>wait</edge_to>
                        </edge>
                        <edge>
                                <edge_name>contains</edge_name>
                                <edge_to>calculated</edge_to>
                        </edge>
                </state>
        </ontology_state>
</state_machine>
```

Fig. 9. XML representation of shopping-cart ontology states – The cart default is empty. A product can be added, its price or final price-&-tax calculated, ending the transaction.

Figure 11 contains the generated model classes.

Figure 12 displays the Shopping cart case study test script. In contrast to the ATM case study, here mock objects are applied (we used the RSpec-Mocks library [18]). The

Feature: Adding to a shopping cart
Scenario: Add items to shopping cart
  Given An empty shopping cart
  When I add 1 item of Product A ($10)
  And I add 2 items of Product B ($20 each)
  And the tax is 8%
  Then the shopping cart contains 3 items
  And the total price is 54$
	Feature: Adding to a shopping cart
	Scenario: Add items to shopping cart
	  Given empty shopping cart
	  When I add <quantity> 1 of Product <name> "A" to shopping cart
	  And I add <quantity> 2 items of Product <name> "B" to shopping cart
	  And tax is <tax> 8% percent
	  Then shopping cart contains <quantity> 3 items

Fig. 10. Shopping-cart – Adding items to a shopping cart. In the left-hand-side one sees a simple Gherkin specification. In the right-hand-side a tagged specification, augmented with modifier tags in bold red within angular brackets, to facilitate code generation.

```
class Shopping_cart
    attr_accessor :products
    attr_accessor :items_per_product
    attr_accessor :tax
    attr_accessor :current_price
    attr_accessor :total_price

    def add(quantity, product)
    end
    def contains
    end
end
class Product
    attr_accessor :name
    attr_accessor :price
    attr_accessor :serial_number
    attr_accessor :part_number
end
```

Fig. 11. Shopping-cart: Extracted model – Ruby generated classes.

mock expectations are met by adding calls to stub objects – in bold red in Fig. 16. The script adds products A and B to empty cart, applies tax and make assertions.

Once the mock expectations were set and the test script is ready, it only remains to run it in a test runner tool (see the screenshot in Fig. 13). This test script can later be reused and re-issued to check correctness of the actual developing implementation.

```
require "test/unit/assertions"
World(Test::Unit::Assertions)

    Given /^empty shopping cart$/ do
            @shopping_cart = Shopping_cart.new
    end
    When /^I add <quantity> (\d+) of Product <name> "(.*?)"
            to shopping cart$/ do |quantity, name|
            product = Product.new
            product.name = name
            @shopping_cart.add(quantity , product)
    end
    When /^I add <quantity> (\d+) items of Product <name> "(.*?)"
                    to shopping cart$/ do |quantity, name|
            product = Product.new
            product.name = name
            @shopping_cart.add(quantity , product)
    end
    When /^tax is <tax> (\d+)% percent$/ do |tax|
            @shopping_cart.tax = tax
    end
    Then /^shopping cart contains <quantity> (\d+) items$/
            do |quantity|
            @shopping_cart.stub(:contains).and_return(3)
            assert quantity == @shopping_cart.contains( )
    end
```

Fig. 12. Shopping-cart: Runnable test script.

```
anton@anton-lap:~/Documents/project/shop$ cucumber features/shop.features
Feature: Adding to a shopping cart

  Scenario: Add items to shopping cart                                          # features/shop.features:3
    Given empty shopping cart                                                   # features/step_definitions/shop_steps.rb:11
    When I add <quantity> 1 of Product <name> "A" that cost <price> 30$ to shopping cart     # features/step_definitions/shop_steps.rb:15
    And I add <quantity> 2 items of Product <name> "B" that costs <price> 50$ to shopping cart # features/step_definitions/shop_steps.rb:22
    And tax is <tax> 8% percent                                                 # features/step_definitions/shop_steps.rb:29
    Then shopping cart contains <quantity> 3 items                              # features/step_definitions/shop_steps.rb:33

1 scenario (1 passed)
5 steps (5 passed)
0m0.002s
```

Fig. 13. Shopping-cart: Running test results – screenshot of running of the above test script with Cucumber. It is a passing test, since all expectations where met by the models, all of the steps in the test script were successfully done.

Lastly for this case study, Fig. 13 is a screenshot resulting from running the generated test script with Cucumber. The steps from the scenario are marked green meaning that the test tool could successfully run and all expectations were met.

5.3   Extended ATM-Account System with a Card

In this case study we extend the ATM-Account system (Sect. 5.1) with a Card, whose ontology is seen in Fig. 14. The respective ontology states are seen in Fig. 15.

```
<?xml version="1.0"?>
<ontology>
<class>
    <name>Card</name>
    <attribute>accepted</attribute>
    <attribute>returned</attribute>
    <attribute>retained</attribute>
  </class>
</ontology>
```

Fig. 14. XML representation of card ontology – This is the third ontology to be added to the two ontologies of the ATM-account system.

```
<?xml version="1.0"?>
<state_machine>
  <ontology_state>
    <class_name>Card</class_name>
    <state> <state_name>insertion</state_name>
      <edge><edge_name>password_entered</edge_name>
            <edge_to>validation</edge_to> </edge>
    </state>
  </ontology_state>
  <ontology_state>
    <class_name>Card</class_name>
    <state> <state_name>validation</state_name>
      <edge> <edge_name>valid</edge_name>
            <edge_to>validated</edge_to> </edge>
    </state>
  </ontology_state>
  <ontology_state>
    <class_name>Card</class_name>
    <state> <state_name>validation</state_name>
      <edge><edge_name>not_valid</edge_name>
        <edge_to>rejected</edge_to>
      </edge>
    </state>
  </ontology_state>
  <ontology_state>
    <class_name>Card</class_name>
    <state> <state_name>validated</state_name>
      <edge><edge_name>operation_completed</edge_name>
            <edge_to>returned</edge_to> </edge>
    </state>
  </ontology_state>
</state_machine>
```

Fig. 15. XML representation of card ontology states – The states are: insertion, validation, validated.

Feature: Account Withdrawal
   Scenario: Successful withdrawal from an account
       Given Account has a balance of <balance> $100
       When <amount> $20 are withdrawn from ATM
       Then the account balance should be <balance> $80
   Scenario: Account has sufficient funds
       Given the account balance is <balance> $100
       And the **card is valid**
       And the ATM contains <amount> $500
       When the Account request cash <amount> $20
       Then the ATM should dispense cash <dispense_cash> $20
       Then the account balance should be <balance> $80
       And the **card should be** <returned> **"returned"**
   Scenario: Account has insufficient funds
       Given the account balance is <balance> $10
       And the **card is valid**
       And the ATM contains <amount> $50
       When the Account Holder withdraw <amount> $20 from ATM
     Then the ATM should Print <message>"there are insufficient funds"
       And the account balance should be <balance> $10
       And the **card should be** <returned> **"returned"**
   Scenario: Card has been disabled
       Given account
       And the **card is not valid**
       And the ATM contains <amount> $500
       When the Account request cash <amount> $20
       Then the **card should be** <retained>**"retained"**
     And the ATM should Print <message>"the card has been retained"

Fig. 16. Scenarios involving a card in the ATM-account system – The card may be valid (in bold green color) and it is returned, but still there may be sufficient or insufficient funds for a withdrawal operation. If the card is not valid (in bold red color, within the last scenario), it is retained. This scenario is based upon a user story found in http://dannorth.net/whats-in-a-story/ (Color figure online).

The next system input is the set of scenarios, now involving the Card, seen in Fig. 16.

In the next Fig. 17, one can see the Ruby code generated by KODEGEN with the testing steps.

The work order in a test script (such as in Fig. 17) is as follows: objects are created in the *Given* part; messages are printed in the *When* part; verification occurs in the *Then* part. A metaclass is used to save a message to be verified in the *Then* part. This is an example of a Ruby meta-programming feature to dynamically add methods to (yet) non-existing model classes and later test the right interaction with them.

Finally, Fig. 18 displays a screenshot of the respective testing run.

```
require "/usr/lib/ruby/vendor_ruby/cucumber/rspec/doubles"
require "test/unit/assertions"
World(Test::Unit::Assertions)

Given /^Account has a balance of <balance> \$(\d+)$/ do |balance|
    @account = Account.new
    @account.balance = balance                          end
When /^<amount> \$(\d+) are withdrawn from ATM$/ do |amount|
    atm = ATM.new
    atm.withdraw(amount)                                end
Then /^the account balance should be <balance> \$(\d+)$/ do |balance|
    @account.stub(:balance).and_return(balance)
    assert balance == @account.balance.to_s             end
Given /^the account balance is <balance> \$(\d+)$/ do |balance|
    @account = Account.new
    @account.balance = balance                          end
Given /^the Card is valid$/ do
    @card = Card.new
    @card.valid()                                       end
Given /^the ATM contains <amount> \$(\d+)$/ do |amount|
    @atm = ATM.new
    @atm.contains = amount
    print_manager=mock("Print_Manager")
    metaclass = class << print_manager; self; end
    metaclass.send :attr_accessor, :text
    def print_manager.print(text)
            @text = text            end
    @atm.print_manager=print_manager                    end
When /^the Account request cash <amount> \$(\d+)$/ do |amount|
    @account.request(amount)                            end
Then /^the ATM should dispense cash <dispense_cash> \$(\d+)$/ do |dispense_cash|
    @atm.stub(:dispense_cash).and_return(dispense_cash)
    assert dispense_cash == @atm.dispense_cash.to_s end
Then /^the card should be <returned> "(.*?)"$/ do |returned|
    @card.stub(:returned).and_return(returned)
    assert returned == @card.returned.to_s              end
When /^the Account Holder withdraw <amount> \$(\d+) from ATM$/ do |amount|
    account_holder = mock( "Account_Holder ")
    account_holder.stub!( :withdraw ).with( amount ) do
            @atm.withdraw( amount )     end
    account_holder.withdraw( amount )                   end
Then /^the ATM should Print <message>"(.*?)"$/ do |message|
    @atm.print_manager.print( message )
    assert @atm.print_manager.text == message           end
Given /^account$/ do
    @account = Account.new                               end
Given /^the card is not valid$/ do
    @card = Card.new
    @card.not_valid()                                   end
Then /^the card should be <retained>"(.*?)"$/ do |retained|
    @card.stub(:retained).and_return(retained)
    assert retained == @card.retained.to_s              end
```

Fig. 17. KODEGEN generated Ruby code with testing steps – Mock objects for concepts (as *printer* and *account_holder*) not appearing in the system ontologies are stressed in bold red (Color figure online).

```
Feature: Account Withdrawal

  Scenario: Successful withdrawal from an account    # ATM.features:2
    Given Account has a balance of <balance> $100     # step_definitions/ATM_steps.rb:5
    When <amount> $20 are withdrawn from ATM          # step_definitions/ATM_steps.rb:10
    Then the account balance should be <balance> $80  # step_definitions/ATM_steps.rb:15

  Scenario: Account has sufficient funds              # ATM.features:6
    Given the account balance is <balance> $100                  # step_definitions/ATM_steps.rb:20
    And the Card is valid                                        # step_definitions/ATM_steps.rb:25
    And the ATM contains <amount> $500                           # step_definitions/ATM_steps.rb:30
    When the Account request cash <amount> $20                   # step_definitions/ATM_steps.rb:42
    Then the ATM should dispense cash <dispense_cash> $20        # step_definitions/ATM_steps.rb:46
    Then the account balance should be <balance> $80             # step_definitions/ATM_steps.rb:51
    And the card should be <returned> "returned"                 # step_definitions/ATM_steps.rb:51

  Scenario: Account has insufficient funds                                 # ATM.features:14
    Given the account balance is <balance> $10                             # step_definitions/ATM_steps.rb:20
    And the Card is valid                                                  # step_definitions/ATM_steps.rb:25
    And the ATM contains <amount> $50                                      # step_definitions/ATM_steps.rb:30
    When the Account Holder withdraw <amount> $20 from ATM                 # step_definitions/ATM_steps.rb:56
    Then the ATM should Print <message>"there are insufficient funds"      # step_definitions/ATM_steps.rb:64
    And the account balance should be <balance> $10                        # step_definitions/ATM_steps.rb:15
    And the card should be <returned> "returned"                           # step_definitions/ATM_steps.rb:51

  Scenario: Card has been disabled                              # ATM.features:22
    Given account                                               # step_definitions/ATM_steps.rb:69
    And the card is not valid                                   # step_definitions/ATM_steps.rb:73
    And the ATM contains <amount> $500                          # step_definitions/ATM_steps.rb:30
    When the Account request cash <amount> $20                  # step_definitions/ATM_steps.rb:42
    Then the card should be <retained>"retained"                # step_definitions/ATM_steps.rb:78
    And the ATM should Print <message>"the card has been retained" # step_definitions/ATM_steps.rb:64

4 scenarios (4 passed)
23 steps (23 passed)
0m0.015s
```

Fig. 18. Account withdrawal test run for the four scenarios – A screenshot showing that all the steps in all the four scenarios were passed, after a few interactive improvement iterations.

6 Discussion

The KODEGEN tool, as applied to the case studies described in this work, clearly demonstrate the feasibility of the approach, for applications in their scale range. Thus, the concrete realization of a specification into a running test script is done through KODEGEN. We have opened and resolved a series of specific issues resulting into an evolution of the tool itself.

Next we discuss some of the characteristics of KODEGEN.

6.1 KODEGEN Characteristics

KODEGEN is written in Java, and the source code with the discussed examples can be obtained here [12].

KODEGEN embodies quite a significant knowledge as a set of rules to handle common patterns and idioms when dealing with inputs. For example, during the test script generation, an object under test is recognized according to the ontology and by its appearance in the specification. Thereafter, the actions performed in the following steps are related implicitly or explicitly to this object under test. We continue growing this set as we use the tool for different domains and input sizes.

A significant step taken in the tool evolution was the modification of the Gherkin syntax through the introduction of <tags>, needed to fill certain gaps between ontologies and executable specification.

Mock object libraries are not necessarily mandatory – as only concepts not essential for the system ontologies need to be implemented by mocks. But mock objects may be used to pass tests, to enable the system developer to test the model integrity.

6.2 Future Work

Among issues still open to investigation is the extent of KODEGEN automation: will it remain a useful quasi-automatic tool? Or will the automation gap be safely and significantly covered, approaching the efficiency and reliability of current compilers?

In this work the tools produce code in Ruby which is more concise than, e.g., C#/Java. One can also use specific language features to improve the produced scripts, e.g., using partial classes in C# to separate expectations from the test script. Will a certain language assume a definitive role for KDE?

The case studies in the current work still are of limited scope. Can we extend the current tool and techniques to industrial production of large scale software systems? We are also building a GUI based tool that will better support the iterative human aided process needed for growing the models for a large project.

Given a set of ontologies, how to determine the amount of scenarios needed to develop a consistent and stable system?

Finally, a most important issue is the ability to overcome gradual, ad hoc and localized improvements, to reach a stage of generalized techniques that are independent of specific applications.

6.3 Main Contribution

The main contribution of this work is the usage of code generation as a fast implementation means to check system design while still in the highest *Runnable Knowledge* abstraction level.

References

1. Adzic, G.: Test Driven .NET Development with FitNesse. Neuri, London (2008)
2. Adzic, G.: Bridging the Communication Gap: Specification by Example and Agile Acceptance Testing. Neuri, London (2009)
3. Adzic, G.: Specification by Example – How Successful Teams Deliver the Right Software. Manning, New York (2011)
4. Beck, K.: Test Driven Development: By Example. Addison-Wesley, Boston (2002)
5. Boehm, B.W.: Software engineering economics. IEEE Trans. Softw. Eng. **10**, 4–21 (1984)
6. Brown, A.W.: Model driven architecture: principles and practice. Softw. Syst. Model **3**, 314–327 (2004). doi:10.1007/s10270-004-0061-2
7. Calero, C., Ruiz, F., Piattini, M. (eds.): Ontologies in Software Engineering and Software Technology. Springer, Heidelberg (2006)

8. Chelimsky, D., Astels, D., Dennis, Z., Hellesoy, A., Helmkamp, B., North, D.: The RSpec Book: Behaviour Driven Development with RSpec, Cucumber, and Friends. Pragmatic Programmer, New York (2010)
9. Exman, I., Llorens, J., Fraga, A.: Software knowledge. In: Exman, I., Llorens, J., Fraga, A. (eds.) Proceedings of SKY 2011, 2nd International Workshop on Software Knowledge (2010)
10. Exman, I., Yagel, R.: ROM: a runnable ontology model testing tool. In: Fred, A., Dietz, J.L.G., Liu, K., Filipe, J. (eds.) Knowledge Discovery, Knowledge Engineering and Knowledge Management, pp. 271–283. Springer, Heidelberg (2012)
11. Freeman, S., Pryce, N.: Growing Object-Oriented Software, Guided by Tests. Addison-Wesley, Boston (2009)
12. KODEGEN – the tool (2013). https://github.com/AntonLitovka/KODEGEN
13. Moq – the simplest mocking library for .NET and Silverlight (2012). http://code.google.com/p/moq/
14. North, D.: Introducing Behaviour Driven Development. Better Software Magazine (2006). http://dannorth.net/introducing-bdd/
15. NUnit (2012). http://www.nunit.org
16. Pan, J.Z., Staab, S., Assmann, U., Ebert, J., Zhao, Y. (eds.): Ontology-Driven Software Development. Springer, Heidelberg (2013)
17. Parreiras, F.S.: Semantic Web and Model-Driven Engineering. John Wiley and IEEE Press, Hoboken (2012)
18. RSpec mocks library (2013). https://github.com/rspec/rspec-mocks
19. Smart, J.F.: BDD in Action Behavior-Driven Development for the Whole Software Lifecycle. Manning, New York (2014)
20. SpecFlow – Pragmatic BDD for .NET (2010). http://specflow.org
21. Wynne, M., Hellesoy, A.: The Cucumber Book: Behaviour Driven Development for Testers and Developers. Pragmatic Programmer, New York (2012)
22. Yagel, R.: Can executable specifications close the gap between software requirements and implementation? In: Exman, I., Llorens, J., Fraga, A. (eds.) Proceedings of SKY 2011 International Workshop on Software Engineering, pp. 87–91. SciTePress, France, (2011)

Using Context for Search, Browse and Recommendation in Software Development

Paulo Gomes[1,2]($\boxtimes$), Bruno Antunes[1,2], and Bárbara Furtado[1,2]

[1] CISUC - DEI, Faculdade de Ciências e Tecnologia
da Universidade de Coimbra, Coimbra, Portugal
[2] DEI - Pólo II, Universidade de Coimbra, Coimbra, Portugal
{pgomes,bema}@dei.uc.pt, bfurtado@student.dei.uc.pt

Abstract. Finding information is a critical task in software development, especially when is hard to remember every piece of information that developers need to do their job. The increase in the average size of projects and in the number of technologies used makes the task of finding the right information or the right piece of code a nightmare for most developers. Tools that help developers be more efficient in finding this information need to be developed. Not only they need to be more efficient, but they also need to be more powerful, making the job of the developer easier. In this paper, we describe the SDiC (Software Development in Context) system, which implements three types of information finding: search, recommendation and browsing. The main innovation in SDiC is using the developer's context to make more efficient these mechanisms, enabling not only a more accurate answer, but also a more efficient way in leading with the information overload. *abstract* environment.

Keywords: User context · Search · Recommendation · Browsing · Software development

1 Introduction

Nowadays, developers need to work with software projects that are increasing in size and complexity [1]. To be able to deal with the amount of information in an IDE (Integrated Development Environment) developers need tools that can cope with huge amounts of highly interconnected code. Usually developers tend to focus in specific parts of the code, when trying to solve specific issues, which makes the task context very important in solving issues. Much of these tasks imply the navigation in complex code, trying to find the correct parts of the code to modify [2].

The need of information is common to most of the professions nowadays, especially to the knowledge workers. There are three different knowledge needs:

- Information that one knows, but that one does not remember right now.
- Information that one needs, but do not know it apriori, and one needs to find it.

© Springer-Verlag Berlin Heidelberg 2015
A. Fred et al. (Eds.): IC3K 2013, CCIS 454, pp. 276–289, 2015.
DOI: 10.1007/978-3-662-46549-3_18

– Information that one does not know that it is important and that should be known.

The first two levels are usually solved using search mechanisms. These are also information needs more simple to solve. But the last level is very hard to solve and needs a different mechanism(s) to solve this issue. These mechanisms can be: recommendation or browsing. Either way, these are two different ways to suggest new information to the user that might be relevant for what the user is doing. Recommendation works by having the system taking the initiative, while in browsing it is the user that takes the initiative.

In this paper, we describe the SDiC (Software Development in Context) system, which uses search, recommendation and browsing to provide information to the software developer. The main distinctive feature of SDiC is the use of the developer's context to provide better results. The context in SDiC is seen taking into account search, recommendation and browsing, but in layers as seen in Fig. 1. The personal layer deals with the work of the developer at the IDE level. The project layer deals with the knowledge at the team or project level, where a team of people are working in the same project. The organization level works at the organization level, and the community level is a much broader level of knowledge, one that transcends the organization limits. In this work we focus on the personal layer.

The next sections describe the SDiC system with regard to: the architecture, the knowledge base, the context model, the context-based search, the context-based recommendation, the context-based browsing and the weight learning. We conclude with some experimental work, conclusions and future work.

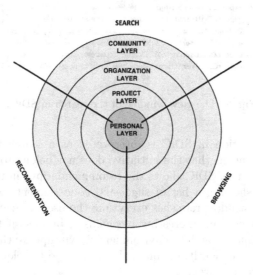

Fig. 1. The context model in SDiC.

2 SDiC Overview

SDiC[1] is a plug in for Eclipse[2] that constantly monitors what the user is doing so that it creates and maintains a history of the main actions that the user makes in the IDE. The visible part of SDiC comprises several views in Eclipse that allow the user to search for code, get code recommendations or browse the code that the user is working with.

The search functionality is triggered by the user using the view in Fig. 2. This view allows the user to search for code elements (classes, interfaces or methods) that satisfy the user query and that are near or related with the context in which the user is working on. Using the context enables the user to have more accurate search results for the information needed.

Fig. 2. The view window of the search in SDiC.

The recommendation in SDiC is interpreted as a search without a query (see Fig. 3), which means that the initiative does not belong to the user, but to the system. In this case, SDiC shows the recommendations in the same window as the search, by showing a list of suggested code elements when there is no search query. This window refreshes everytime the user changes the location in terms of focus. This is an interesting mechanism, because if the user has the search window visible in the IDE environment, it will update the contents of the suggested elements frequently, enabling the user to go to relevant code, even if s/he do not perform a search.

---

[1] http://sdic.dei.uc.pt.

[2] http://www.eclipse.org.

Fig. 3. The view window of the recommendation in SDiC.

The browsing mechanism opens a new view and allows the user to browse the code using the context to show or hide code elements that are not strongly related with the user navigational context (see Fig. 4). The size of the circles representing different code elements are bigger depending on their role regarding the user context, with the less important elements dimed in relation to the important ones. The user can navigate in the code graph by expanding and collapsing the various nodes.

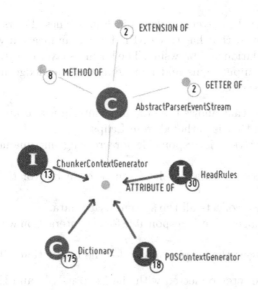

Fig. 4. The view window of the browsing in SDiC.

The conceptual architecture of SDiC is shown in Fig. 5, where the four main modules are: the context-based search, responsible for performing the search; the context-based recommendation, responsible for performing the recommendation; the knowledge base that stores all the knowledge needed by the system; and the context model that stores and manages the user context. These modules are going to be described in more detail in the next sections.

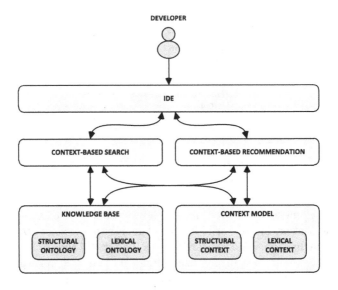

Fig. 5. The conceptual architecture of SDiC.

SDiC is organized in layers as shown in Fig. 6. These layers are: the presentation/interface layer, that has three APIs for communication with the Eclipse: search, recommendation and browsing. The business layer is responsible for doing all the complex computations and managing the knowledge in the system. It comprises several services:

- Monitor Service: that allows the system to interpret and analyse all the user interaction data that is gathered from Eclipse.
- Context Service: that is responsible for creating and maintaining the user context.
- Explorer Service: accesses the Java source code files of the user using the Eclipse JDT.
- Collector Service: collects all the system usage data.
- Knowledge Manager: that is responsible for the interaction with the knowledge base.
- Database Manager: that is responsible for the interaction with the database.

These modules are interconnected with the Eclipse JDT and Eclipse Platform. The data layer comprises two main parts: the knowledge base where all the knowledge is stored in the form of ontologies; the database that stores more simple data, such as user data and permissions.

SDiC monitors the user interactions with Eclipse, managing the interface views of SDiC and managing the communication with the server. In the next sections we will describe the most important parts of SDiC: the knowledge base, the context model, the search, the recommendation, the browsing and the weight learning mechanisms.

3   Knowledge Base

The Knowledge Base used in SDiC stores all the knowledge used by system. It comprises four different parts: the structural ontology that stores the code structural representation; the lexical ontology that stores the lexical knowledge extracted from class and method names; the user context, which will be described in detail in Sect. 4; and the Lucene index that stores all the linguistic terms needed to be searched. Both the structural and the lexical ontologies are described in the next sections. In this section, it will also be described the process of creating the knowledge base and we all also provide an example of how a java code element is represented in these ontologies.

3.1   Ontologies

The complete Ontology used in SDiC is shown in Fig. 7 and comprises two parts: the structural ontology and the lexical ontology.

The structural ontology represents the source code elements and comprises classes that represent concepts common in Java code, such as: Type, Class, Interface and Method. It also represents the relations that exist between source code

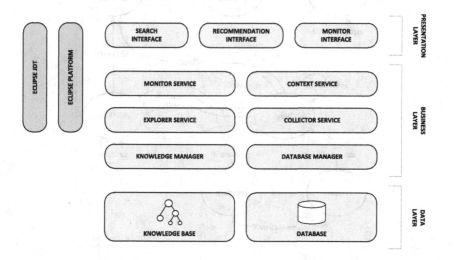

Fig. 6. The architecture of SDiC.

elements, such as: attributeOf, subclassOf, implementationOf, parameterOf, and others. This model allows the explicit representation of the source code structure, including the relations between source code elements, which are very important to reason in this domain.

The lexical ontology stores the information about linguistic terms that is extracted from the classes, interfaces and method names. These names are parsed and interpreted to extract the words that compose their names. SDiC not only extracts words, but it also extracts the relation associated With between the words that are in the same name. This relation represents the co-occurrence between terms, which can also be interpreted as semantic relatedness or semantic proximity in linguistics [3]. This information is then stored in the lexical ontology and terms are associated with structural elements as well.

3.2   Knowledge Base Creation

The process of creating and maintaining the knowledge base, in particular the structural and lexical ontologies is illustrated in Fig. 8. The process begins by getting the source code files from the IDE workspace of the user, which are then parsed and transformed in an Abstract Syntax Tree (AST) of the code. Then the source code elements are extracted from the AST going through two subprocesses: reference disambiguation and term extraction. In the first, the result is the extraction of the elements for the structural ontology, while the second extracts the terms and respective relations for the lexical ontology.

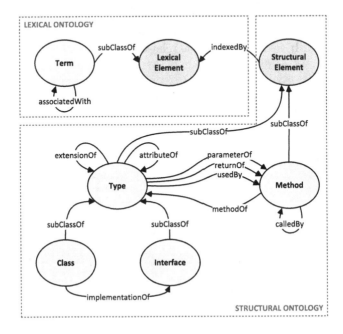

Fig. 7. The Ontologies used in SDiC.

SDiC constantly monitors the IDE and in special the interactions between the user and the source code. If the user adds, removes or updates a source code file, then the knowledge base gets updated in regard to the changed elements.

Another important aspect of the knowledge base is that the system indexes all the lexical elements in a search engine (Apache Lucene[3]). This search engine indexes not only the extracted lexical elements (for example: 'database' and 'manager'), but also the structural elements that originated them (following the previous example: 'DatabaseManager').

3.3   Example

We now present an example of how the system works regarding information extraction and knowledge base creation. Taking as example the source code in Fig. 9, the resulting parsing tree (the AST tree) can be seen in Fig. 10. The resulting elements extracted from this AST are shown in Fig. 11. Notice that the lexical

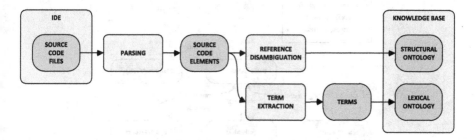

Fig. 8. The process used in the creation of the knowledge base.

```
1  package example.database;
2
3  import example.database.model.*;
4
5  public class DatabaseManager implements IDatabaseManager
6  {
7      [...]
8
9      public void addProduct(Product product)
10     {
11         [...]
12     }
13
14     public Product getProduct(String id)
15     {
16         [...]
17     }
18
19     public boolean updateProduct(Product product)
20     {
21         [...]
22
23         Product currentProduct = getProduct(id);
24
25         [...]
26     }
27
28     [...]
29 }
```

Fig. 9. The Java source code used in the example.

---

[3] http://lucene.apache.org.

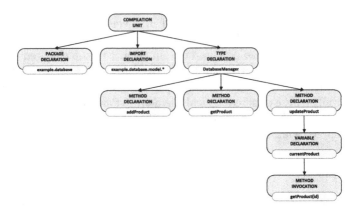

Fig. 10. AST generated for the example source code.

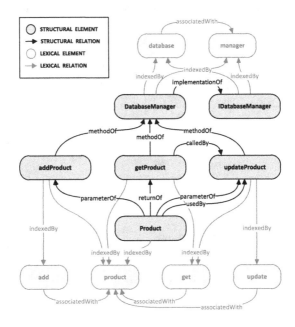

Fig. 11. Ontology elements extracted and created from the example source code.

elements are extracted from the structural element names using the Camel-Case rule (http://en.wikipedia.org/wiki/CamelCase). The underscore character is also used to split words, and the system also deals with acronyms.

4   Context Model

The context model of SDiC tries to represent the actual work focus of the user, in terms of source code being changed or developed. The context model is created using the interactions between the user and Eclipse, which are constantly being monitorized by SDiC. The context model comprises two parts: the structural part and the lexical part, which is to say, the code elements that the user interacted with and their corresponding terms.

The structural context model represents the structural elements and relations that the user interacted with. An interaction can be: opening the element, activate an element, edit an element and close an element. Following these actions, SDiC creates the respective structural element along with an associated degree of interest associated (DOI), which is a number. The DOI increases if the interaction is an open, activate or edit action, and is decreased if it is a close action. The DOI is an important concept in SDiC, since it represents the actual interest of the user in a specific structural element. DOIs are dynamic, in the sense that they increase or decrease in time, not only with user actions, but also with the natural passage of time. There is a decay function that decreases the DOI of all elements in time.

A similar process occurs for structural relations and for the lexical context. The lexical context is based on the lexical terms of the structural elements with which the user interacts with. For the interested reader, the context model used in SDiC, as well as its building process, are described in more details in [4].

5   Context-Based Search

One of the main functionalities of SDiC is search, which is far more accurate than the Eclipse search, because it uses the context of the user to rank the retrieved elements. The context-based search process has two main phases: the retrieval and the ranking.

The retrieval process takes as input the search query and uses Lucene to retrieve a list of relevant source code elements. These elements have a retrieval score associated, which is the score given by Lucene scoring function. This score is a standard score in the information retrieval area, which is based on the TF-IDF metric (term frequency and inverse document frequency). This would be the expected score if one wants to use a standard keyword search engine.

The second phase, which is the distinctive phase in SDiC, is the use of a ranking algorithm that takes into account the user context model to rearrange the way the search results are ranked. The ranking algorithm uses a similarity metric based on a weighted sum of three aspects: the retrieval score that comes from Lucene, the structural score that results from the user structural context, and the lexical score that results from the user lexical context. The structural context measures the similarity between the search result and the user structural context. The lexical score measures the similarity between the search result and the user lexical context. For the interested reader, a more detailed description of the context-based search process is provided in [5].

6   Context-Based Recommendation

The context-based recommendation is a functionality of SDiC in which the system suggests source code elements that might be relevant to the user. The relevancy concept in this case is defined as code elements in which the user might be interested in navigating to or inspecting. The process is similar to the search process, with the difference that there is no query. In this case, the most important part of the structural context is used to retrieve code elements. This retrieval takes into account two aspects: the elements in the user context that are more important and the time elapsed since these elements were accessed.

After the retrieval of the code elements that might be relevant to the tasks that the user is performing, the system ranks these elements. The ranking process takes into account four different aspects:

- The interest score: this score originates from the retrieval process and represents the score of the elements retrieved taking into account the interests in the context model.
- The time score: this score also comes from the retrieval process and represents the elements retrieved based on the elapsed time.
- The structural score: this score reflects the structural relevance of the element in relation to the context model.
- The lexical score: this score reflects the lexical relevance of the element in relation to the context model.

The formula to compute the final ranking score is a weighted sum, with each score having an associated weight. For the interested reader, a more detailed description of the context-based recommendation process is provided in [6].

7　Context-Based Browsing

The context-based browsing is a functionality that aids the developers to explore and understand the code elements and how they are connected. It enables developers to see the source code as a graph with the structural elements (classes, interfaces and methods) as nodes and the relations between these elements as links. Figure 12 shows two browsing situations, where in situation 1 we selected to hide the ReturnOf relation and the resulting visualization is presented in 2.

The context model is used to limit the nodes to be visualized, which can be seen in the example of Fig. 4. In this figure, it shows a visualization of the code elements in the context of the developer (class AbstractParserEventStream and ChunkerContextGenerator). Both classes are used to focus the code visualization showing which are the most important code elements connected to them.

Browsing has several important features such as:

- Enables the visualization of several different nodes and relations.
- Aggregates nodes to deal with information overflow, using the context model as an intelligent way to deal with this.
- Enables search within the nodes being shown in the visualization problem.
- Filters node and relation types according to the user needs.
- Provides several different ways of navigation in the graph.
- Provides a visual timeline so that the user can go backwards in the visualization history.

8   Weight Learning

One of the characteristics of SDiC is the use of several weigths to control the different parts of the ranking metrics. These weights are different for each user, making the system personalizable. This is a huge advantage, since each developers have a different way to navigate the source code. In order to have this functionality, SDiC uses a weight learning function that uses the developer search and recommendation results selections to fine tune the ranking weights. The basic idea is to increase the weights of the components contributing positively to the selected, and decrease the ones that are contributing negatively to the selected result.

This functionality is very important to fine tune the results of the search and recommendation processes. Figure 13 shows the monitorization of the weights related to the search process. There are 4 important parts of this figure:

1. The number of search results by type of access, search window or search view, which also gives the total number of searches performed.
2. The evolution of the average rankings of selected search results.

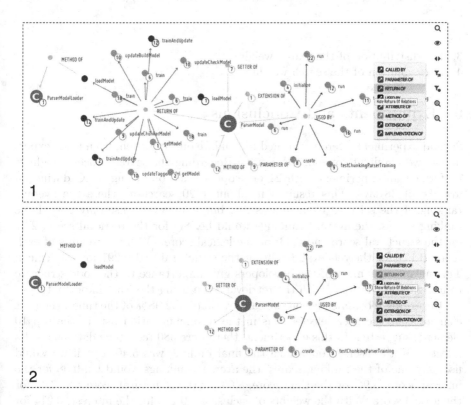

Fig. 12. A screenshot of two browsing situations where a filter was applied to situation 1 and resulted in situation 2.

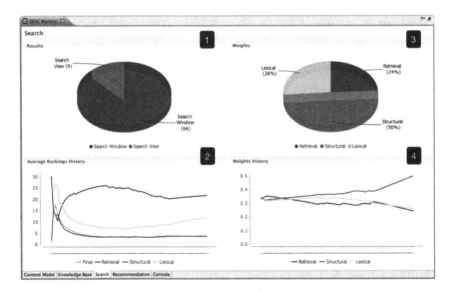

Fig. 13. A screenshot of the monitoring of the weight learning process for the search process.

3. The distribution of the search weights.
4. The evolution of the search weights.

9   Experiments and Conclusions

Several experiments were performed to validate our approach, from these experiments we highlight some of the results. Regarding the search process, we have performed an experiment with 21 developers, each one using SDiC during an average of 38 days. This resulted in about 1120 searches, where the average ranking of the selected element was 2.4, but if we would use only one of the ranking scores, the average rankings would be: 8.0 for the retrieval score, 2.31 for the structural score and 5.43 for the lexical score. With the weights of each score: 0.177 for the retrieval, 0.529 for the structural and 0.291 for the lexical. This means that most of the developers give importance to the code structure to navigate in the code, with the retrieval score being the least important.

Regarding recommendation, we can say that in 52.98 % of the times, the next element that the user selected was in the recommendation list, in the top 10 elements suggested. In this experiment, there were 380 recommendations results selected by the users, from which the final ranking was 5.40. But if we would use only one of the ranking scores, the average rankings would be: 9.08 for the interest score, 8.96 for the time score, 8.03 for the structural score and 7.41 for the lexical score. With the weights of each score: 0.224 for the interest, 0.214 for the time, 0.297 for the structural and 0.264 for the lexical. These results do not allow any conclusion, contrary to the search results.

As mentioned in the introduction, SDiC enables users to find code elements related to the task that developers are performing based on the context of the developer. Combining search, recommendation and browsing provides the user with three powerful, but complementary ways of finding information. While search can be used to find information that we already know, but we do not remenber. Browsing and recommendation empowers developers with mechanisms that are able to provide the developer new knowledge about the code.

Future work on SDiC will be on the other context layers: project, organization and community layers. This will enable a more broad usage of knowledge and will also allow a bigger return in terms of gains for the developer. Another important direction for the future work is the use of semantic search to improve the search mechanism.

Acknowledgements. Bruno Antunes was supported by a FCT scholarship grant SRFH/ BD/43336/2008, co-funded by ESF (European Social Fund).

References

1. Robillard, M., Walker, R., Zimmermann, T.: Recommendation systems for software engineering. IEEE Softw. **27**(4), 80–86 (2010)
2. Murphy, G.C., Kersten, M., Findlater, L.: How are Java software developers using the elipse IDE? IEEE Softw. **23**(4), 76–83 (2006)
3. Harris, Z.: Distributional structure. Word **10**(23), 146–162 (1954)
4. Antunes, B., Cordeiro, J., Gomes, P.: Context modeling and context transition detection in software development. In: Proceedings of the 7th International Conference on Software Paradigm Trends (ICSOFT 2012) (2012)
5. Antunes, B., Cordeiro, J., Gomes, P.: Context-based search in software development. In: Proceedings of the 7th Conference on Prestigious Applications of Intelligent Systems (PAIS 2012) of the 20th European Conference on Artificial Intelligence (ECAI 2012) (2012)
6. Antunes, B., Cordeiro, J., Gomes, P.: An approach to context-based recommendation in software development. In: Proceedings of the 6th ACM Conference on Recommender Systems (RecSys 2012) (2012)

Knowledge Management and
Information Sharing

ESA Knowledge Management Agenda

Roberta Mugellesi Dow[1], Damiano Guerrucci[1],
Raul Cano Argamasilla[2(✉)], Diogo Bernardino[3],
and Siegmar Pallaschke[2]

[1] European Space Agency, Robert Bosch Strasse 5, Darmstadt, Germany
{roberta.mugellesi.dow,damiano.guerrucci}@esa.int
[2] Terma GmbH, Europaplatz 5, Darmstadt, Germany
raul.cano.argamasilla@esa.int, espall@t-online.de
[3] Serco GmbH, Robert Bosch Strasse 7, Darmstadt, Germany
diogo.bernardino@esa.int

Abstract. During the last years, ESA has gathered an extensive experience in Knowledge Management (KM). As a knowledge-intensive organisation, ESA needs to implement an efficient management of its expertise and know-how. To this purpose, an ESA corporate KM initiative was conducted during the last few years and consisting of several pilot projects. The selected approach was built on pillars summarized in these four keywords: "integrated" (interconnection of its parts), "pragmatic" (concrete solutions compatible with the corporate culture), "business" (linked to the core business) and "open" (input and best practices gathered inside and outside the Agency). This paper presents the activities carried out within this set of pilots and the relations among them. The pilot projects presented in the paper are: the KM Portal called iKnow, the Competency Management Tool, the Expertise Directory, the Knowledge Capture and Handover process, the KM Officer and Lessons Learned Harmonization. The future of KM in ESA is a challenge aiming at proposing integrated solutions in an environment with different cultures and several existing individual solutions. The achievement of it will certainly provide a better leverage for the institutional KM.

Keywords: ESA · Knowledge Management · Portal · Competency Management · Knowledge Capture · Lessons Learned

1 Introduction

Knowledge Management (KM) is a key pillar for an efficient evolution of knowledge-intensive organisations such as the European Space Agency (ESA). ESA is in charge of promoting, developing and operating space missions (www.esa.int) and it does that from its six establishments located in France (ESA Headquarters), UK (ECSAT), the Netherlands (ESTEC), Germany (ESOC), Spain (ESAC) and Italy (ESRIN). The coexistence of research, development and operation programmes makes ESA a multidisciplinary knowledge-intensive organization.

ESA has recognised the crucial role of KM as a tool not only to primarily preserve and share information and know-how, but also to help to guarantee its own cost-effective and qualitative evolution. In order to achieve this objective, a corporate

© Springer-Verlag Berlin Heidelberg 2015
A. Fred et al. (Eds.): IC3K 2013, CCIS 454, pp. 293–310, 2015.
DOI: 10.1007/978-3-662-46549-3_19

strategic approach to KM, in support to both the transversal and the local nature of the Agency's operations, has been defined and implemented.

This paper introduces the knowledge processes and tools which have been launched in different areas and identifies key challenges and opportunities to move ESA's knowledge agenda forward in a phased manner.

2 Background

The KM topic has gained momentum at ESA thanks to a number of KM support initiatives undertaken within the last decade. Two examples [5] are quoted for illustration:

- Long duration projects [10]: Knowledge preservation is not only required due to staff mobility but also in the light of long duration satellite missions, such as the European space mission Rosetta. Initial analysis for the project started in 1986, the launch took place in 2004 and the approach to the comet 67P/Churyumov-Gerasimenko will occur in 2014. Knowledge preservation is an important issue for Rosetta and the first steps towards KM were initiated already in 1999 to be incorporated since the beginning into the mission.
- Internal ESA reviews concerning efficiency: In 2003 an internal review was started concerning the engineering approach at ESOC. One area of concern was the effective transfer and communication of vital information and expertise across projects. The review called into existence a dedicated working group which was one of the driving forces in ESA KM domain.

Major steps have been achieved in the area of KM in ESA:

- At a general level managers and employees are conscious of the importance of KM, not only as a learning and development issue but also as strategic and business related;
- A wide set of KM and communication technologies has already been developed;
- Important KM practices such as lessons learned are to some extent applied to space missions;
- Operative knowledge is shared within several individual Divisions.

However, some outstanding issues are still not solved:

- Even if the importance of KM is recognised, the culture of sharing and re-use is not yet completely adopted in people's mind and in business processes;
- The institutional knowledge base risks to shrink as individuals retire or move to different projects/programmes and mitigation actions are only partially in place;
- New staff are immersed into new programs and expected to start working without a substantial introduction to the ESA history and processes;
- Knowledge-explicit categorisation is missing and the evolution of competencies reflecting the objectives is developed in the heads of managers without adopting a systematic process and appropriate tools.

3   ESA KM Strategy

The ESA KM strategy [7] focuses on three major objectives:

- Capture, preserve and help evolving knowledge across missions and projects in order to increase efficiency, minimize the risks and avoid loss of expertise;
- Facilitate knowledge sharing to increase collaboration, synergies and, eventually, innovation;
- Set up methodologies and tools for people to find, organize, and share knowledge.

The first objective focuses on securing the gathered experience (i.e. Lessons Learned, Best Practices) and making it available to the community so that the proper knowledge is available at the appropriate place and time within work processes. Maintaining the existing knowledge provides the foundations to evolve with less effort towards the new knowledge needed for future endeavours. There is a large amount of knowledge generated during the day-to-day work from formal documents, reports, presentations to the less formal knowledge, such as information exchange between colleagues that represents a significant and valuable part of the knowledge generated and accumulated. Actually, in most cases this knowledge remains in a tacit form, retained in the heads of the staff. When not formalized or made explicit, this knowledge naturally tends to be dispersed and eventually loses its value for ESA and its stakeholders.

The second objective focuses on facilitating the sharing of knowledge in order to increase the collaboration between individuals and teams. This heavily contributes to create a more fertile environment for continuous learning and for the search for innovative solutions. An effective knowledge sharing depends on how efficiently and effectively the knowledge is managed internally and how quickly it capitalizes on the skills and experiences gathered in the different functional areas of the organization.

Findings from interviews and brainstorming sessions with staff indicate that they need information technology to act on:

· *Search capability*. Documents such as reports or other types of information should be electronically searchable. Staff wants to be able to find a particular knowledge product with keywords and to be able to trust that it is the current version.

. *Ease of use*. Staff wants knowledge to be well organized, to some extent summarized, easy to locate and easy to retrieve.

. *Access*. Staff wants to be able to access the information from the office or home.

. *Sharing*. Staff wants to be able to exchange easily with colleagues information and knowledge products.

Box 1. Staff expectations.

The third objective focuses on helping people to access the information and resources they need to complete their tasks by providing the right KM tools, resources and methodologies. The way to find out which elements could build this objective was mostly the direct input of the people and brainstorming. Box 1 offers an insight on this.

The localised initiatives conducted in ESOC were the key contributor for introducing knowledge management ESA-wide. In view of maximising the KM positive impact on the users and minimise the effects of a too sudden methodological change, a phased approach consisting of two phases was adopted, taking into account the actual needs of the users and having pilot projects before any solution proposed for application at corporate level. Moreover, if a local solution was already found available, the idea was to study its possible integration into a larger system without imposing structural changes to localised successful processes.

4   Overview of KM Pilot Projects

The two phases approach was the following. The first phase consisted in the analysis of the ESA as-is situation in terms of KM, in order to propose a set of pilot projects that would build the corporate KM system. A set of thirteen pilots were envisaged. They were:

For Organisation/processes: KM Governance, Review Process (with Concurrent Design Methodology), Lessons learned, Project Map and KM Officer;
For Technology/IT Tools: Competence Appraisal, KM Portal, Expert Directory, Unique Search Engine and E-Learning;
For Staff: Knowledge Capture plus Handover and Knowledge Sharing Incentives;
For Communication: KM Awareness Campaign.

The second phase consisted in the effective implementation of the pilot projects selected amongst those identified in the previous phase. The framework of KM selected projects and activities [3] which have been undertaken and launched consist of the seven projects listed below and in the next sections. A short summary of the objectives and goals is provided below for each pilot, whereas a more detailed description will follow in the subsequent chapter.

In parallel to pilot projects, transversal activities were carried out with the goal to be informed of the state-of-the-art platforms and methodologies which would enable ESA to have a KM serving efficiently its goals. To that purpose, the participation in KM groups (e.g. KM for Space - Google Group, Knowledge Managers from CNES and NASA), the contribution to the investigation of new methodologies (via collaborating in paper studies or creating their own research) and benchmarking new ideas for implementation with companies working on KM tools and methodologies was reinforced.

4.1   Pilots Main Objectives

iKnow, the KM Portal. The focus of this pilot was to improve the Knowledge Management Portal (called iKnow) already existing in response to the feedback from the user community. The areas identified for improvements were: interface look & feel and ease of use, performance, functionality extension by adding video transcriptions, search engine integration and mail to post mechanism.

Expert Directory. The aim of this pilot was to create a network between the ESA professionals, contractors and industry. The way to achieve that was the development of a platform that, interfacing with the competency management database, would provide a catalogue of people with their expertise.

Competency Management Tool. The scope of this pilot was to consolidate a previously drafted competency management process and design a tool to support it. The result of this activity was the presentation of a competency management process (supported by a tool), which covers the following objectives:

- Guarantee the availability of the current and future ESA competencies (knowledge, skills, and abilities) in regard to the ESA tasks that constitute ESA (critical) strategic capabilities and services.
- Provide a management support process and tool for identifying competency gaps, training opportunities and development plans on both short and long terms.
- Support the definition of the strategic objectives of the directorate.
- Allow an efficient maintenance and up-to-date overview of the related competencies.

Knowledge Capture and Handover. This pilot was selected to assess and propose possible mechanisms to be used to guarantee an effective capture of knowledge, tacit and explicit, in case of staff leaving for retirement or other assignment. The scope of the pilot is also to define a Handover Procedure.

The steps that were done:

- Analysis of KC events with respect to methodology and visual aspects.
- Proposal for capturing of experience.
- Proposal of a new way to present the KC events.
- Proposal of a Handover sequence.

KM Officer. This pilot was born with the objective to define the role, responsibilities and the functioning model of the KM Officer that could become the pivot of the knowledge management processes inside and outside the projects.

The actions taken were:

- Collection and assessment of the possible KMO tasks.
- Summarize the aspects of integration into the project.

Lessons Learned. The objectives of this pilot project were the analysis of mechanisms and tools for the collection of Lessons Learned (LL) and also the dissemination of the Best Practices and Lessons Learned into the project database (with respect to efficiency and anomaly aspects) and across projects. To this purpose, the following has been achieved:

- Analysis of current ESA LL systems together with a comparison.
- Analysis concerning the collecting, managing and archiving of LL (compliant to the formal ESA process) with emphasis on description, transferability and sustenance.

KM Awareness. The scope of the KM Awareness Campaign was to promote a Knowledge Management culture and to communicate in a smart way the KM initiatives. The deliverables have been:

- Communication plan to choose topics to communicate, the media and the tone of voice.
- Video and posters to promote KM.

4.2 Pilots Pillars

The individual pilots provide powerful solutions to the scope they have been designed for. Moreover, there is a more holistic view to regard the project as the integration of its parts. One of the principles on which the corporate KM project was built was the "Integrated" approach that stands for the whole project has a meaning when observed as a unity instead of as many subsystems.

Each of the pilots is connected in some form or another to the rest (Fig. 1) in a way that if this connection is not available, the functionality of each of them will be significantly decreased.

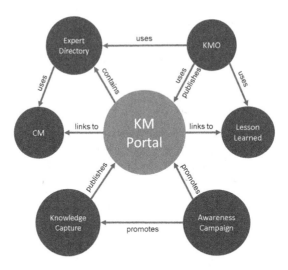

Fig. 1. Integration among pilot projects.

Just to quote some examples. The Competence Management tool provides the input to the Expert Directory. At the same time, the Knowledge Management Officer can use this information to either request help for a project or to propose a Knowledge Capture event based on the available experts. The KM team takes care of this event and once this is performed and processed, the videos and material would be uploaded to the KM Portal. Finally, through the awareness campaign the people know that the portal includes such information and are inspired to share more.

The "Pragmatic" approach is currently the main way of reaching KM goals in ESA. The resources available are very limited and only by implementing concrete and simple solutions, the main goal of each pilot can be reached. The adoption of this approach is often characterised by the need of simplifying or tailoring the solution in parts that shall not limit the scope of the pilot. The simplification taken are tracked so to be later considered in case of a future corporate complete solution.

Every time a pilot is started, the "Business" approach has driven the activities by identifying first a concrete local interest without forgetting the need of a scalable design to make it suitable for future extension at corporate level. The local interest ensures the reaching of a very good level of maturity since derived by a real case study with a clear and short time need. Later on the collected real data, it will facilitate the leverage raising it suitable at corporate level.

The last, but not the least, pillar is to be "Open". The KM team shall always work in collaboration with users and minimise the impact to existing processes, especially to the successful ones. Best practises are key input to any KM activities, since are essential for addressing real and practical needs and later for ensuring a well-received support.

5   Detailed KM Pilot Projects

This section describes on a deeper detail the activities related to the pilots which have a larger reach, that is to say, affecting the biggest possible amount of users. Hence, the pilots KM Officer and Awareness are not further described in here.

5.1   iKnow, the KM Portal

iKnow was born to be the backbone of all the information at ESOC, more or less like an enhanced multimedia library, with forums, events, news and all relevant materials. Moreover, as the user is the that creates most of the content, this portal is inspired by a "Web 2.0" approach for the knowledge management in ESOC.

The information that can be accessed from iKnow can be of very different nature, like relevant news, collaborative articles or any document and file indexed by the search engine or a link to an existing tool. One of the main characteristics of iKnow is that every piece of content (article, forum post, etc.) is assigned according to the group or community that it belongs to and also, to the tags that the author considered more descriptive ("free tagging" principle). So both methods are set by the user. Due to this, the content is categorized by the person that uses it and really knows what it is about, which guarantees that the knowledge is where it is expected to be. In the context of this corporate KM project, the functionalities described in the next sub-sections were added to the platform.

Search Engine. A commercial search engine has been running in a testing phase for a few months and now is implemented both as a standalone operational tool and integrated within iKnow. It is able to index a wide range of repositories without any customization of the tool.

In support to the idea of providing a single entry point to information, the search engine functionality has been fully integrated in iKnow (Fig. 2) so the user can now enter a search request directly on the portal and get the results back from either the portal or all the other repositories indexed by the search engine.

Video Transcriptions. The video functionality was enhanced in iKnow to allow everyone to upload videos and optionally a transcription of them, such that they will be synchronized while the video is running. In combination with the search engine, this is a powerful way of extracting the knowledge from videos. Once the transcriptions are indexed, the information will also appear in the search results.

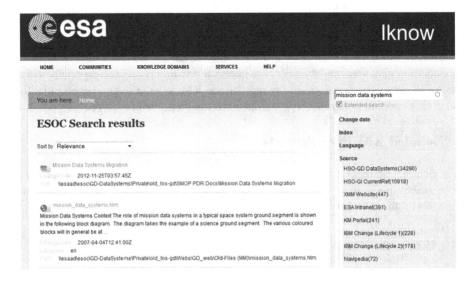

Fig. 2.   iKnow search engine functionality.

Mail2post. This is a new feature in iKnow to facilitate the knowledge sharing by allowing the user to publish contents just by sending an email to a particular address. With this functionality, users can easily share all the valuable information that otherwise would remain buried in the email box.

5.2   Expert Directory

In order to build a network of professionals closer to each other and to facilitate the identification of the necessary competencies, it was decided to create a directory of experts that would be validated by a formal process inside ESA. Therefore, it was agreed that the Expert Directory should use the Competency Management (described below) database as its primary source of data. This database can be considered as a verified dataset by Human Resources. The tool should only be able to explore the database or search through it for a member that has enough competences and present its data. No editing features were needed or had to be developed.

Though this started as an independent pilot, a module was developed and integrated in iKnow. After a successful development, any user could perform searches per competencies or knowledge areas. This module also provides with a view of the employee profile which would help, among other things, to design a training plan.

5.3 Competency Management

The Competency Management (CM) pilot project aims to develop a comprehensive and validated process that provides the organization with a view as well as insight of the current and future required competencies. The process is expected to benefit different stakeholders such as Workforce Management and Training and Development programs.

The major objectives of CM have been reflected in activities included in the process and supported by the tool:

- Provide visibility of current and future competencies with respect to the roles that constitute ESA's strategic capabilities and services in general, and in particular.
- Make individual expertise better accessible within the organization. For this purpose, the focus is on staff specific competencies, i.e. competencies acquired by a person and not necessarily relevant for their current role(s).
- Allow the identification of competency gaps. For example, managers can get an accurate picture of the competency strengths and weaknesses of the staff under their responsibility and therefore assess potential competency gaps, criticality and competency coverage.
- Provide a way for efficient competency maintenance and up-to-date overview, i.e. the workforce needs to be shaped and modelled to cope with desired future directions of the Agency, changes in technology, etc.

The process needs to be supported by a CM Tool structured such as to guide the different actors in the fulfilment of their relevant CM activities. Figure 3 illustrates this process.

At each cycle of the process, an iteration with the KM team will be also considered for addressing future management needs and process improvements.

The platform supports the different steps allowing the different actors to interact through the tool with the common goal of providing the current picture of the organisation and at the same time highlighting the areas where improvements or actions are necessary. The tool can provide to the managers summary reports for an overall analysis of matching between objectives and current assets and identifying possible gaps.

Once the process was consolidated, the main requirements for the tool were clear:

- Maintain a central repository of the CM related data, including both the definition and management of competencies and roles, and the information from the assessment. The repository would store historical information in order to keep track of the evolution over time.
- Support efficient introduction of data by users, manually using forms, and automatically from batch files or other formats.

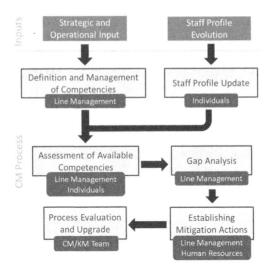

Fig. 3. CM process.

- Offer a web based interface for multiple access with customized dashboard per user role.
- Automate the generation of the reports required as output of the different steps of the process.
- Implement the logic for the analysis of competence gaps, and proposal of mitigation actions.
- Offer a training plan specifying who needs training, what are the competences to be trained and the urgency of training.

Responding to the mentioned requirements, a market analysis was conducted before starting any development. Since the CM process was very likely subject to changes as effect of the pilot project implementation, the analysis suggested to take the approach allowing the highest flexibility and lower cost in performing modifications. In house implementation was selected, also facilitated by the existence in the Agency of a previous suitable tool already developed, though for other purposes (BIRF - Business Intelligence and Reporting Framework - built over Java Enterprise Edition platform). The existing tool based on a multi layered design could be strongly reused in the following generic components:

- The Extraction, transformation and loading (ETL) process responsible for the batch loading of the input data files (e.g. Excel) in the Data Warehouse, performing necessary adaptations.
- The Data Warehouse as a central data storage based on a relational database management system.
- Web browser based clients, which allow users to access the application from any location and device through the network.
- The web interface for providing the user level functionality, and divided itself into three layers:

- Third party components, providing general purpose functionality by integration of open source libraries (e.g. web framework, chart libraries, reporting library, etc.).
- Data analysis and reporting framework, providing generic application level components supporting all the required functionality (e.g. dashboards, report generator, security, data querying, etc.).
- CM customization, implementing the adaptation of the reporting framework to the CM process (e.g. adaptors, data model, user interface layout, etc.).

The design of the tool is taking into account future integration within other ESA platforms, in order not to duplicate information and realise better synergies. In this case, examples are: Workforce Management software, eLearning platforms, or Expert Directory. A validation of this pilot is being carried through one department of the organization.

5.4 Knowledge Capture

Knowledge Capture activity has been followed up during the past years with more than a dozen of events in which two methods were considered: the expert debriefing and the interview. The background and its underlying characteristics are described in the quoted reference [6]. These two methods are not oriented to factual results of project phases but to personal experience gained by a staff member either at the termination of a task or at the end of a position.

In brief words, the knowledge capture procedure consists of four steps: the review of the status (knowledge, expertise), the planning, the conduct of the expert debriefing/interview and the documentation of the event.

The expert debriefing is a kind of forum/workshop in order to allow the expert to explain the specialized knowledge for a selected range of topics to a larger group of participants while the other method, the interviews, is applied for the description of complex subjects to a reduced audience. The characteristics of the two methods are summarized in Table 1.

Table 1. Characteristics of Knowledge Capture methods.

	Expert debriefing	Interview
Type	Workshop	Script
Audience	Larger group of experts, colleagues within knowledge area	Interview team only
Interaction	Face-to-face interaction with audience	No direct interaction, interview script to be followed
Moderation	Required	Depending on subject
Recognition of Merit	Direct in public	Indirect

Knowledge Capture Procedure. The knowledge capture procedure consists of following steps:

Step 1: Review of status with an inquiry. The objective is to assess the important knowledge subjects of the leaving staff as well as the knowledge demand required by the group and the successor. For the identification of the essential knowledge items to be captured different viewpoints (e.g. different role) should be adopted in addition to the chronological review of the supported projects. The quality aspects could be another perspective such as best and worst practices, contribution of the success and mistakes leading to a possible failure.

Step 2: Plan for debriefing and interview. This includes the sequence of knowledge subjects for the debriefing/interviews and its structure.

Step 3: Conduction of the Debriefing/Interviews with the goal of capturing the tacit knowledge and facilitating its documentation. The various viewpoints mentioned for the first step above have to be adopted for the conduct of the expert debriefing and the interviews as well.

Step 4: Documentation (including the transcription) of the event.

Knowledge Capture Events. Knowledge capture events are video-recorded, can last from 20 up to 120 min. and comprise several sub-topics followed directly by a question and answer session. The sub-topics should not explain explicit knowledge (easily found in text books) but should go more into the direction of experience or knowledge not easily found in literature.

Some more explanations have to be added for the better understanding of the capture events at ESOC.

- The duration for the expert debriefing is limited to two hours. This seems to be adequate as the expert debriefing is additional to the standard handover procedure. Furthermore about four to five subjects are selected. Another reason for the short duration lies in the assumption that with increasing duration the interest in participation decreases.
- The choice of the four subjects depends on the relevance of the gained experience for the department/organisation. Normally, two topics concerning specialized knowledge and two ones concerning experienced knowledge are selected.
- The group of the participants is decided from case to case depending on the topics and the envisaged subsequent discussions. In case of detailed discussions the group of participants should be limited as the intensity of the discussion decreases with increasing participation.

Capturing Experience. In the context of knowledge capture, the handling of the experience was also investigated [8]. In everyday life, experience is understood as the individual occurrence which affects one personally and which leads one to draw subjective and life-like conclusions and not a universally valid understanding. It is a subjective discovery and perception.

Since more than a decade the value of storytelling for the transfer of knowledge by experience has gained in importance since stories can better describe the complexity of

Table 2. Overview of narrative methods.

Type of story	Description	Application
Learning (interpret the past)		
Report on experience (real, authentic)	Report on one's own experience for the benefit of the colleagues	Sharing of experience
Story construction (Report of experience, however, constructed)	Real stories and anecdotes (content and message) are mixed and formed to a newly-constructed story	Propagation of a desired message to the staff. (Lessons learned, best practices)
Learning history	Reporting of essential projects by various members with subsequent analysis	Discovery of implicit relations
Motivation (shape the future)		
Springboard Story	Short stories (real or fictitious) facilitating a mental jump	Creation of understanding/ acceptance of an upcoming change

reality than any theoretical explanations. Table 2 lists the various storytelling options. The application of storytelling in knowledge management has become greatly in demand and various companies have tested and use this method. Knowledge by experience is also considered as a central resource for the conduction of organizational processes: it cannot easily be discovered with the help of the appraisal results and requires additional interviews with the leaving staff, the superior and the group.

Table 3. Summary of enhancements.

Step	Knowledge Capture procedure	Enhancements concerning experience
Inquiry	Assessment and identification of critical knowledge areas	(a) Narrative interviews (how expertise was developed and applied) (b) Analysis in order to identify structures through associations (c) Deeper interviews
Planning of event	Sequence of knowledge subjects	
Conduct of event	Assessment and identification of critical knowledge areas	(a) Story without ppt, message to be mentioned at the end (b) Visualisation of comments provided by audience (c) Reflection, feedback oriented dialog, no discussion with arguments
Prepare documentation		

The aim of the narrative interviews is to explore how the experience was developed and how it was applied. Of course, the outcome of the interviews needs to be analysed with the help of associations for identifying the underlying experience. Structured interviews could follow for the better understanding of the experience. The four steps (as mentioned above for the capture procedure) will also be followed in the case of storytelling. Whereas the planning (step 2) will remain unchanged, the actual conduct of the event (step 3) will include narrative methods. Narration cannot be compared with explanations (rational descriptions), being more feedback-oriented dialogues. The audience needs to conduct its own reflection for drawing the conclusions. Table 3 outlines the changes for the various steps of the capture procedure.

5.5   Lessons Learned

Lessons Learned stands for the systematic collection, evaluation and compression of experiences, developments, failures, successes and risks encountered within the course of a project. In essence, Lessons Learned is equivalent to reviewing the conduct of a project and looking for improvements. The conclusions derived from positive and negative project experiences should help to avoid the repetition of mistakes and the wastage of time for subsequent projects. Unfortunately, these experiences are not always recorded systematically due to missing time, conflicting priorities, missing preparedness or lack of error culture to report on errors openly.

The goal of the pilot was to analyze and to improve the collection and dissemination mechanisms of the lessons learned process already in place in the Agency. The pilot work was based on a wide range of input from the various sectors as well as textbooks and publications. The various recommendations were grouped in categories such as: extended workshop, description & categorization, dissemination & transferability and sustainability (see Table 4). Particular emphasis was given to the description and the transferability aspects since these are closely linked as the transferability is facilitated by a correct and suitable description of the Lessons Learned.

Three-level Structure. The structure of the documented Lessons Learned follows very often logical aspects which might not be optimal for the reader. Barbara Minto [2, 4] proposes a three level structure (see Fig. 4) which is based on pyramid thinking. The content is presented in three levels where the amount and depth of information increases with the next lower level. Hence, the decision concerning the relevance of the particular Lessons Learned can be taken at an early stage.

The structure was tailored for optimizing the search, preparation and re-use of Lessons Learned and experience.

Classification and Keywords. The methodology used for the classification of lessons learned is important for the archiving and for future retrieval. Emphasis is also put on the classification as it is needed for application to other projects. Keywords are very important for the later association and application. The following paragraph lists the range of possible keywords going from the domain part to the management part with different perspectives.

For the domain-related part of the classification the corresponding taxonomy can be used as a basis. In addition to the domain (subject) related part categories can be

Table 4. Sequence of steps.

Step	Description	Comment
Extraction, Generation Proposal, Description	After reviews, at anomalies or ad hoc proposals	Collection without filter (question of culture)
		Guidelines for expressing Lessons Learned
Assessment, Verification, Validation	Verification by nominated experts, domain responsibles and review board	
Storage, Archiving	Incorporation into documentation/ standards	Categorization
	Structured according to the key processes and knowledge domains	
Distribution, Promotion	IT tools (virtual classrooms, etc.)	Ambassador as part of project
	F2F methods (public presentations, training events, audits)	
	F2F exchange between project	
Application	Integration into core business	Re-use
	Linking Lessons Learned with prevention mechanism	
Maintenance	Periodic reviews for the identification of potentially obsolete Lessons Learned	Sustainability with progressing time
Reporting	Annual report	

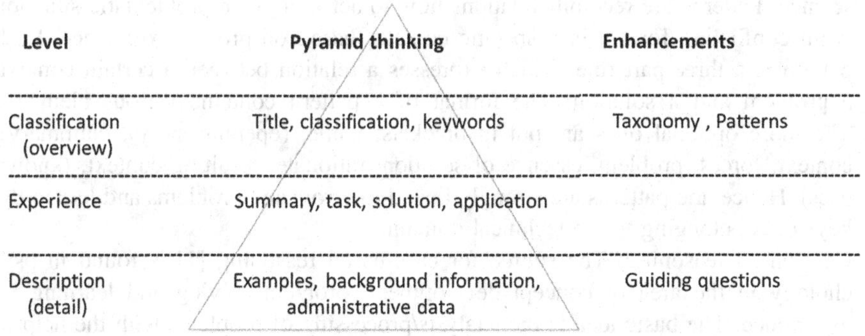

Level	Pyramid thinking	Enhancements
Classification (overview)	Title, classification, keywords	Taxonomy , Patterns
Experience	Summary, task, solution, application	
Description (detail)	Examples, background information, administrative data	Guiding questions

Fig. 4. Three-level Structure.

formed for other areas such as management and communication. In case Lessons Learned refer to errors these could be broken down in: ignorance of symptoms, handling of side-effects, assessment of exponential development (as IT) or rigidity in planning and goal. These errors could also be seen from the perspective of intention distinguishing between un-intentional (blunder) to intentional actions (procedures, knowledge).

For the archiving direction primarily the area of categories and keywords is important. For the retrieval direction (for linking the current problem with one of the archived Lessons Learned) pattern and case-based reasoning become of additional interest (see **Applicability and Re-use**).

Description. There are at least three major goals to be considered for the description of Lessons Learned. One goal is the reduction of the description to the essential (key) points, the other is the provision of the required information/background for being able to reproduce the concluded Lessons Learned and the third relates to the correct expression.

Nina Plum developed in her doctoral thesis [9] a structure of guiding questions for Lessons Learned. The guiding questions should inspire for a stepwise reconstruction of the occurrence and for a better description of the context. The guiding questions are broken down into two parts relating to the description of the situation and to the recommendation.

Applicability and Re-use. The retrieval of previous Lessons Learned applicable to a current problem is easier through a good and concise classification and keyword system. In addition to this system, patterns and case-based reasoning can support the retrieval of applicable Lessons Learned. The idea of patterns and case-based reasoning is built on the assumption that experts tend to use the experience of previous similar cases for solving the current problem. Hence, the Lessons Learned should not only be stored according to domain-specific keywords but should also be structured based on a problem-oriented perspective. Here the two options of pattern and case-based reasoning are presented.

- Pattern: The idea of patterns was introduced by Christopher Alexander (1977) in the field of architecture and has been adapted for various disciplines including computer science. Patterns are recommendations how to act in a given problematic situation (with conflicting forces) in a specific context, based on proven experience. Each pattern is a three-part rule, which expresses a relation between a certain context, a problem and a solution. The format of a pattern contains various elements. The more optional ones are put in brackets: name, (opening story), (summary), context, forces, problem, essence of solution, rationale, resulting context, (known uses). Hence, the patterns are strongly linked to errors and problems and less to the keywords belonging to the technical domain.
- Case-based reasoning: The source for case-based reasoning [1] is found in psychology in the area of concept development, problem solving and learning by experience. The basic idea is the analysis/processing of problems with the help of solutions already applied in earlier situations (see Fig. 5). A repository contains the various solutions to earlier problems. In the case of a new problem one tries to match the current one with one of the earlier cases. Of course, the required steps are:
  - Retrieve a similar case (or more);
  - Reuse and revise those for establishing an approach to the current situation;
  - Retain the solution and update the repository.

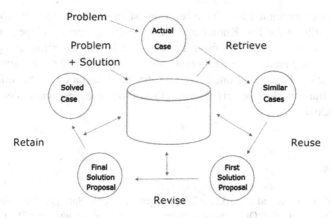

Fig. 5. Case-based reasoning.

It has to be mentioned that the case-based reasoning is well suited for the maintenance of the Lessons Learned system as the repository will be updated with each new problem encountered (see Fig. 6).

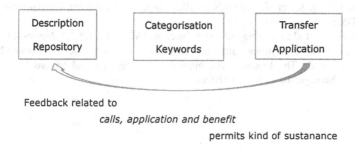

Fig. 6. Maintenance of Lessons Learned.

6   Conclusions

Achieving the knowledge management objectives is pivotal to the Agency's efforts to reach its goal of cost effectiveness and risk minimization. Knowledge management is an ESA-wide initiative that requires strong leadership and commitment at every level, effective incentive mechanisms, ownership by all departments, offices, units, and active participation by all staff members. Moreover, it requires a change in how ESA knowledge is perceived and supported.

The benefit of implementing the KM framework of projects presented in this paper will result in an overall improvement of processes and mechanisms for capturing, sharing and storing information, leading to an increased operational efficiency and facilitating innovation.

The activities presented here have set a milestone in the history of KM for ESA since this constitutes the first Knowledge Management system developed at corporate level. There are two remarkable facts: the first one is that the pilots comprising the system are covering areas of very different nature, demonstrating the big spread of KM in an organization; the second one is the fact that the integration of the different parts into a single framework is perceived more efficient by managers and easier to be used by the individuals.

References

1. Bodendorf, F.: Data and Knowledge Management, 2nd edn. Springer, New York (2006)
2. Bunse, Ch., et al.: SoftQuali–An integrated approach for software quality improvement (1999)
3. Guerrucci, D., et al.: Technological aspects of the KM system in ESA. In: 4th International Conference on KM, Toulouse SpaceShow 2012, Toulouse (2012)
4. Minto, B.: The Pyramid Principle, 1st edn. Prentice Hall, Upper Saddle River (1991)
5. Mugellesi Dow, R., et al.: A knowledge management initiative in ESA/ESOC. J. Knowl. Manage. **10**(2), 22–35 (2006)
6. Mugellesi Dow, R., et al.: Managing knowledge for spacecraft operations at ESOC. J. Knowl. Manage. **14**(5), 659–677 (2010)
7. Mugellesi Dow, R., et al.: Towards an ESA corporate knowledge management. In: IAC 2011, D5,2,1, Cape Town, SA (2011)
8. Pallaschke, S., et al.: Capturing of experience. In: IAC 2012, D5,2,3, Naples, IT (2012)
9. Plum, N.: Structure of guiding questions for Lessons Learned, Doctoral Thesis (2006)
10. Zender, J., et al.: The Rosetta video approach: an overview and lessons learned so far. J. Knowl. Manage. **10**(2), 66–75 (2006)

Multilevel Self-organization in Smart Environment: Approach and Major Technologies

Alexander Smirnov, Nikolay Shilov$^{(\boxtimes)}$, and Alexey Kashevnik

SPIIRAS, 39, 14th Line, St. Petersburg, Russia
{smir,nick,alexey}@iias.spb.su

Abstract. Efficient operation of complex distributed systems such as smart environments requires development of self-organisation mechanisms. However, uncontrolled self-organization can often lead to wrong results. The paper proposes solving this problem through the "top-to-bottom" configuration principle. A reference model for a smart environment member is proposed. The approach is illustrated via a museum smart environment case study. It is then extended with introducing a recommending system based on the developed smart environment architecture.

Keywords: Smart environment · Multi-level self-organization · Service-oriented architecture · Recommending system

1 Introduction

The expanding capabilities of mobile devices enables their usage in the growing number of human activities. However, such trend requires a significant increase of information sharing. Smart environments are aimed to assist in solving this problem. They assume presence of a number of physical devices that use shared view of the resources and services provided by them [1].

In order for such systems to operate efficiently, they need self-organisation mechanisms. The network is self-organised in the sense that it autonomically (without external control) monitors available context in the network, provides the required context and any other necessary network service support to the requested services, and self-adapts when context changes. These systems are particularly robust, since they adapt to changes, and are able to ensure their own survivability [2]. The process of self-organisation of a network assumes creating and maintaining a logical network structure on top of a dynamically changing physical network topology. This logical network structure can be used as a scalable infrastructure by various functional entities like address management, routing, service registry, media delivery, etc. The autonomous and dynamic structuring of components, context information and resources is the essential work of self-organisation [3].

The key mechanisms supporting self-organising networks are self-organisation mechanisms and negotiation models. The following self-organisation mechanisms are usually selected [7]: intelligent relaying; adaptive cell sizes; situational awareness; dynamic pricing; intelligent handover.

© Springer-Verlag Berlin Heidelberg 2015
A. Fred et al. (Eds.): IC3K 2013, CCIS 454, pp. 311–325, 2015.
DOI: 10.1007/978-3-662-46549-3_20

The following negotiation models can be mentioned [8]:

- Different forms of spontaneous *self-aggregation*, to enable both multiple distributed services/agents to provide collectively and adaptively a distributed service, e.g. a holonic (self-similar) aggregation.
- *Self-management* as a way to enforce control in the ecology of services/agents if needed (e.g. assignment of "manager rights" to a service/agent.
- *Situation awareness* – organization of situational information and their access by services/agents, promoting more informed adaptation choices by them and advanced forms of stigmergic (indirect) interactions.

The analysis of literature related to organizational behaviour & team management has showed that the most efficient teams are self-organizing teams working in the organizational context (Fig. 1). For example, social self-organisation has been researched by [4, 5]. However, in this case there is a significant risk for the group to choose a wrong strategy preventing from achieving desired goals. For this purpose, self-organising groups/systems need to have a certain guiding control. This consideration produces the idea of multi-level self-organization.

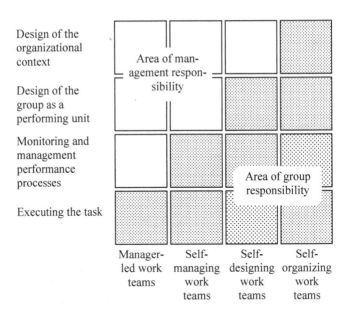

Fig. 1. Authority of work group types (adapted from [6]).

The multilevel self-organisation has not been addressed yet in research. This approach would enable a more efficient self-organisation based on the "top-to-bottom" configuration principle, which assumes conceptual configuration followed by parametric configuration (Fig. 2).

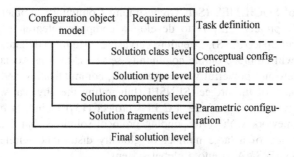

Fig. 2. Multi-level configuration.

The paper extends the 2013 "Knowledge Management and Information Sharing" International Conference contribution [9] introducing the further developments in the area of recommending system implementation (Sect. 6). The next section presents some related work followed by the description of the proposed approach. Section 4 describes the developed reference model of self-organising smart environment member. The case study related to the development of smart museum service is given in Sect. 5. Section 6 extends the case study introducing a recommending system based on the developed approach. Main results are summarised in the conclusion.

2 Related Work

The approaches to creating systems of autonomous elements are currently being widely developed in the areas of context-dependent decision support systems (sponsored by DARPA), forming self-contextualized networks (IST-2004-2.4.5 Ambient Networks, FP 6), creation of self-organising systems (ICT-2007.1.1 Self-optimisation and self-configuration in wireless networks, FP 7) and other.

DARPA ITO Project S3: Scalable Self-Organizing Simulations [10] addresses development and distribution of Scalable Simulation Framework (SSF) and the SSFNet Internet modelling tools.

Another DARPA sponsored project Self-Organizing Sensor Networks [11] assumes that self-organizing sensor networks may be built from sensor nodes that may spontaneously create impromptu network, assemble the network themselves, dynamically adapt to device failure and degradation, manage movement of sensor nodes, and react to changes in task and network requirements. Reconfigurable smart sensor nodes enable sensor devices to be self-aware, self-reconfigurable and autonomous.

The Ambient Networks EC FP6 project [3] is addressing these challenges by developing mobile network solutions for increased competition and cooperation in an environment with a multitude of access technologies, network operators and business actors. It offers a complete, coherent wireless network solution based on dynamic composition of networks that provide access to any network through the instant establishment of inter-network agreements. The concept offers common control functions to a wide range of different applications and access technologies, enabling the integrated, scalable and transparent control of network capabilities.

The vision of SOCIETIES (Self Orchestrating CommunIty ambient IntelligEnce Spaces, EC FP7 project, [12] is to develop a complete integrated solution via a Community Smart Space (CSS), extending pervasive systems to dynamic communities of users. CSSs will embrace online community services, such as Social Networking, and thus offer new and powerful ways of working, communicating and socialising.

Tangible results of the project SENSEI (Integrating the physical with the digital world of the network of the future, EC FP7 project [13]) include a highly scalable architectural framework with corresponding protocol solutions that enable easy plug and play integration of a large number of globally distributed wireless sensor and actuator networks (WS&AN) into a global system.

One more EC FP7 project SOCRATES (Self-optimization and self-configuration in wireless networks [14]) -investigates the application of self-organization methods, which includes mechanisms for self-optimization, self-configuration and self-healing, as a promising opportunity to automate wireless access network planning and opti-mization, thus reducing substantially the Operational Expenditure (OPEX) and improving network coverage, resource utilization and service quality. Fundamental drivers for the deployment of self-organization methods are the complexity of the contemporary heterogeneous access network technologies, the growing diversity in offered services and the need for enhanced competitiveness.

3   Service-Oriented Approach

The proposed approach is based on the idea of smart environment where all partici-pating devices are represented via services [15]. The service-oriented architecture (SOA) is a step towards information-driven collaboration. This term today is closely related to other terms such as ubiquitous computing, pervasive computing, smart space and similar, which significantly overlap each other [16].

The proposed service-oriented approach to efficient multilevel self-organisation of services in the smart environment assumes information actualization in accordance with the current situation. An ontological model is used in the approach to solve the problem of service heterogeneity [17]. This model makes it possible to enable inter-operability between heterogeneous services due to provision of their common semantics [18]. Application of the context model makes it possible to reduce the amount of information to be processed. This model enables management of informa-tion relevant for the current situation [19]. The access to the services, information acquisition, transfer, and processing (including integration) are performed via usage of the technology of Web-services.

Figure 3 represents the generic scheme of the approach. The main idea of the approach is to represent the smart environment members by sets of services. This makes it possible to replace the organisation of the smart environment with that of distributed service network. As it was mentioned the configuration is done based on the "top-to-bottom" configuration principle, which assumes conceptual configuration fol-lowed by parametric configuration. The information between levels is transferred through guiding from the upper level to the lower level.

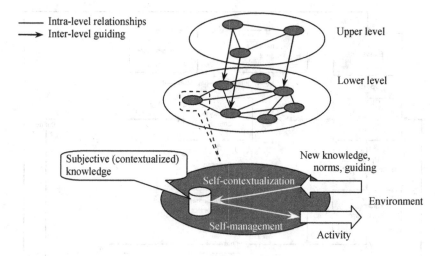

Fig. 3. Generic scheme of the approach.

For the purpose of interoperability, the services are represented by Web-services using the common notation described by the application ontology. Depending on the problem considered the relevant part of the ontology is selected forming the abstract context that, in turn, is filled with values from the sources resulting in the operational context. The operational context represents the constraint satisfaction problem that is used during self-organization of services for problem solving.

4 Reference Model of Smart Environment Member

The proposed reference model of the multilevel self-organization is presented in Fig. 4. Below, its main components are described in detail.

Smart environment member (service, agent, sensor, etc.) is an acting unit of the multilevel self-organization process. It has **structural and parametric knowledge**, and **profile**. It is characterized by such properties as self-contextualization, self-management, autonomy, and proactiveness.

Structural Knowledge is a conceptual description of the problems to be solved by the **smart environment member**. This is the member's internal ontology. It describes the structure of the member's **parametric knowledge**. Depending on the situation it can be modified (adapted) via the **self- management** capability. It also describes the terminology of the member's **context** and **profile**.

Parametric knowledge is knowledge about the actual situation defining the smart environment member's **behavior**. Its structure is described by the member's **internal ontology**, and the parametric content depends on the **context**.

Context is any information that can be used to characterize the situation of an entity where an entity is a person, place, or object that is considered relevant to the interaction between a user and an application, including the user and applications themselves [20]. The context is purposed to represent only relevant information and knowledge from

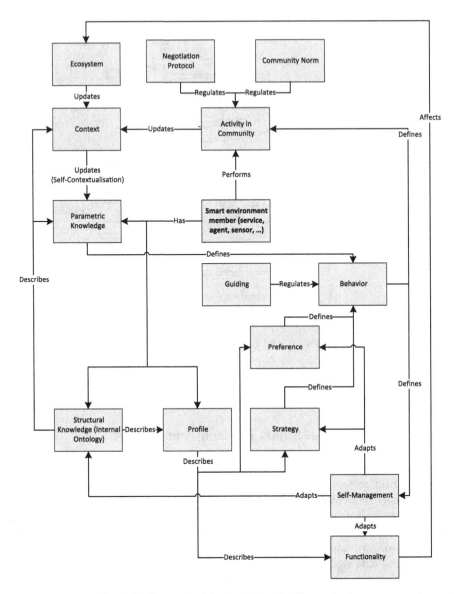

Fig. 4. Reference model of multi-level self-organisation.

the large amount of those. Relevance of information and knowledge is evaluated on a basis how they are related to a modelling of an ad hoc problem. The context is represented in terms of the smart environment member's **internal ontology**. It is updated depending on the information from the member's **ecosystem** and as a result of its **activity in the community**. The context updates the member's **parametric knowledge**, which in turn defines the member's behaviour. The ability of a system (smart environment member) to describe, use and adapt its behavior to its context is referred to as **self-contextualization** [21]. The present research exploits the idea of self-contextualization to autonomously adapt behaviors of multiple members to the context of the current situation in order to provide their services according to this context and to propose context-based decisions. To achieve this purpose the smart environment members have to be context/situation – aware and context-adaptable.

Ecosystem is the surroundings of the smart environment, that may interact with its members. The ecosystem affects the members' **context**. The **smart environment member** can affect the ecosystem if it has appropriate **functionality** (e.g., a manipulator can change the location of a corresponding part).

Functionality is a set of functions the **smart environment member** can perform. Via it the **member** can modify its **ecosystem**. The member's functionality can be modified in certain extent via the **self-management** capability. The functionality is described by the member's **profile**.

Profile describes the smart environment member's **functionality** in terms of the member's **internal ontology** and in a way understandable by other members of the smart environment.

Self-Management is a smart environment member's capability to modify (reconfigure) its **internal ontology, functionality, strategy**, and **preferences** in response to changes in the **ecosystem**.

Behavior is the smart environment member's capability to perform certain actions (**activity in community** and/or **functionality**) in order to change the own state and the state of the **ecosystem** from the current to the preferred ones. The behavior is defined by the member's **preferences** and **strategies**, as well as by the **guiding** from a higher level of the self-organization.

Guiding is a set of principles and/or rules coming from a higher level of self-organization to direct the **behavior** and achieve rational outcomes on a lower level of self-organization.

Preference is a smart environment member's attitude towards a set of own and/or environmental states and/or against other states. The preferences are described by the member's **profile** and affect the member's **behavior**.

Strategy is a pre-defined plan of actions rules of action selection to change the smart environment member's own state and the state of the **ecosystem** from the current to the preferred ones. The strategy is described by the member's **profile** defines the member's behavior.

Activity in community is a capability of the **smart environment member** to communicate with other members and negotiate with them. It is regulated by the **negotiation protocol** and **community norms**.

Negotiation protocol is a set of basic rules so that when smart environment members follow them, the system behaves as it supposed to. It defines the **activity in community** of the members.

Community Norm is a law that governs the smart environment member's **activity in community**. Unlike the **negotiation protocol** the community norms have certain degree of necessity ("it would be nice to follow a certain norm").

5 Case Study: Smart Museum Service

The tourism is becoming more and more popular. People travel around the world and visit museums and other places of interests. They have a restricted amount of time and usually would like to see many museums. In this regard a system is needed, which would allow assisting visitors (using their mobile devices), in planning their museum attending time and excursion plans depending on the context information about the current situation in the museum (amount of visitors around exhibits, closed exhibits, reconstructions and other) and visitor's preferences. This could be done using personal mobile devices, which have Internet connection and possibility to show appropriate information to visitors.

Mobile devices interact with each other through the smart environment. Every visitor installs a smart environment client to his/here mobile device. This client shares needed information with other mobile devices in the smart environment. As a result, each mobile device can acquire only shared information from other mobile devices. When the visitor registers in the environment, his/her mobile device creates the visitor's profile (stored in a cloud and containing long-term context information of the visitor). The information storage cloud (not computing, which is distributed among the services of the smart environment) might belong to the system or be a public cloud. The only requirement is providing for the security of the stored personal data. The profile allows specifying visitor requirements in the smart environment and personifying the information and knowledge flow from the service to the visitor. Each time the visitor appears in the smart environment, the mobile device shares information from the visitor's profile with other devices.

Visitor context accumulates and stores current information about the visitor in the smart environment (current visitor context). It includes: visitor location, museum reaching times for the visitor, current weather (e.g., in case of rain it is better to attend indoor museums), visitor role (e.g., tourist, school teacher), information about closed at the moment museums or exhibits.

To get external information for different system modules, the services of four types are used:

- Positioning service (calculates current indoor and outdoor positions of the visitor based on raw data provided by visitor mobile device);
- Information service (provides visitor mobile device with needed information about exhibits, e.g., Wikipedia, Google Art Project, museum internal services, etc.);
- Current situation service (provides information about the current situation in the region, e.g., weather, GIS information, traffic information);
- Museum/exhibition (provides information related to the museum and exhibits, e.g., holidays, closed exhibits).

The proposed ontological scheme for the case study is presented in Fig. 5. Each visitor has a mobile device, which communicates with mobile devices of other visitors (shares own information to them and gets needed information), uses different services for getting and processing information, accesses and manages the visitor's profile, and processes information and knowledge stored in visitor context.

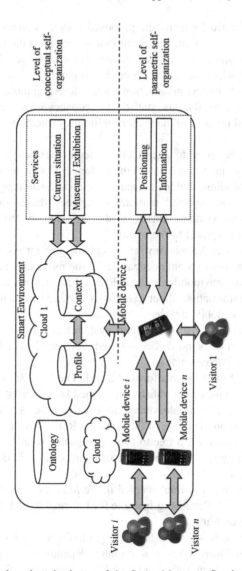

Fig. 5. Ontology-based scheme of the Smart Museums Service case study.

Visitor's profile and context are stored in the cloud, which allows visitors to access them from any internet enabled devices (when the visitor changes his/her mobile device it is needed only to install the appropriate software to use the new device). The conceptual level self-organization takes place in the cloud sending resulting information to the users' devices.

The visitor context is formed based on the interaction process between the visitor's mobile device and different services through the smart environment (parametric level self-organization). The context is the description of the visitor's task in terms of the ontology taking into consideration the current situation in the museums. Visitor's task in the proposed approach is a list of museums the visitor would like to attend.

The following scenario for using the proposed system is considered. The visitor arrives to a region. His/her mobile device finds the museums the visitor is going to attend in this region (stored in the visitor's profile). The mobile device generates the context, which describes the current situation of this region. It negotiates with different services to extract information about interesting museums (working time, closed museums, closed exhibitions, statistical occupancy of interesting museums for the next few days) and propose to the visitor preliminary interested museums attending plan.

When the visitor is going to attend the museum (next day), the mobile device updates the context by current situation in the region, e.g.: weather (in case of rain it is better to postpone attending outdoor museums), traffic situation on the roads, current museum occupancy, and expected museum occupancy (based on negotiation with mobile devices of other visitors). Based on this information, the corrected museum attending plan can be proposed to the visitor.

When the visitor enters the museum, an acceptable path for visiting museum rooms is built based on the museum room occupancies at the moment. Using location service and Wi-Fi infrastructure the mobile device calculates the visitor's location and shares it with other devices. Information about exhibits is acquired from the service and displayed on the visitor's mobile device.

The intelligent museum visitor's support system has been implemented based on the proposed approach. Maemo 5 OS – based devices (Nokia N900) and Python language are used for implementation.

It is based on the Smart-M3 platform [22, 23]. The key idea in Smart-M3 is that devices and software entities can publish their embedded information for other devices and software entities through simple, shared information brokers. The understandability of information is based on the usage of the common RDF-compatible ontology models and common data formats. It is a free to use, open source solution available under the BSD license [24]. Communication between software entities is developed via Smart Space Access Protocol (SSAP).

The system has been partly implemented in the Museum of Karl May Gymnasium History [25] located in St. Petersburg Institute for Informatics and Automation Russian Academy of Science building.

The visitor downloads software for getting intelligent museum visitors support. Installation of this software takes few minutes depending on operating system of mobile device (at the moment only Maemo 5 OS is supported). When the visitor runs the system for the first time the profile has to be completed. This procedure takes not more than 10 minutes. The visitor can fill the profile or can use a default profile. In case of default profile the system can not propose preferred exhibitions to the visitor.

Response time of the Internet services depends on the Internet connection speed in the museum, number of people connected to the network, and workload of the services. The average response time does not exceed one second.

An example museum attending plan is presented in Fig. 6. It consists of five museums: The State Hermitage, the Kunstkamera, the Museum of Karl May Gymnasium History, St. Isaac Cathedral, the Dostoevsky museum.

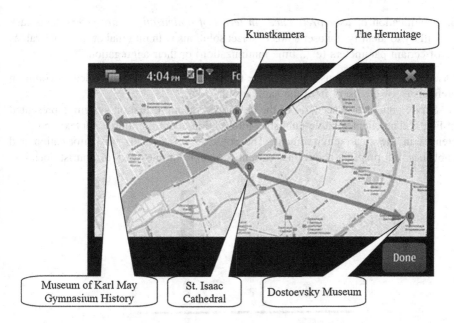

Fig. 6. A smaple of museum attending plan in a visitor mobile device in the center of St. Petersburg.

6 Smart Environment-Based Recommending System

The further development of the described above prototype is aimed at introducing features of group recommending systems. Generation of feasible museum attending plans that takes into account explicit and tacit preferences requires strong IT-based support of decision making so that the preferences from multiple users (accumulated in the system and/or obtained from social networks) could be taken into account [26–28]. Group recommending systems are purposed to solve this problem. Recommendation/recommend-ing/recommender systems have been widely used in the Internet for suggesting products, activities, etc. for a single user considering his/her interests and tastes [29], in various business applications [e.g., 30, 31] as well as in product development [e.g., 32, 33].

The preference revealing can be interpreted as identification of patterns of the solution selection (decision) by a user from a generated set of solutions. The ability to automatically identify patterns of the solution selection allows to sort the set of solutions, so that the most relevant (to user needs) solutions would be in the top of the list of solutions presented to the user.

Currently, three major tasks of identification of user preferences are selected:

1. Identification of *user preferences based on solutions generated for the same context*. In this case, the problem structure is always the same, however its parameters may differ.
2. Identification of user *preferences based on solutions generated for different contexts*. This task will be more complex then the first one since structures of the problem will be different.

3. Identification of *user preferences in terms of optimization parameters*. This task will try to identify if a user tends to select solutions with minimal or maximal values of certain parameters (e.g., time minimization) or their aggregation.

Based on the identified user groups, the user preferences can be revealed as common preferences of the users from the same group.

Service interaction scenario of the developed recommendation system is presented in Fig. 7. The client module publishes information about the tourist context and preferences in the smart environment. The context service reads this information and publishes information about the current situation in the area of tourist location.

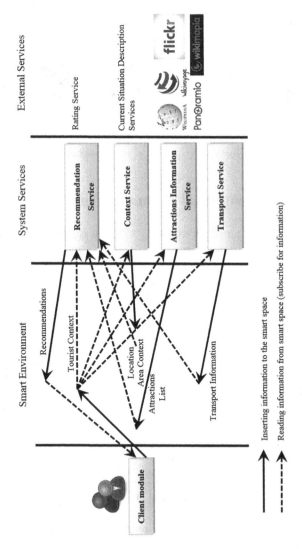

Fig. 7. Tourist attraction recommendation system services interaction based on smart environment technology.

The attraction information service publishes the list of attractions that meets the tourist context and preferences. The recommendation service reads information about attractions, tourist preferences, and context from the smart environment, gets information from the rating service, and generates a list of most appropriate at the moment for the tourist attractions related to the his/her interests. Based on information from the transport service, the recommendation service analyzes which attractions and in which order can be proposed to the tourist for visiting. Then, the recommendation service publishes this information in the smart environment and it becomes accessible for the client module, which presents it to the tourist.

The recommendation service reads information about attractions, tourist preferences, and context from the smart environment, gets information from the rating service, and generates a list of most appropriate for the tourist at the moment attractions related to the his/her interests. Based on information from the transport service, the recommendation service analyzes which attractions and in which order can be proposed to the tourist for visiting. Then, the recommendation service publishes this information in the smart environment and it becomes accessible for the client module, which presents it to the tourist.

The recommendation system allows the tourist to browse the attractions' descriptions for making a decision about visiting them. In this case, the client module publishes the information that the tourist is interested in an attraction. The attraction service gets this information, searches for descriptions of this attraction, and publishes links to information sources with these descriptions to the smart environment. The recommendation service analyzes these information sources, user preferences, ratings made by other tourists for these sources and shares description (or descriptions) that are better for the tourist at the moment with the client module.

7 Conclusions

The paper presents a reference model and proposes major technologies for a multi-level self- organization in a smart environment. The proposed reference model enables a more efficient self-organisation based on the "top-to-bottom" configuration principle, which assumes conceptual configuration followed by parametric configuration. The service-oriented architecture makes it possible to replace the organisation of the smart environment with that of distributed service network. Application of the approach is illustrated via a museum smart environment case study.

The approach has some limitations. In particular, it requires the corresponding services (transportation & museums) to be available in the smart environment. At the moment, some of such services have been developed as prototypes and wrappers for existing third-party services have been developed. For development of a working application, this issue has to be kept in mind. Besides, the functioning of the client's application requires almost permanent Internet connection.

Acknowledgements. The research presented is motivated by a joint project between SPIIRAS and Nokia Research Center. Some parts of the work have been sponsored by grants # 12-07-00298, # 12-07-00302, # 13-07-13159, and # 13-07-12095 of the Russian Foundation for

Basic Research, project # 213 of the research program "Intelligent information technologies, mathematical modelling, system analysis and automation" of the Russian Academy of Sciences, and project 2.2 "Methodology development for building group information and recommendation systems" of the basic research program "Intelligent information technologies, system analysis and automation" of the Nanotechnology and Information technology Department of the Russian Academy of Sciences.

References

1. Smirnov, A., Kashevnik, A., Shilov, N., Oliver, I., Boldyrev, S., Balandin, S.: Profile-based context aware smart spaces. In Proceedings of the XII International symposium on problems of redundancy in information and control systems, pp. 249–253 (2009)
2. Serugendo, G.D.M., Gleizes, M.-P., Karageorgos, A.: Self-organisation and emergence in MAS: an overview. Informatica 30, 45–54 (2006)
3. Ambient Networks Phase 2. Integrated Design for Context, Network and Policy Management, Deliverable D10.-D1, 2006. http://www.ambient-networks.org/Files/ deliverables/D10-D.1_PU.pdf. Accessed 18 April 2007
4. Hofkirchner, W.: Emergence and the logic of explanation. an argument for the unity of science. Acta Polytech. Scand. Math. Comput. Manage. Eng. Ser. 91, 23–30 (1998)
5. Fuchs, C.: Globalization and Self-Organization in the Knowledge-Based Society. TripleC 1 (2), 105–169 (2003). http://triplec.uti.at
6. Hackman, J.R.: The design of work teams. In: Lorch, J.W. (ed.) Handbook of Organizational Behavior, p. 1987. Prentice Hall, Upper Saddle River (1987)
7. Telenor. R&D Report, Project No TFPFAN, Program Peer-to-peer computing. http://www. telenor.com/rd/pub/rep03/R_17_2003.pdf. Accessed 18 April 2007
8. De Mola, F., Quitadamo, R.: Towards an agent model for future autonomic communications. In: Proceedings of the 7th WOA 2006 Workshop From Objects to Agents. http://sunsite. informatik.rwth-aachen.de/Publications/CEUR-WS/Vol-204/P07.pdf. Accessed 25 March 2013
9. Smirnov, A., Shilov N., Kashevnik, A.: Multilevel self-organization in smart environment: service-oriented approach. In: Proceedings of International Conference on Knowledge Management and Information Sharing (KMIS 2013), pp. 290–297 (2013)
10. ITO Project S3. Scalable Self-Organizing Simulations, 2000. http://www.dimacs.rutgers. edu/Projects/Simulations/darpa/. Accessed 25 March 2013
11. Institute for Rexconfigurable Smart Components. Self-Organizing Sensor Networks. http:// www.eng.auburn.edu/users/lim/sensit.html. Accessed 25 March 2013
12. Waterford Institute of Technology. Self Orches-trating CommunIty ambiEnt IntelligEnce Spaces, 2013. http://cordis.europa.eu/search/index.cfm?fuseaction=proj.document&PJ_ RCN=11486081. Accessed 25 March 2013
13. SENSEI. Integrating the physical with the digital world of the network of the future. http:// www.sensei-project.eu/. Accessed 25 March 2013
14. SOCRATES. Self-Optimisation and self-ConfiguRATion in wirelEss networkS. http://www. fp7-socrates.org/. Accessed 25 March 2013
15. Johannesson, P., Andersson, B., Bergholtz, M., Weigand, H.: Enterprise modelling for value based service analysis. In: Stirna, J., Persson, A. (eds.) The Practice of Enterprise Modeling. LNBIP, vol. 15, pp. 153–167. Springer, Heidelberg (2008)

16. Balandin, S., Moltchanov, D., Koucheryavy, Y.: Smart spaces and next generation wired/wireless networking. In: Balandin, S., Moltchanov, D., Koucheryavy, Y. (eds.) Smart Spaces and Next Generation Wired/Wireless Networking. LNCS, vol. 5764. Springer, Heidelberg (2009)

17. Smirnov, A., Kashevnik, A., Shilov, N.: Ontology matching in context-driven collaborative recommending systems. In: Proceedings of International Conference on Knowledge Management and Information Sharing (KMIS 2012), pp. 139–144 (2012)

18. Uschold, M., Grüninger, M.: Ontologies: principles, methods and applications. Knowl. Eng. Rev. 11(2), 93–155 (1996)

19. Dey, A.K.: Understanding and using context. Pers. Ubiquitous Comput. J. 5(1), 4–7 (2001)

20. Dey, A.K., Salber, D., Abowd, G.D.: A conceptual framework and a toolkit for suppor-ting the rapid prototyping of context-aware applications. In: Moran, T.P., et al. (eds.) Context-aware Computing. A Special Triple Issue of Human-Computer Interaction, vol. 16, pp. 229–241 (2001)

21. Raz, D., Juhola, A.T., Serrat-Fernandez, J., Galis, A.: Fast and Efficient Context-Aware Services. Willey, London (2006)

22. Honkola, J., Laine, H., Brown, R., Tyrkkö, O.: Smart-M3 information sharing platform. In: The 1st Inernational Workshop on Semantic Interoperability for Smart Spaces (SISS 2010) in Conjunction with IEEE ISCC 2010 (2010)

23. Smart-M3 at Wikipedia. http://en.wikipedia.org/wiki/Smart-M3. Accessed 25 March 2013

24. Smart-M3 at Sourceforge. http://sourceforge.net/projects/smart-m3. Accessed 25 March 2013

25. The Museum of Karl May Gym-nasium History. http://www.spiiras.nw.ru/modules.php?name=Content&pa=shshowpa&pid=8. Accessed 25 March 2013 (in Russian)

26. McCarthy, K., Salamo, M., Coyole, L., McGinty, L., Smyth, B., Nixon, P.: Group recommender systems: a critiquing based approach. In: IUI 2006: Proceedings of the 11th International Conference on Intelligent User Interfaces. pp. 267–269 (2006)

27. Wang Y., Chan, S.C.-F. and Ngai, G.: Applicability of demographic recommender system to tourist attractions: a case study on TripAdvisor. In: Proceedings of 2012 IEEE/WIC/ACM International Conferences on Web Intelligence and Intelligent Agent Technology, pp. 97–101. IEEE (2012)

28. Zhang, S., Zhang, R., Liu, X., Sun, H.: A Personalized trust-aware model for travelogue discovering. In: Proceedings of 2012 IEEE/WIC/ACM International Conferences on Web Intelligence and Intelligent Agent Technology, pp. 112–116. IEEE (2012)

29. Garcia, I., Sebastia, L., Onaindia, E., Guzman, C.: A group recommender system for tourist activities. In: Di Noia, T., Buccafurri, F. (eds.) E-Commerce and Web Technologies. LNCS, vol. 5692. Springer, Heidelberg (2009)

30. Hornung, T., Koschmider, A., Oberweis, A.: A Recommender system for business process models. In: Proceedings of the 17th Annual Workshop on Information Technologies & Systems, WITS 2009. Electronic resource. http://ssrn.com/abstract=1328244. Accessed February 2012

31. Zhena, L., Huangb, G.Q., Jiang, Z.: Recommender system based on workflow. Decis. Support Syst. 48(1), 237–245 (2009). (Elsevier)

32. Moon, S.K., Simpson, T.W., Kumara, S.R.T.: An agent-based recommender system for developing customized families of products. J. Intell. Manuf. 20(6), 649–659 (2009). (Springer)

33. Chen, Y.-J., Chen, Y.-M., Wu, M.-S.: An expert recommendation system for product empirical knowledge consultation. In: The 3rd IEEE International Conference on Computer Science and Information Technology ICCSIT2010, pp. 23–27. IEEE (2010)

Mitigating Barriers to Healthcare Knowledge Sharing in the Collaboration of Traditional and Western Practitioners in Chinese Hospitals

Lihong Zhou[1] and Miguel Baptista Nunes[2(✉)]

[1] School of Information Management, Wuhan University,
Wuhan 430072, Hubei, China
L.zhou@whu.edu.cn
[2] Information School, The University of Sheffield, Regent Court,
211 Portobello, Sheffield S1 4DP, UK
j.m.nunes@sheffield.ac.uk

Abstract. This paper reports on a research study that aimed to mitigate and propose strategies to overcome barriers to patient-centered knowledge sharing (KS) in the interprofessional collaboration of Traditional Chinese Medicine (TCM) and Western Medicine (WM) professionals, who coexist and collaborate in the same hospitals in China. This research adopted a Grounded Theory approach using a Chinese public hospital as case-study, at which 49 informants were interviewed by using semi-structured and evolving interview scripts. 11 KS barriers emerged from significant and prevalent philosophical and professional tensions between the two professional groups. Therefore, to improve KS and mitigate the tensions, three strategies are proposed: (1) formalizing KS processes and exploring effective communication channels; (2) establishing specific inter-professional training schemes and programs; (3) eliminating imbalances of professional power and statues and creating conducive KS environment.

Keywords: Chinese hospitals · Knowledge sharing · Traditional chinese medicine · Western medicine · Patient-centered healthcare

1 Introduction

Different from any other Nation in the world, the Chinese healthcare system uniquely incorporates two entirely different healthcare approaches, namely, Traditional Chinese Medicine (TCM) and Western Medicine (WM). TCM has been a consistent element of Chinese culture [1] and was developed based on the result of the accumulation of experiences and medical practices for over 2300 years [2]. It has been consistently suggested that TCM is not just "folk" medicine, but a highly developed system and science [3]. However, TCM lost the dominant position it had for thousands years over the Chinese public health systems to Western Medicine (WM) at the beginning of the twentieth century. Modern WM, based on the scientific paradigm and evidence-based practices, was developed in Europe and North America after the industry revolution and is largely considered as the main component of today's Chinese medical system, despite its coexistence with TCM [4].

© Springer-Verlag Berlin Heidelberg 2015
A. Fred et al. (Eds.): IC3K 2013, CCIS 454, pp. 326–339, 2015.
DOI: 10.1007/978-3-662-46549-3_21

The coexistence of the two healthcare philosophies and professional communities were initially formulated under a political decision made by Chairman Mao Zedong in early 1950s, immediately after the establishment of the People's Republic of China (PRC). The original purpose of the political decision was to use a reformulated and systematized TCM as a strategic tool to distinguish the new communist China from its superstitious and feudal past as well as to illustrate the Chinese cultural heritage. Despite the political nature of the decision, many researchers have claimed that it created conditions for a complementary relationship with WM [5, 6]. This relationship was unexpectedly very successful, since it unites and synergizes the two types of professionals working cooperatively against a number of diseases deemed to be un-treatable solely by WM doctors [6]. The interprofessional collaboration of TCM and WM healthcare professionals gradually emerged as the central basis to the provision of healthcare services in today's Chinese hospitals.

However, the two professional communities, which sometimes operate in the same building, do not really co-exist harmoniously in the national healthcare system [7]. This co-existence arose from the initial political decision, but it became very quickly apparent that simply putting the two communities together and expect them to work collaboratively was not without problems. In fact, each community have integral and very distinctive medical beliefs, diagnose and treatment methodologies. This careless integration of the two generated disbeliefs, distrust and disregard between the two communities and resulted in the problems of coexistence in Chinese hospitals today [7]. In any case, regardless of any disagreements, dispute and problems of co-existence, it is politically decided that the two communities have to collaborate.

Since 2006, with the implementation of the patient-centered healthcare policy, an additional layer of political requirements was forced upon the TCM and WM collabo-ration. That is, the needs, requirements and benefits of patients must be constantly ensured and carefully protected throughout the processes of TCM and WM collaboration [8, 9].

The provision of patient-centered healthcare service relies on effective and suffi-cient communication and knowledge sharing (KS) [10, 11]. Nonetheless, and in reality, TCM and WM professionals do not necessarily actively and voluntarily communicate and share knowledge with each other [12]. In fact, there are barriers hindering the two types of professionals from actively engaging in KS [12, 13].

Despite public awareness of the issues that emerged from the TCM and WM coex-istence and a continuing debate on philosophical superiority, the KS problem between TCM and WM professionals has not been politically recognized and academically investigated. This paper presents, criticizes and discusses the barriers to patient-centered KS between TCM and WM professionals. In addition, this paper proposes and discusses actionable strategies that can be employed by hospital management to improve inter-professional communication and KS in TCM and WM collaboration.

2 Literature Review

2.1 Duality and Complementarity of TCM and WM

Through several decades of exploration and negotiation, TCM and WM practitioners in Chinese hospitals have gradually accumulated and formed complementary relationships.

In order to thoroughly explain the complementary relationships, it is necessary to understand the basic beliefs, base philosophies, and diagnosis and treatment methods of the two types of medicine.

TCM emphasizes on the integrity of the human body as whole and its close relationship with the environment [2]. According to the study of Ma [14], traditional Chinese healing practice is intended to enhance the immune system of human body, antiviral effects, anti-inflammation, balance of mind and body, aches and pain relief, and cholesterol reduction. There are four main categories of Chinese medicine treatments, namely herbal medicine (oral intake and external use), heat therapy (moxibustion and cupping), massage (oriental massage, Gua Sha and magnets) and acupuncture [15].

Conversely, WM employs a scientific attitude in treating patient problems [16]. It is believed that achievements from intensive and evidence-based fundamental scientific research have brought WM to an unchallengeable dominant position in world health care as well as in China [17]. In fact, and despite the plurality, in the Chinese healthcare system WM takes the primary position being complemented by TCM as an alternative healthcare therapy. It is widely accepted in China that WM is more effective in the acute stage of many diseases and works much faster than TCM in treating these acute diseases [14]. However, it is also acknowledged that WM creates more adverse side effects [18]. Nevertheless, healing herbs, acupuncture, massage and other health methods from TCM may be more appropriate in health promotion, prevention, treatment, and rehabilitation. Moreover, TCM may be used as a last resort, when Western medicine is either too toxic or unable to provide any further expected benefit [19].

The main difference of Chinese traditional medicine in relation to its Western counterpart is its adoption of a holistic concept of healing, which emphasizes the integrity of the human body as a whole and its close relationship with the environment [2]. In contrast, WM doctors are more interested in localized diseases or illnesses and the corresponding part of the human body. WM practitioners aim at healing that specific part of the human body rather the more general problems of the patients [16].

Moreover, TCM and WM have entirely different conceptual systems. For TCM doctors, the Yin-Yang theory is an ancient Chinese belief and way of understanding the universe and is the most essential theoretical foundation to the practice of TCM [2]. In contrast to TCM, which is based on Chinese ancient beliefs, WM is based on scientific paradigms and evidence-based research [12] and is a combination of modern science and the art of healing [20].

Furthermore, the two types of healthcare methodology have completely different diagnosis methods. TCM doctors follow the ancient theory of Bian-zheng (distinguishing patterns) [2], which can be generally defined as "the process of identifying the basic disharmony that underlies all clinical manifestation" [21]. To support the processes of Bian-zheng, TCM doctors apply four diagnosis methods to patients, namely "inspection", "listening and smelling", "inquiry" and "palpation" [22]. In fact, the TCM diagnosis mainly relies on the doctors' professional experiences and personal understandings of Bian-zheng [7]. In this case, it is very common for different TCM doctors to produce totally different diagnoses of the same patient [7]. In contrast, WM professionals investigate the problems of patients and make decisions based on the identification of accurate medical evidence and the employment of modern diagnostic technologies, such as x-rays, laboratory tests, and computed tomography (CT) [23].

Finally, TCM and WM professionals have very different treatment approaches to dealing with patient problems. In the TCM methodology, there are four main categories of treatments: herbal medicine (oral intake and external use); heat therapy (moxibustion and cupping); massage (oriental massage, Gua Sha and magnets); and acupuncture [15]. These methods used by TCM doctors are often considered as too unusual by those WM healthcare professionals who are following the doctrine of modern medical science. To them, patient treatments can be simply divided into two categories, namely: medication and surgery [24].

WM is a hard science, whereas TCM is an empirical [soft] science [7]. Even though the two approaches are entirely different, the integration of the two healing beliefs into the Chinese healthcare system constitutes a unique therapeutic plurality, which is believed to be beneficial to patients, and which is only presented in the structure of the Chinese healthcare system.

The advantages and benefits of integrating TCM and WM services into a single healthcare system, as well as the implementation of complementary treatment have become evident. In any case, the complementarity and collaboration of the two types of healthcare professionals should be based on the communication and sharing of technical and patient knowledge with each other.

2.2   Patient-Centered Knowledge Sharing

KS can be simply understood as the behavior of making knowledge available to others [25]. In the healthcare environment, KS is defined as follows:

> "Healthcare knowledge sharing can be characterized as the explication and dissemination of context-sensitive healthcare knowledge by and for healthcare stakeholders through a collaborative communication medium in order to advance the knowledge quotient of the participating healthcare stakeholders." [26]

According to this definition, and considering the patient-centered TCM and WM collaboration, healthcare professionals need to share the following three types of patient knowledge:

- Technical Knowledge includes identification of patient conditions and problems, reasons and objectives of patient care, patient background, agreement to treatment strategy, and explicit patient requirements and needs [27].
- Ethical and Emotional Knowledge is about ethically dealing with patient feelings, emotions, and psychological status; approaches to communicating with, persuading and managing individual patients; and maintaining trusting and collaborative professional-patient relationships [28].
- Social and Behavioral Knowledge is concerned with anticipating how others will behave, perception of patients' implicit requirements, behaviors and reactions, and expectations [28].

Among the three types of patient knowledge, the sharing of technical knowledge is the least problematic, since technical knowledge is easier to share and is usually recorded explicitly in the patient records. Moreover, the two types of healthcare professionals have adopted two entirely different therapeutic systems and each other's philosophical beliefs and technical insights do not seem to matter in the complementary provision of medical service [29, 30]. On the other hand, the ethical and emotional knowledge and the social and behavioral knowledge consist of experiences and perceptions of individual professionals, which are accumulated through processes of dealing and interacting with individual patients. Therefore, when compared with the technical knowledge, these two types of tacit patient knowledge are more difficult and more important to share among healthcare professionals. Thus, this study focuses on these two types of tacit patient knowledge.

3   Research Methodology and Design

3.1   Research Questions

According to the main aim of this study, which is to identify barriers to sharing patient knowledge in TCM and WM collaboration, the following research question was formulated:

> What are the barriers to sharing patient knowledge between healthcare professionals from Traditional and Western medicine in their patient-centered interprofessional collaborations?

In the light of the main research question, three specific research questions were established:

- What are the barriers that hinder the sharing of patient knowledge between TCM and WM healthcare professionals?
- What are the relationships between these barriers?
- What practical strategies can be formulated in order to improve KS?

The research questions were adopted to point a direction to the selection of research methodology, the research design as well as the collection and analysis of data.

3.2   Research Approach and Design

Since there are virtually no empirical studies that have been performed on the communication problems between TCM and WM professionals in Chinese hospitals, this study adopted an inductive approach and aimed at developing a new and contextualized theory. Therefore, a Straussian Grounded Theory (GT) was selected as the main research methodology, since GT is widely recognized as particularly useful for theory generation and development [31]. In addition, in order to allow a theory to emerge from a suitable research context, GT was applied in a social context provided by case-study.

Moreover, considering China is one of the largest countries in the world, with a population exceeding 1.3 billion and with 56 ethnic groups and 34 provinces, it would be virtually impossible to generate a theory that would encompass the whole nation. Consequently, and since this project aimed at generating a first set of insights into this problem, a single case-study design was adopted. A public hospital in the city now city of Xiangyang (Xiangfan at the time of data collection), province of Hubei, was selected for the case-study. This hospital was chosen for two main reasons. Firstly, it provides both WM and TCM services to patients and has done so for several decades. Secondly, the researcher obtained guaranteed and management supported access to the informants and the project.

Furthermore, during the processes of data collection and analysis, it was observed that different departments in the hospital exhibited very different levels of integration of complementary treatments. This study therefore focuses on one specific department, namely the Department of Neurosurgery. This department has a proven history of using WM and TCM compound treatments for rehabilitating patients after craniotomies.

Semi-structured interviews were adopted as the data collection tool. Moreover, as required by the GT theoretical sampling strategy, interview participants were sampled by the emerging theory and interviewed using evolving interview question scripts. Overall, 46 informants were interviewed in a total number of 49 interviews. These informants were 27 healthcare professionals, 7 TCM professionals, 1 chief hospital manager, 1 hospital ICT manager, 1 TCM professor at local university, 1 healthcare politician in local government, and 8 patient relatives and carers.

As required by GT, the processes of data collection and analysis were operationalized interactively. That is, immediately after each individual interview, the collected data were transcribed and analyzed. The analysis of data collected adopted two essential GT analytical tools, namely, coding (open, axial and selective) and constant comparative analysis. Consequently, data collection and analysis coexisted until the theoretical saturation was achieved, that is, until no new open codes emerged from the data analysis. The final theory saturated with 11 KS barriers.

4 Research Findings

4.1 Process of Sharing Patient Knowledge

Through the processes of data analysis, it became clear that the interprofessional collaboration of TCM and WM professionals is considered as fundamental to the treatment of neurosurgical patients, since "more than half of our [neurosurgical] patients are using TCM treatments" (Interview WMD 2.72). As described by the interview informants, when dealing with patient problems, WM is employed as the primary methodology and was always used in the first instance. TCM methods are implemented as a complementary approach and are usually considered as more effective at the post-craniotomy and rehabilitation stages.

"Patient usually has some problems after the brain surgeries. These problems may lead to some serious sequelae. For these problems, patients can use TCM herbal medicine and acupuncture to assist rehabilitation after surgeries. TCM is not usually used before surgeries." Interview WMD 20.15

Interprofessional collaboration and KS usually occur in consultation sessions, which are usually requested by a neurosurgeon, when a patient condition is perceived to be better treated by TCM doctors. The nurse in charge usually initiates the process at the request of the neurosurgeon and contacts the TCM doctors directly to make an informal enquiry. If the TCM doctor agrees his/her commitment, the neurosurgeon initiates a consultation note as a formal invitation for collaboration. The consultation note records a very brief description of all procedures and medical decisions that are made during the consultation session. After this consultation session, WM and TCM professionals never meet again to discuss that particular patient, unless in the case of emergencies. The consultation note must be signed by doctors from both sides and documented in the patient records.

As perceived, these consultation sessions could be a relatively good communication channel for KS, since they require the presence of professionals from both teams and to work collaboratively and interactively on specific health problems of a specific patient. However, the data collected reflect that these meetings in reality cannot be considered as a good communication channel and is fraught with barriers that hinder interprofessional communication and the sharing of patient knowledge.

4.2   KS Barriers

Philosophical Barriers. The data collected show that WM and TCM have completely different conceptual, philosophical and methodological systems. These fundamental differences in the philosophical roots of the two types of medicines have resulted in significant barriers to the sharing of patient knowledge. Specifically, five barriers emerged and were identified in the data analysis as follows:

Different Conceptual Systems. The KS problems between TCM and WM professionals are rooted in the basic concepts and beliefs of the two types of medicines. The data analysis revealed that, apart from a unified purpose to resolve patients' problem, the provision of TCM and WM services are based on two entirely divergent systems, including differences in philosophical views, theoretical foundations, treatment and diagnostic approaches. This finding confirms that findings of the literature review.

> "They [TCM doctors] have a totally different theoretical system, which we [WM professionals] do not understand. [...] Undeniably, there are a number of conflicts between the two theoretical systems, but their [TCM] methods are effective. Nevertheless, WM is probably more effective and as a WM doctor, I believe in our system. They believe in theirs. There are clear conflicts." Interview WMD 9.25

These conceptual differences could cause conflicts of understandings of patient problems and requirements, and result in conflicts in actions aimed at solving patient problems and achieving patient requirements. These conflicts could hinder processes of interprofessional communication and prevent activities of sharing patient knowledge.

Different Terminology Systems. Upon the completely divergent conceptual and methodological systems, TCM and WM healthcare professionals have entirely different systems of terminology and use very different professional terms and jargon to describe and explain patient problems and requirements.

> "[WM and TCM] have two terminological systems. Maybe both of them have an identical purpose, but how they express the purpose is entirely different." TCM 15.35

Differences in terminology make KS particularly difficult, since TCM and WM professionals cannot understand each other's language. Patient knowledge shared by one side probably cannot be correctly received and fully comprehended by the other side. Therefore, the terminology difference is a KS barrier.

Conflicts of Philosophical Beliefs. During the interviews, TCM and WM healthcare professionals showed a consistent lack of belief in each other's practices. Many interviewed WM professionals not only expressed that WM is "purely scientific and superior to TCM" (Interview WMN 14.15), but also showed strong disbelief, distrust, disagreement and even discrimination against TCM. In fact, TCM philosophy and methodology was often harshly criticized as "unscientific" (Interview WMD 1.64) and useless "superstition" (Interview WMN 14.17). On the other side, TCM doctors strongly disagree that TCM is considered as inferior to WM. Many TCM doctors defended their methodology is a "solid medical methodology" (Interview TCM 4.9), which consists of a systematic and consistent set of diagnostic and treatment methods and which has been accumulating and revising through an evolution of thousands of years. Moreover, many TCM interviewees disagree with some of the WM beliefs and methods, which they asserted are not always appropriate and which sometimes have adverse effects on patients' conditions.

Evidently, the philosophical conflicts augments conflicts of opinions and perspectives of the two types of professionals, and have created a climate of distrust, disregard, and unwillingness to communicate in the two communities.

Inadequate Interprofessional Common Ground. The data analysis identified a lack of interprofessional common ground, which can be theorized as a knowledge base of overlapping interests and shared conceptual understandings. The research findings show that the lack of interprofessional common ground could result in philosophical conflicts and disagreements with each other's views and opinions, enhance untrusting relationships between the two medical communities, and thus is identified as a KS barrier.

"Communication, if without a knowledge basis, is impossible. For me, I can easily communicate with WM doctors, because I nearly learnt all WM knowledge. But if WM doctors do not learn TCM, they will never accept our philosophy." Interview TCM 6.72

Insufficient interprofessional education and training: The inadequacy in interprofessional common ground, as indicated in the research findings, is probably resulted by a lack of interprofessional education in Chinese healthcare HE. Specifically, the healthcare HE structure in China consists of TCM education and WM education, as two almost entirely isolated systems with very limited programs, courses and modules designed and included focusing on the convergent areas of TCM and WM.

"We [WM professionals] only have a very basic understanding about TCM, actually very superficial. We only learnt something like the palpation, nothing else. Almost nothing learnt in medical school." Interview WMN 14.29

Moreover, the data analysis identified an absence of focus and exercises on hospital interprofessional training on the areas of convergence aiming at establishing mutual understandings between the two professional communities. Consequently, it is evident that, due to the insufficient interprofessional education and training, TCM and WM practitioners do not have a sufficient common ground to facilitate necessary interprofessional communication and KS.

Professional Barriers. Apart from the KS barriers emerged from the substantial divergences of TCM and WM philosophy, some professional issues were emerged as barriers to interprofessional KS. Specifically, the data analysis identified six professional barriers.

Asymmetrical Decisional Power. The data collected exhibit evidences of substantial asymmetries of positional power and professional standing of the two medical communities.

"If neurosurgical patients need acupuncture treatments, neurosurgeons would initiate a consultation note and telephone us. Then we go to treat patient with acupuncture. [...] In this process, we do not have decision power. For example, this patient clearly needs TCM treatment, but we cannot do anything about it, because neurosurgeons need to make this decision, not us." Interview TCM 16.17

As shown in data, for instance the quotation above, when collaborating with TCM doctors, WM practitioners have a relatively higher professional standing and almost complete control over patients. Comparably, TCM doctors possess a relatively lower professional standing and hold less power. Therefore, TCM doctors are most likely to maintain a passive position, avoid any confrontations and to follow instructions, instead of actively and voluntarily proposing their ideas, opinions and suggestions. For them, even if they intend to share knowledge, they have very little power or influence to have their views recognized. Therefore, it is a significant KS barrier need to be carefully resolved.

Overwhelmingly High Workload. As both witnessed in the field and reflected in data, both types of practitioners were extremely busy and had very high workloads. A number of interviewees, therefore, informed that they are more concerned with "take care of patient [solving patient's immediate problems]" (Interview TCM 15.45), rather than contributing time in interprofessional communication and KS. This also emerged as a KS barrier, since processes of sharing patient knowledge could be largely neglected.

"[In the consultation] usually they do not ask many questions, and we do not talk that much. We all are very busy. As long as we can treat the patient, that is all right. We all are too busy to actually sit down and to have a deep conversation." Interview TCM 37.63

Rigid Problem-oriented Collaboration Approach. As identified in the data analysis, the sharing of patient knowledge is constrained and hindered by the adoption of an overly rigid problem-oriented approach to collaboration. In this approach, as long as those patient problems can be resolved, interprofessional communication and KS would be considered as not really important and as something that can probably be ignored, for instance a TCM and a WM informant stated that:

"(In WM and TCM collaboration) we do not need to know TCM theory and method. We just want them (TCM doctors) to help us to solve patients' problems." Interview WMD 48.12

"The reason why neurosurgeons invite us to join a consultation is that they want us to solve their problems. I don't think they are trying to understand TCM or how we think of the patient." Interview TCM 4.81

Evidently, this approach to collaboration is not an encouraging mechanism for sharing any form of knowledge.

Inefficient Communication Channels. As discussed in Sect. 4.1, KS occur in consultation sessions, which could be perceived as useful vehicle for exchanging patient knowledge. However, as a communication channel, these meetings can only play a very limited role in real KS between the two professional groups. In reality, as expressed by a number of informants, the meetings last usually "no more than 10 or 20 min" (Interview WMN 7.119), in which "the diagnosis of the patient is presented by a WM doctor and then usually we [the visiting TCM doctor and the neurosurgeon in charge] need to have a brief discussion" (Interview TCM 4.92). This is of course not conducive to in-depth interprofessional discussions. Thus, the consultation meeting in fact becomes a formal handover of patients and not a vehicle for the exchange of patient knowledge.

Absence of Explicit KS Requirements from Hospital Management. As shown in the data collected, even though the hospital management has been repetitively emphasized on integrating KS concepts and practices into the provision of healthcare services, "no specific requirements or guidelines have been formulated which explicitly demand interprofessional communication and KS" (Interview WMD 20.13). Therefore, professionals from both medical teams probably perceive that communication and KS are optional, not compulsory, and not important.

> "If there are have some kind of regulations that WM and TCM teams need to adequately communicate and KS, practitioners are forced to do this. But, we do not have these requirement. It is like if you [a WM practitioner] do not talk with TCM doctors for ten years, no one would care about that and no one would criticise you. There is no supervision." Interview WMD 1.83

Imbalanced Management Support. As reflected in the data, the hospital management provides more attention and support to WM departments, whereas the TCM department is not only less supported, but also could be discriminated by the hospital management and viewed as "secondary in the hospital" (Interview WMD 23.17). There are two reasons as point out in the data analysis: firstly, nearly all power figures in the hospital management team have WM backgrounds and hence would attach more attention and support on WM departments; secondly, and more critically, WM departments are much more financially profitable when compared with the TCM department. It is important to note that financial profitability became particularly important to the survival of hospitals in China after the implementation of Market Economy Policy, which determined that all hospitals are themselves responsible for all hospital operation expenses.

The imbalanced hospital support has exacerbated the already existing philosophical conflicts, encouraged interprofessional competition, augmented imbalances of power and distanced professional standings. Thus, professionals from both communities are not motivated and even unwilling to communicate and share knowledge with each other.

5 Discussion and Conclusion

With further distillation and conceptualization of the findings, it became clear that the identified KS barriers have resulted in two types of interprofessional tensions, namely philosophical tensions, and professional tensions, which then emerged as the centers of the KS problems between TCM and WM practitioners.

- Philosophical tensions are caused by the substantial divergence in philosophies, theoretical grounds and conceptual systems of TCM and WM. These tensions have resulted in conflicts of opinions and perspectives, which in turn have created a climate of distrust, disregard, and unwillingness to communicate in the two communities. Additionally, the philosophical tensions are resulted by a lack of interprofessional common ground to facilitate communication and KS. The lack of interprofessional common ground is caused by lacking of interprofessional education in the Chinese healthcare education and by lacking of interprofessional training in the hospital environment.
- Professional tensions result from the substantial asymmetries of power and professional standings between the two medical communities. The data analysis clearly revealed that neurosurgeons have relatively higher professional standings and have almost dominant power over patients. Therefore, they often explicitly instruct and regulate TCM doctors on what to do with the patient. Comparably, TCM doctors have lower professional standings and hold relatively less power. Therefore, TCM doctors are most likely to maintain a passive position when collaborating with neurosurgical practitioners, avoid any confrontations and to follow instructions, instead of actively and voluntarily proposing their ideas, understandings and suggestions.

Moreover, the conceptualization of the research findings included an analysis of the cause-consequence relationships between individual barriers. The result of the analysis can be illustrated in a concept map as shown in Fig. 1.

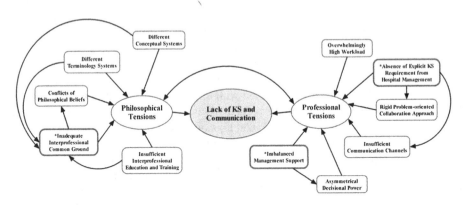

Fig. 1. Philosophical and Professional Tensions as Main Barriers to KS between TCM and WM practitioners.

As shown in Fig. 1, KS barriers are causes to philosophical tensions and professional tensions as two conceptual centers of the emergent theory. Furthermore, both types of tensions are interrelated and reinforce each other. Hence, to improve KS and communication between TCM and WM practitioners, efforts need to be put on mitigating and resolving the two types of interprofessional tensions. More specifically, to effectively resolve the two tension, it is necessary to examine individual KS barriers and establishing actionable strategies to mitigate the effect of each barrier.

In addition, as shown in Fig. 1, three KS barriers are marked with "*", namely, inadequate interprofessional common ground, imbalanced management support, and absence of explicit requirement from hospital management. These barriers are inter-linked with others as either causes, or consequences. In this case, strategies should be developed targeting at these barriers. As reflected from the research findings, the following three strategies should be adopted and implemented by the hospital management:

1. To develop and reinforce the interprofessional common ground, the hospital management should establish very specific interprofessional training schemes and programs. For both types of healthcare professionals, these programs and sessions could increase mutual understanding, acceptance of each other's philosophy and beliefs, could enhance a better understanding of each other's professional terminology and, more importantly, effectively put in place a common ground to enable, facilitate and motivate interprofessional communication and KS.
2. In order to relieve the professional tensions, explicit management strategies should be formulated and implemented aiming at equally supporting TCM and WM communities, eliminate imbalances of power and professional standings and foster a harmonious hospital environment, which could be more conducive for interprofessional collaboration and communication.
3. It is necessary to formalize the process of interprofessional collaboration and formally define activities and processes of sharing patient knowledge. Moreover, there is a need to explore new communication channels and tools to facilitate the process of sharing patient knowledge, for instance and as reflected in data, patient records and consultation notes could be much better used and explored. Finally, as also identified during the process of data collection in the field, the hospital was under the processes of designing and implementing a new Information System. Therefore, there is the opportunity to create new communication platforms that can be developed within the hospital intranet and support better communication and KS.

Finally, it needs to be highlighted that these strategies must be fully supported by hospital managers and leaders in both medical communities, who should realize that the collaboration of TCM and WM is not just a political imperative, but may bring tangible benefits to patient welfare, through mutual trust between these complimentary medical communities.

Acknowledgements. This paper is supported by the National Natural Science Foundation of China (Project No. 71203165) and the Wuhan University Research Grant (Project No. 2012GSP076).

References

1. Wong, T., Wong, S., Donna, S.: Traditional Chinese medicine and western medicine in Hong Kong: a comparison of the consultation processes and side effects. Hong Kong Med. J. **45**, 278–284 (1993)
2. Cheng, J.: Review: drug therapy in chinese traditional medicine. J. Clin. Pharmacol. **40**, 445–450 (2000)

3. Hyatt, R.: Chinese Herbal Medicine: Ancient Art and Modern Science. Wildwood House Limited, London (1978)
4. Chi, C.: Integrating traditional medicine into modern health care systems: examing the role of Chinese medicine in Taiwan. Soc. Sci. Med. **39**, 307–321 (1994)
5. Fruehauf, H.: Chinese medicine in crisis. J. Chin. Med. **61**, 1–9 (1999)
6. Taylor, K.: Divergent interests and cultivated misunderstanding: the influence of the west on modern Chinese medicine. Soc. Hist. Med. **17**, 93–111 (2004)
7. Liu, L.: Considering Traditional Chinese Medicine (in Chinese). Guangxi Normal University, Guilin (2003)
8. Zhong, H.: The patient-centered care and hospital marketing strategies. Manage. Obs. **10**, 234–235 (2009). (in Chinese)
9. Hu, G.: Take patient as the centre and provide premium services. Jiangsu Healthc. Ind. Manage. **20**, 49–50 (2009). (in Chinese)
10. Steward, M.: Towards a global definition of patient centered care: the patient should be the judge of patient centered care. Br. Med. J. **322**, 444–445 (2001)
11. Maizes, V., Rakel, D., Niemiec, C.: Integrative medicine and patient-centered care. Explore **5**, 277–289 (2009)
12. Zhou, L., Nunes, M.: Identifying knowledge sharing barriers in the collaboration of traditional and western medicine professionals in Chinese hospitals: a case study. J. Libr. Inf. Sci. **44**, 238–248 (2012)
13. Sun, C.: Research in communication of traditional Chinese medicine and western medicine doctors. J. Nanjing Univ. Tradit. Chin. Med. **4**, 6–9 (2003). (in Chinese)
14. Ma, X.: Between two worlds: the use of traditional and western health service by Chinese immigrants. J. Community Health **24**, 421–437 (1999)
15. Sherman, K., Cherkin, D., Eisenberg, D., Erro, J., Hrbek, A., Deyo, R.: The practice of acupuncture: who are the providers and what do they do? Ann. Fam. Med. **3**, 151–158 (2005)
16. Dally, A.: The Trouble with Doctors: Fashions, Motives and Mistakes. Robson Books, London (2003)
17. Unschuld, P.: Medicine in China: a History of Ideas. The University of California Press, Los Angeles (1985)
18. Kaptchuk, T.: The Web that Has No Weaver: Understanding Chinese Medicine. McGraw Hill, New York (2000)
19. Chen, C.: Medicine in Rural China. University of California Press, Los Angeles (1989)
20. Warrell, D., Cox, T., Firth, J.: Oxford Textbook of Medicine. Oxford University Press, New York (2005)
21. Maciocia, G.: The Foundations of Chinese Medicine: A Comprehensive Text for Acupuncturists and Herbalists. Churchill Livingstone, London (1989)
22. Wang, X., Qiu, H., Liu, P., Cheng, Y.: A self-learning expert system for diagnosis in traditional Chinese medicine. Expert Syst. Appl. **26**, 557–566 (2004)
23. Fitzgerald, F.: Physical diagnosis versus modern technology - a review. West J. Med. **152**, 377–382 (1990)
24. Goldman, L., Ausiello, D.: Cecil Medicine: An Expert Consult Title Online + Print. Saunders Elsevier, Philadelphia (2008)
25. Ipe, M.: Knowledge sharing in organizations: a conceptual framework. Hum. Resour. Dev. Rev. **2**, 337–359 (2003)
26. Abidi, S.: Healthcare knowledge sharing: purpose, practices, and prospects. In: Bali, R., Dwivedi, A. (eds.) Healthcare Knowledge Management: Issues, Advances, and Successes. Springer, New York (2007)
27. Smith, R.: Information in practice. Br Med J **313**, 1062–1068 (1996)

28. Fennessy, G., Burstein, F.: Role of information professionals as intermediaries for knowledge management in evidence-based healthcare. In: Bali, R., Dwivedi, A. (eds.) Healthcare Knowledge Management: Issues, Advances, and Successes. Springer, New York (2007)

29. Guo, F.: Discussion of the theories of the integration of traditional Chinese medicine and western medicine. J. Exp. Clin. Med. **5**, 408–409 (2006). (in Chinese)

30. Yang, Y.: Discussion of the integration of traditional chinese medicine and western medicine. China Foreign Med. J. **3**, 84–86 (2005). (in Chinese)

31. Strauss, A., Corbin, J.: Basics of Qualitative Research: Techniques and Procedures for Developing Grounded Theory. Sage Publications, London (1998)

A Risk Diagnosing Methodology Web-Based Platform for Micro, Small and Medium Businesses: Remarks and Enhancements

Luís Pereira[1(✉)], Alexandra Tenera[1,2], João Bispo[1],
and João Wemans[3]

[1] Faculdade de Ciências e Tecnologia (FCT),
Universidade NOVA de Lisboa (UNL), Lisbon, Portugal
pl10336@campus.fct.unl.pt, abt@fct.unl.pt,
jbixpo@gmail.com
[2] Department of Mechanical and Industrial Engineering,
UNIDEMI, Lisbon, Portugal
[3] WS Energia, Lisbon, Portugal
wemans@ws-energia.com

Abstract. The indisputable necessity to innovate brings to companies vital responsibilities such as the inevitable errand of having consistent innovation and risk management procedures along their projects. This work presents a risk diagnosing methodology (RDM) web-platform, that can provide to Small and Medium Enterprises (SMEs) the ability to identify and self-assess risks associated with their innovative project's portfolio, from the project's idealization to its commercialization. The creation of this tool attempts to encourage SMEs on the systematic use of risk management approaches, in order to increase project's successful rates. As groundwork, this paper also includes a bibliographical review and comparative analysis of other existing risk management tools and models available to SMEs. The present article also uncovers first empirical results available from exploratory research done and reveals ongoing research efforts regarding risk perception and potential bias correction.

Keywords: Risk assessment · Innovation management · SMEs · Decision tools

1 Introduction

During the last decades the world's economy has undergone a process of deep re-establishment [28], moving into a fast-changing, knowledge-based economy in a global scale [29] which dragged businesses into a daily struggle to survive/prevail in a new difficult and challenging economic environment [6]. To prevail in this global competition and overcome the rapid technology changes and product variety expansion, the development of an integrated capability to innovate is becoming a predominant strategy for SMEs [5]. Innovation is an inexhaustible motive force for socio-economic development, which makes it a key factor to measure national competitiveness [4]. Innovation can be seen as the action or process of creating a new method or idea [2] which also includes its exploration and commercialization [17]. The use of a formal and

© Springer-Verlag Berlin Heidelberg 2015
A. Fred et al. (Eds.): IC3K 2013, CCIS 454, pp. 340–356, 2015.
DOI: 10.1007/978-3-662-46549-3_22

systematic process in the development of new products/services, has been considered a decision factor of the project's success or failure [9], in which the new product development (NPD) innovative approach is one of the most wide known and used to formally support the innovation project processes among SMEs.

Since the creation of something new is the essence of an innovation, this process necessarily involves risk, and consequently early risk identification and management is specially vital and required in innovative SMEs [24]. Nonetheless, efforts to develop empirical models, metrics and tools to accurately assist SMEs in the risk management of innovative projects still need development [1]. Bibliographical review also shows that SMEs operate in the same environment as their larger counterparts, but their attitude towards risk grandly differs, being that SMEs chief executives don't always recognize the need to escalate the importance of risk identification and its minimization [23] SMEs also have inevitable limitations regarding internal availability of resources, restraining the company's ability to engage in innovative activities [7], because they easily became fully occupied solving short-term operational problems, which leads to a lack of attention to their long-term strategy and to disregard risk management importance in their initiatives, remaining stuck in a permanent operational problem solving [26], Knowing that the percentage of existing SMEs around the world stands over 97 % [3], developing studies regarding the risk assessment and management in innovative projects for SMEs is then critical.

Therefore, this paper seeks to present a solution to systematically support management risk practices in SMEs, in order to satisfy their need to attain useful and pragmatic approaches to manage risks in their portfolio of ideas development and innovative projects. We will then begin this article by focusing on the presentation and discussion of some of the most relevant risk tools and models currently available to SMEs. Secondly, we will propose a web-based integrated risk perception and response tool, designed to SMEs and start-up enterprises, which will provide an early stage risk assessment throughout a web-based platform – the Spotrisk®, along with further validation tests and available results.

2   Assessing Risks in Innovative SMEs

2.1   Current SMEs Risk Support Tools and Models

SMEs are usually characterized by the central role of their owners, high multiplicity of one's duties and close employee identification [19]. Often, managing directors engage in the overall risk assessment without sharing and discussing it with team members [10]. Consequently, enterprises in their starting phase often underestimate risks, ignoring them or just having only one risk strategy for bearing the risks [10].

Furthermore, SMEs typically do not have the resources to acquire specialists for each enterprise's position nor in administration functions such as risk management [18]. Also, SMEs usually do not tend to use specific techniques to identify or manage risks and as its related bibliography is limited and still in an early stage of development [13].

Moreover, due to limitations regarding infrastructure, management, technical expertise, intellectual and financial resources, SMEs are far from adopting a proactive approach towards risk [12], despite its critical importance regarding their sustainability and results.

As groundwork to the present research and for the aftermost risk tool comparative analysis, several existing risk models and tools were previously analyzed. A list of the most relevant cases found is presented on Table 1, which includes a brief description along with the identified advantages and disadvantages of each one of them.

All the cited tools use a subjective risk approach based on uncertainties and nearly all identifies, prioritizes and address risks. However none of the tools is suitable per se to SMEs, because they are either too expensive or too complex, compelling the company to take up extensive efforts and time that most SMEs' managers consider not to be a binding activity.

From Table 1 analysis, Risk Diagnosing Methodology (RDM) emerges as an important model that, if adapted to a platform that reduces the existing complexity, time expenditure and facilitator's function as originally required, could dwell as a strong risk supporting solution for SMEs and start-up enterprises.

3   The Spotrisk® Tool

The development of the Spotrisk® tool was made possible with the contribution of a group of researchers that included Jimme Keizer, one of RDM's creators. Spotrisk® has a RDM framework basis that provides SMEs with an adapted system of risk assessment and response. The tool seeks to allow enterprises to diagnose thoroughly and methodically both internal and external risks that an innovative project generally faces, formulating the type of risk management strategy to be established.

The tool was developed on a web-based platform, therefore universally accessible, that automates the RDM's "Risk identification" and "Risk Response development and control" phases through a web integrated system, as summarized in Fig. 1.

The interaction with the organization is based on the RDM's "Risk assessment" phase, where the manager and each project team member need to answer to a standard risk questionnaire, in order to put forward the project's risk profile analysis and check its progress in time.

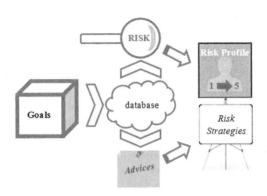

Fig. 1. Spotrisk® global structure.

Table 1. Main existing risk management support tools and models.

Name/Year	Functionalities	Advantages	Disadvantages
Iris Intelligence 2005	Organizational integrated risk management within wide business practices.	Integration with Microsoft Office; User-Friendly; Cloud based possibility; Complete.	Very Expensive; Designed to manage only organizational risks.
RiskCloud 2003	Risk identification and assessment with personal support.	Cloud based (no installation required); Visual & User-friendly; AS/NZS/ISO31000:2009; Complete.	Expensive; Difficult access; Organization analysis and not specific project analysis.
ProjectFuture 2003	Project's quantitative and qualitative risk calculation and identification software.	List of possible risks, effects, causes and responses; Risks associated with tasks; Possibility to evaluate severity of risks associated with different dates.	Limited number of Projects; Expensive; Software installation required.
RiskyProject 2002	Project planning, scheduling, quantitative risk analysis, and performance measurement.	Add-In association with Microsoft Project; Possibility to regulate the risk tolerance.	Complex; No risk identification; Software installation required; Dilatory and slow processes.
SME-at-Risk 2002	Service that aims to provide a comprehensive understanding of the risk management basics.	Provides a vast know-how basis; Shortens the access to existing articles; Shares information.	Lack of management; No functionalities in terms of tool; Only provides literature review.
RDM 2002	Methodology that through a series of interviews and a checklist questionnaire collects results, allowing a company to diagnose and manage project's risks.	Diagnoses thoroughly and systematically the project's risks; Develops technological, organizational and business approaches; Formulates suitable risk output strategies.	Needs a risk facilitator; Time expenditure; Complex; Difficult access to the methodology.

3.1 Goal Oriented Questionnaire

The Spotrisk® web-based platform is centered on a goal-oriented questionnaire in order to help identify common potential risks of product innovation projects in the main domains proposed on the RDM's approach: technology, market, finance and operations [14].

The questionnaire brings forward a selection from RDM's reference list with potential risk issues in the innovation process. A deep analysis of the whole reference list was previously designed for that and the selected issues were introduced and standardized in the Spotrisk's questionnaire (see Appendix the selected risk issues considered). After some debate between researcher and development team some issues were added, such as question number 7 from Idea Stage or question number 3 from Feasibility Stage, setting up a total of 35 critical issues.

In order to manage the project portfolio better, the RDM uses the approach, known as the "Innovation Funnel" created in the early nineties [8]. This approach, based on the conceptual model of Wheelwright and Clark [27] consisted originally of six stages in which projects are defined. However, in the Spotrisk® risk approach, only the four stages shown in Fig. 2 were considered, while the remaining original stages "Post Launch Evaluation" and "Rollout Contender" were taken out, being the most important aspects of these phases incorporated in the issues of the Launch Stage.

As it can be seen in the Appendix, the selected critical issues resorted in the goal oriented questionnaire were distributed through this four key staged process, from the initial conception of the idea, to the launching of the new product/service, elapsing through feasibility and capability phases to safeguard the potential and readiness of the project.

The critical questions were rendered into positive statements of goal objectives, meaning that each objective, if realized within a project, will translate it as a safe project. Each goal/objective on the questionnaire needs to be responded individually standing on three different parameters assessment:

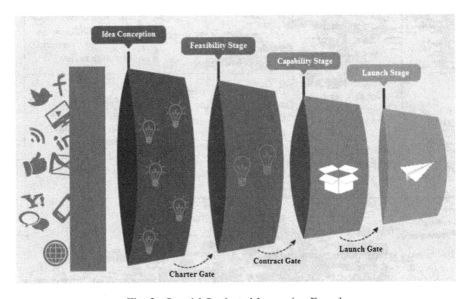

Fig. 2. Spotrisk® adapted Innovation Funnel.

- *Level of Implementation* – represents how much of the specific goal the project already has or the level of certainty that it will be realized; i.e. the strength of the statement's truth, within the project's reality.
- *Capacity to Influence* – represents the ability of the project team to guarantee the accomplishment of the project's goal, within the time and resource limits.
- *Severity of the Consequences* – represents the potential level of negative impact on the project's performance by not attending to the specified goal.

For each parameter considered, an answer is required on a Likert five-point-scale, as for "Very low" representing the lowest reflection of the analysis and "Very high" being the highest. Each response represents a numeric quantity to be used in the risk profile calculation (#), being that the first two variables behave according to a "the higher, the better" logic, unlike the third variable which behaves in the opposite sense (see Table 2).

The questionnaire results are then conducted into a database, where each goal is categorized into a risk class, returning from the data base the respective categorization: "Safety", "Low Risk", "Medium Risk", "High Risk" or "Failure", as exemplified in Fig. 3.

Table 2. Qualitative risk metrics.

Level of implementation		Capacity to influence		Severity of the consequences	
Answer	#	Answer	#	Answer	#
V.Low	1	V.Low	1	V.Low	5
Low	2	Low	2	Low	4
Medium	3	Medium	3	Medium	3
High	4	High	4	High	2
V.High	5	V.High	5	V.High	1

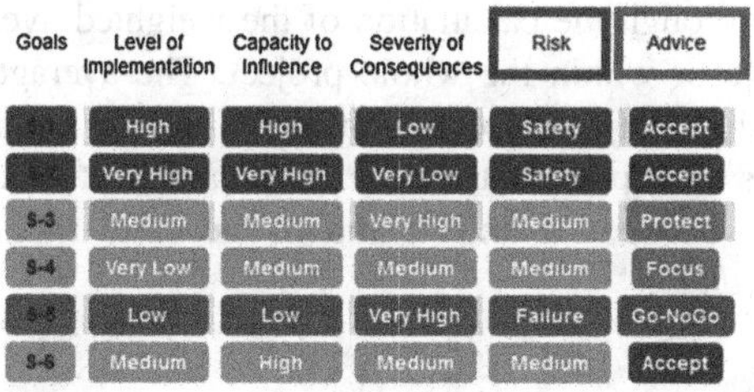

Fig. 3. Print screen of the capability stage assessment on a certain project.

3.2 Results Module

The project risk profile calculation is inspired once more by the work developed by Jimme Keizer and his team [14]. Likewise, every risk is classified (according to the three parameters) into three different groups by the following decision rules:

- ("*"): Scores are 1 or 2, with 1 being "very risky.
- ("0"): Scores are 4 or 5.
- ("m"): Scores of three.

Meanwhile, the database formulates the risk management strategies to respond to the assessed risk goals. Each goal is analyzed with a criterion that relates the number of answers below a given value, then generating an advice (specified in Table 3).

For each goal, one of five broad advices can be generated and, for each advice specific actions will be later formulated [21].

Moreover, apart from the calculation of objective's risk classes and the generation of advices, the platform is also able to calculate the average risk *per* stage or the project's general risk profile (as exemplified on Fig. 4).

Table 3. Correspondent parameter conditions for the generated broad advices.

Advices	Accept	Focus	Acquire	Protect	Go/No-Go
Condition	Each of the parameters: ≤ 2	Level of Implementation ≤ 2	Capacity to Influence: ≤ 2	Severity of the consequences: ≤ 2	At least 2 of 3 parameters: ≤ 2

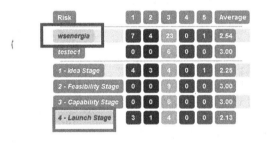

Fig. 4. Project's risk profile.

This is performed through the calculation of the weighted average of the assessed goals within each stage or within the whole project. The average risk profile attains values from 1 to 5, being that a project is considered as "*Excellent*" if the risk profile stands bellow "2"; gets the representativeness of "*Viable*" if it stands between 2 and 3; as "*Risky*" if it stands between 3 and 4; and in the case it stands above "4" it is considered "*Impracticable*".

3.3   Benchmarking Discussion

Small businesses, in which inevitable limitations of internal resources ultimately constraint the ability to engage in innovative activities [22], will barely gather conditions to obtain a risk facilitator, which will attain the RDM risk management approach. In this context, Spotrisk® grants the enterprise the possibility to earn feedback from a global network of its users, biding the information collected. So, the services provided through the platform may stand automatically for the risk facilitator's role.

Each submitted project takes a tended place in the database, contributing to data collection from which every user can compare his project. Each project can then be compared in terms of overall risk ranking, stage status or goal assessment and scaled up with the overall average of existing projects, within a specific user or with a specific project. For example, a project manager may wish to compare his project with a specific company's type of project (e.g. Google's) that it could be possible if the stated profile was created. Thereby, it is possible to incrementally enrich an assessed project by submitting it to a global comparison and working as an innovation network.

The continuing use of the platform will provide further information to the database, working as a bilateral delivery that will bring feedback to platform administrators regarding incremental improvements, layout suggestions, new specific project advices and other issues.

4 Tool Test and Assessment

4.1 Tests Description and Main Results

The Spotrisk® tool was developed by a Portuguese SME, developer of products and services in the solar photovoltaic industry, who shared the exposed difficulties of lack of risk management practices over their projects. Thus, the first validation test was carried out in six projects within the company through a series of individual evaluations made by each of the six members inserted in the respective project. The evaluated projects, despite pertaining within the same company, were able to reach different areas and components, such as operational, research & development and financial departments, providing a preliminary test regarding the universal content of the goal oriented questionnaire.

The first results obtained from internal tests showed that the purpose of a web integration of a tool could directly and effectively support the use of risk management practices, and that the easy access and little time expenditure involved could be the most suitable approach towards SMEs.

Furthermore, conferences and workshops could be held for a wide range of purposes, whereas they have mostly been for helping communities or groups of individuals structuring problematic and supporting action plans or in the decision processes. Yet, a common purpose of these thematic gatherings is also to come up with ideas and visions that would be a suitable base for the strategy development processes carried out by a given community [25]. Therefore, these meetings are learning and creatively substantiated, producing outcomes from each participant contribution. Consequently, a risk management workshop was carried out for start-up enterprises and SMEs, in Madam Park, a Portuguese start-up incubator located in Almada, Portugal. The aim was to externally test the Spotrisk® tool, and check their potential coverage and utility for other organizations outside the energy cluster. 14 participants in total were involved in the event. As seen in Table 4 the participants covered several selected areas such as start-up incubator representatives, SMEs managers, risk academic experts and R&D project researchers.

Table 4. Participants' professional domains.

Main Domain	Participants
Academic	4
SMEs Enterprises	6
Start-up Enterprises	4

The workshop opened with a clarification of its main purpose, followed by a brief individual introduction of each of the participants, a brainstorm session, as well as several presentations regarding the main difficulties felt in different projects on the initial phase. The Spotrisk® tool was then presented in the end. Each participant used the platform to evaluate a particular project that was either held by them or in which they had been inserted, during their professional life. Then all the 14 projects were compared and the risk profiles were analyzed, collecting diverse project risk profile average results. The lowest and highest results were respectively 1.74 and 3.82, which led to a discussion regarding the reasons for the values found in each project. It was concluded, for example, that the lowest result of 1.74 was, in fact, due to the nature of the project analyzed (it attracted very safe conditions from several investors and institutions).

In the end a debate took place, where some observations regarded the value Spotrisk® could bring to SMEs, start-up enterprises and R&D projects, as well as some improvements and suggestions about the platform's performance.

4.2   Spotrisk® Evaluation

During the workshop, each participant was given a small survey to set down some considerations on the most interesting aspects, as well as suggestions and ideas regarding the improvement of the tool they were presented to. They were also asked to fill a small evaluation table, in order to assess a few specific aspects. Table 5 summarizes the main assessment values obtained from the 14 workshop participants. Here, platform's features such as "Usability", "Comprehension", "Appearance"; "Potential Utility" and "Overall Appreciation", were individually evaluated in writing, using a scale of 1 (very bad) to 5 (excellent).

Table 5. Spotisk® workshop assessment results.

Item	Mean	Min	Max
Usability	3,4	2	4
Comprehension	3,5	3	4
Appearance	3,5	3	5
Potential Utility	4,6	4	5
Overall Appreciation	3,9	3	5

The gathered results suggest that the strongest feature of the platform is the potential utility to users, while the aspect in need of more improvement is the usability associated with the navigation in the web-platform.

Additional feedback was also brought into the assessment, driven by the participants' awareness in the process of answering the goal- oriented questionnaire. Participants stated that the questionnaire provided them the possibility of contemplating risks and events that they would never have consciously considered before. These observations suggest that the simple action of answering the questionnaire per se provides the user with an important awareness of some critical risks inherent to a project. Therefore, this risk-assessment tool can be a ideal for start-up incubators (as it was stated by this segment), for it brings important awareness to individuals who normally were never exposed to the given situations.

Furthermore, these endeavored developments and consequent results from the tool's presentation and discussion were exposed on a recent international conference [21]. The tool was actively received and discussed with the audience and consequent reactions pointed out the need to expand further the tool test efforts further, in order to obtain complementary validation results. However, the main board of discussion in both events regarded the potential bias associated with the control of the subjectivity of the answers. Subsequently, it was inferred that the integration in the tool of a model to correct the attained results should be found and implemented.

The process of tracking this inference drove the work's research into endeavors targeted to pursue bias correction related to the answers given [20]. This allowed for a specific study conducted by Gary King from Harvard University to be found. In this study, he and his team attempted to measure response category incomparability in surveys, due to linguistic imprecision or cultural bias, and to correct it. The measurements were performed through individual self-assessments and through hypothetical scenarios described in short vignettes, in order to correct the self-assessments without sophisticated statistical techniques [16]. The proposed solutions were to find anchors and attach the response categories to some standard or anchor [15]. Gary King and his colleagues, in collaboration with the World Health Organization, developed an approach called Anchoring Vignettes to provide such a standard. When using anchoring vignettes, the corresponding answers provided were used to adjust people's self-assessment of a situation or concept, and thus interpersonal comparable measurements were created. Vignette questions are questions about hypothetical situations or scenarios that interviewees are asked to evaluate [16].

As an example, it would be useful to recall the work performed by Gay King and his team. They studied an individual's opinion, across different countries, concerning their own political participation and their respective governments' assistance and efficiency. To that same purpose, a survey was carried out addressing the question "How much say do you have in getting the government to address issues that interest you?" [11]. The options were given in a likert five point scale from 1 (Low political efficacy) to 5 (High political efficacy). A positive scenario (vignette n°1), represented by a character named "Alison", and a negative scenario (vignette n°2), represented by a character named "John", was created.

Like the self-report assessment, the interviewees were asked to rate the degree of political efficacy for each of the presented vignettes – Alison and John - on the same scale as used for the self-report. Moreover, the vignettes were written in the manner that Alison experienced more political efficacy than John. So let's suppose that two hypothetical interviewees – R1 and R2 – have different assessments regarding their self-report and vignettes. Let's suppose R1performs a self-assessment of 2 on the likert scale, and assesses Alison and John's vignettes as 1 and 3, respectively, having the self-assessment on the middle. On the other hand R2 performs the self-assessment as 3, but assesses Alison and John's vignettes as 4 and 5, respectively.

By direct comparison of the two self-assessments it is suggested that political efficacy for the second interviewee is higher than the first. However, comparing the two vignettes' evaluations shows that the respondents have very different response scales. By looking at the second interviewee it is observed that he experiences less political efficacy than John, while the first interviewee experiences more political efficacy than John. Therefore, by adjusting the self-assessment using the vignette answers it is possible to reverse conclusions, where in fact the first interviewee experiences higher political efficacy than the second. Hence, with this method it is suggested that it is possible to re-assign and re-define the self-assessment answers relative to the vignettes, reducing potential existing bias [16].

In the study endeavored by Dr. King and his team, the political participation and efficiency of interviewees from two countries Mexico and China, was assessed, including 430 interviewees from China and 551 interviewees from Mexico. The gathered results resulted in a distinctive perception of Chinese people as having a much more active participation on their democratic decisions than Mexicans do [16].

However, the differences between these countries on political efficiency could hardly be more striking. Back In 2002, when surveys were being completed, citizens of Mexico voted out in an election, closely observed by the international community and widely declared to be free and fair. On the other hand, China is known to be ruled by a government that performs all decisions of national significance, and despite the existence of limited forms of local democracy, nothing resembles the recurring democracy in Mexico. Without the knowledge of these facts, a standard survey would have been seriously misled [16]. Thus, by anchoring vignettes, the respondent's biases were corrected, being able to obtain a more accurate set of information regarding these countries' reality.

In order to reduce the bias of Spotrisk's checklist questionnaire, adequate vignette questions had to be added. Adding the vignette's data to the collected data regarding the perception of the interviewee to the Spotrisk's questionnaire, it is believed to be possible to re-define the self-assessment answers relative to the vignette answers. As soon as a vignette peer (two vignettes) is available and assessed by the interviewees, it should be possible to acknowledge their view of the matter and their perception of a project's specifications and needs, and thus their project's risk profile.

Each vignette scenario had to be cautiously considered. The existing domains affect a certain matter (such as value proposition or target market) and for each domain were created 2 vignettes. Each goal is assessed according to 3 different parameters, Level of implementation, Capacity to influence and Severity of the consequences, which divides each goal in 3 different questions. Each question is analyzed regarding both the self-assessment and the vignettes peer, performing 3 analyses in each parameter examination. Hence, in order to simplify the discourse, let's call a singular analyzes of a

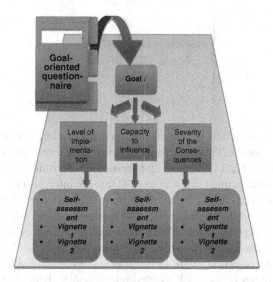

Fig. 5. Assessments made by anchoring vignettes.

singular parameter, through a single project scenario, as a "unitary response". Consequently, once the respondent finishes assessing one goal, 3 parameter answers are collected along 3 different scenarios (vignette n°1, vignette n°2 and the self-assessment), generating 9 unitary responses, as illustrated on the Fig. 5.

The vignette's descriptions play an important role within this method, which leads to a careful attention regarding its elaboration. Anchored vignettes are rated according to Table 6, being attributed the same ratings for each group of vignettes throughout the whole questionnaire, without any knowledge by its respondents.

In order to adequately anchor each vignette, each scenario needed to possess a rigorous description according to concise criteria along the whole questionnaire. The criterion used in describing each vignette was based in consistent management guidelines and standardized models of project management, innovation management and new product development.

Furthermore, a non-parametric estimator should be applied to generate a single unbiased variable C, which recodes and mingles the vignettes and self-assessment questions into a specific value, being a scalar value for some observations and multiple values for others [16]. Within this non-parametric approach, the recoded variable is assigned as 1 if the self-assessment is below Vignette 2, 2 if equal to Vignette 2, 3 if between Vignette 2 and Vignette 1. In a general way may be the self-assessment response and may the variables from z_1 to z_j be the corresponding number of a j vignette response for a single respondent, represented on Eq. (1) [15].

Table 6. Anchored values of vignettes.

	Parameter 1	Parameter 2	Parameter 3
Vignette 1	High (4)	High (4)	High (2)
Vignette 2	Low (2)	Low (2)	High (4)

$$\begin{cases} 1 & \text{if } y < z_1 \\ 2 & \text{if } y = z_2 \\ 3 & \text{if } z_1 < y < z_2 \\ & \cdots \\ & \cdots \\ 2J + 1 & \text{if } y > z_J \end{cases} \tag{1}$$

The remaining issue is how to generalize C, in order to admit tied and inconsistently ordered vignette responses. This is performed by first checking which of the conditions from the right side of Eq. (1) are true and then by condensing C with the vector of responses that are held as true [16]. Values of C that are intervals (vectors), rather than scalar, represent an inconsistent case in which more than one condition is true. In these cases it's not possible to distinguish without further assumptions which condition held true is the most accurate one [16]. Furthermore on Table 7 are represented all 13 possible combinations that can result from two vignette responses and a self-assessment.

Besides the bias correction extant in this method, it is provided the possibility of a more tangible approach of the goal oriented questionnaire, being that by reading the vignettes, respondents are more eager to understand questions as intended and to answer them appositely. However certain details are probed, such as question wording and the order of self-assessment and vignettes. Hopkins and King [11] developed improvement studies regarding anchoring vignettes where they show that if vignette questions are performed first, followed by self-assessments, it primes respondents to define the response scale in a common way. In this case, priming is not a bias to be avoided but means of better communicating the question's purposes [11]. Also it is shown that each vignette peer should be randomly displayed in the course of answering vignettes, due to the fact that, in case it doesn't, respondents can effortlessly identify the

Table 7. Possible vignette combinations.

Ex.	Responses	1 $y<z_1$	2 $y=z_1$	3 $z_1<y<z_2$	4 $y>z_2$	5 $y>z_2$	C
1	$y<z_1<z_2$	1	0	0	0	0	{1}
2	$y=z_1<z_2$	0	1	0	0	0	{2}
3	$z_1<y<z_2$	0	0	1	0	0	{3}
4	$z_1<y=z_2$	0	0	0	1	0	{4}
5	$z_1<z_2<y$	0	0	0	0	1	{5}
6	$y<z_1=z_2$	1	0	0	0	0	{1}
7	$y=z_1=z_2$	0	1	0	1	0	{2,3,4}
8	$z_1=z_2<y$	0	0	0	0	1	{5}
9	$y<z_2<z_1$	1	0	0	0	0	{1}
10	$y=z_2<z_1$	1	0	0	1	0	{1,2,3,4}
11	$z_2<y<z_1$	1	0	0	0	1	{1,2,3,4,5}
12	$z_2<y=z_1$	0	1	0	0	1	{2,3,4,5}
13	$z_2<z_1<y$	0	0	0	0	1	{5}

superior and inferior case scenario, through the logic of display order and not by vignette content, thus denouncing the perception and assessment within the method.

Hence, with the collected literature review it can be assumed that surveys may sometimes need correction. And since this work focuses itself on risk assessment through a questionnaire, a model based on Anchoring Vignettes to adjust the potential biases will continue to be developed and tested for further application on the presented tool and similar tools that use surveys or questionnaires.

5 Conclusive Remarks

This paper sought to present a tool to help filling the gap in SMEs risk management practices, proposing a useful and pragmatic approach to assess risks of innovative projects, provided on a RDM based risk appraisal on a web platform. The proposed Spotrisk® tool is intended to provide an integrated and early stage risk perception and response tool, designed to SMEs. However, recent assessment test results indicate that the tool can be also relevant for start-up enterprises. Through the collected results it can be expected that the simple action of answering the developed questionnaire per se may provide the user with an important awareness of critical risks inherent to a project, showing strong potential. On the other hand, this can also be an important tool for start-up incubators, due to the fact that the start-up enterprises associated are promptly the ones with less notion and tangibility with the market. Therefore, besides making an efficient risk assessment and generating factual risk strategies to be tailed, the platform can also bring important awareness to individuals who, in most of the cases, were never exposed to striving situations and events within an innovative project.

These first empirical results provide us the means to a favourable integration of an universal tool to support innovative projects development on SMEs, regarding that the analysed sample embraced several distinct areas. Yet, it remains a pending concern to extend universally the evidence to all the goal-oriented issues integrated in the platform. Also, it stands as future research to further extend the list of specific advices, linking that to a group of concrete actions that would respond to risks accordingly, so that SMEs might be driven thoroughly and systematically to suitable risk management practices.

Further investigation is also currently being carried out, regarding the analysis of risk perception, alongside additional research efforts to correct potential bias inherent to answers given in the questionnaire. At this stage of investigation, anchoring specific vignettes - the presented pursued solution - seems adequate in reducing the potential existing bias, at least to some degree. Nonetheless, a tangible application of this model and further results are still in the course.

Hopefully these tools and improvements will cause a positive impact by reducing the cost of projects, raising success rates, along with the induction of a higher number of successful innovative projects into market.

Acknowledgements. The authors gratefully acknowledge the funding by QREN, FEDER and PorLisboa, Ministério da Ciência, Tecnologia e Ensino Superior, FCT, Portugal, under grants WS NPT QREN - LISBOA-01-0202-FEDER-011999.

Appendix

Spotrisk's Goal Oriented Questionnaire

1.Idea Stage

1. The idea has a clear business proposition: operational, cost, product, customer or resource leadership.
2. The idea has "springboard potential" (i.e. good prospects to become products or services).
3. The idea has a value proposition with unique points, clear for buyers and partners.
4. The idea is based on a solid market research.
5. The project team has listed all the characteristics that the intended client seeks in the product/service.
6. The target market is well defined and there are clearly described channels.
7. There is a proposal for an effective action plan including eventual contingencies.
8. The team has clearly identified channels to access external knowledge and skills regarding technology, marketing and management.
9. Outsourcing solutions have been identified and are available.
10. The idea is free of eventual property rights disputes.
11. Possible ideas under development from competitors have been described.
12. There is a clear list of competitors by market segment.

2.Feasibility Stage

1. The team possesses the critical competences to develop, produce and market the intended product/service.
2. Partners will deliver in time, with all the specifications as agreed upon.
3. Organization and relations within the team members and partners are clear and goal oriented.
4. The product/service will meet all requirements in terms of licenses, safety, environment, regulations, or others.
5. The company is ready to provide future after sales services.
6. The product/service will satisfy demands and expectations from stakeholders and external bodies/agencies.
7. Financial resources are guaranteed to develop the product/service.
8. The product/service can be delivered with prices that are acceptable to buyers.
9. The product/service will contribute to the long term financial position of the company.

3.Capability Stage

1. There is a clear production/supply process to provide a reliable product delivery.
2. Future scaling up of process has been clearly addressed and described.
3. Prototypes of the product/service have been tested to reach clear pre-defined criterion.

4. Schedule and costs are realistic and achievable.

5. Sales projections for the new product/service are based on consistent data.

6. There is contingency plan to correct schedule and cost deviations along the project.

4.Launch Stage

1. There is an action plan to react to competitors' response to the introduction of the product/service.

2. The roll out of the product/service will happen as planned without information leaks.

3. There is a plan to increase and protect the barriers that the new product/service will create against competitors.

4. The key opinion makers are identified and assured.

5. There is a clear process to measure the product acceptance and marketing & sales.

6. There is a clear strategy to spread the marketing information through multiple channels.

7. A clear ratio of cost/income will be monitored during the launch processes.

8. A financial budget and monthly burn-rate thresholds are clearly defined.

References

1. Aleixo, G.G., Tenera, A.B.: New product development process on high-tech innovation life cycle. World Acad. Sci. Eng. Technol. **58**, 794–800 (2009)

2. Black, J.: Oxford Dictionary of Economics. Oxford University Press, New York (2003)

3. Brancia, A.: SMES risk management: an analysis of the existing literature considering the different risk streams. In: 8th AGSE International Entrepreneurship Research Exchange, pp. 225–239 (2011)

4. Di, Y.: Study on the mechanism of project integrated risk management for technological innovation project. In: International Conference on Information Management, Innovation Management and Industrial Engineering, vol. 2, pp. 197–201 (2010)

5. Ebrahim, N.A., Ahmed, S., Taha, Z.: SMEs; virtual research and development (R&D) teams and new product development: a literature review. Int. J. Phy. Sci. **5**(7), 916–930 (2010)

6. Emmenegger, S., Laurenzini, E., Thönssen, B.: Improving supply-chain-management based on semantically enriched risk descriptions. In: International Conference on Knowledge Management and Information Sharing, pp. 70–80 (2012)

7. Freel, M.S.: Patterns of innovation and skills in small firms. Technovation **25**, 123–134 (2005)

8. Ganguly, A.: Business Driven Research & Development. Palgrave Macmillan, New York (1999)

9. Griffin, A.: PDMA research on new product development practices: updating trends and benchmarking best practices. J. Prod. Innov. Manage **14**(6), 429–458 (1997)

10. Henschel, T.: Risk Management Practices Of SMEs: Evaluating and Implementing Effective Risk Management Systems. Schmidt (Erich), Berlin (2008)

11. Hopkins, D., King, G.: Improving anchoring vignettes: designing surveys to correct interpersonal incomparability. Public Opin. Q. **74**(2), 1–22 (2010)

12. Janney, J.J., Dess, G.G.: The risk concept for entrepreneurs reconsidered: new challenges to the conventional wisdom. J. Bus. Ventur. **21**, 385–400 (2006)

13. Jayathilake, P.M.B.: Risk management practices in small and medium enterprises: evidence from Sri Lanka. Int. J. Multi. Res. **2**(7), 226–234 (2012)

14. Keizer, J.A., Halman, J.I., Song, M.: From experience: applying the risk diagnosing methodology. J. Prod. Innov. Manag. **19**, 213–232 (2002)

15. King, G., Wand, J.: Comparing incomparable survey responses: evaluating and selecting anchoring vignettes. Polit. Anal. Adv. Access **15**(1), 46–66 (2006)

16. King, G., Murray, C.J., Salomon, J.A., Tandon, A.: Enhancing the validity and cross-cultural comparability of measurement in survey research. Am. Polit. Sci. Rev. **98**(1), 191–207 (2004)

17. Massa, S., Testa, S.: Innovation and SMEs: misaligned perspectives and goals among entrepreneurs, academics, and policy makers. Technovation **28**, 390–407 (2008)

18. Matthews, C.H., Scott, S.G.: Uncertainty and planning in small and entrepreneurial firms; an empirical assessment. J. Small Bus. Manage. **33**(4), 34–52 (1995)

19. McKiernan, P., Morris, C.: Strategic planning and financial performance in UK SMEs: does formality matter. Br. J. Manag. **5**, 31–41 (1999)

20. Pereira, L., Tenera, A., Wemans, J.: Insights on individual's risk perception for risk assessment in web-based risk management tools. Procedia Technol. **9**, 886–892 (2013)

21. Pereira, L., Tenera, A., Bispo, J., Wemans, J.: A risk diagnosing methodology web-based tool for SME ' s and start -up enterprises. In: International Conference on Knowledge and Information Sharing, vol. 6, pp. 308–317 (2013)

22. Rothwell, R.: External networking and innovation in small and medium-sized manufacturing firms in Europe. Technovation **11**(2), 93–112 (1991)

23. Smit, Y., Watkins, J.A.: A literature review of small and medium enterprises (SME) risk management practices in South Africa. Afr. J. Bus. Manag. **6**(21), 6324–6330 (2012)

24. Vargas-Hernández, J.G., García-Santillán, A.: Management in the innovation project. J. Knowl. Manag. Econ. Inf. Technol. **1**(7), 1–24 (2011)

25. Vidal, R.V.: Community facilitation of problem structuring and decision making processes: experiences from the EU LEADER + programme. Eur. J. Oper. Res. **199**, 803–810 (2009)

26. Vos, J., Keizer, J.A., Halman, J.I.: Diagnosing constraints in knowledge of SMEs. Technol. Forecast. Soc. Change **58**, 227–239 (1998)

27. Wheelwright, S.C., Clark, K.B.: Revolutionizing Product Development: Quantum Leaps in Speed, Efficiency and Quality. Free Press, New York (1992)

28. Yang, C., Man-li, Z.: Research on product innovation project risk identification thought. In: International Conference on Information Science and Engineering, vol. 2, pp. 3056–3059 (2010)

29. Yun-hong, H., Wen-bo, L., Xiu-ling, X.: An early warning system for technological innovation risk management using artificial neural networks. In: International Conference on Management Science & Engineering, vol. 14, pp. 2128–2133 (2007)

Elucidating Multi-disciplinary and Inter-agency Collaboration Process for Coordinated Elderly Care: A Case Study of a Japanese Care Access Center

Miki Saijo[1(✉)], Tsutomu Suzuki[2], Makiko Watanabe[3],
and Shishin Kawamoto[4]

[1] International Student Center, Tokyo Institute of Technology, Tokyo, Japan
msaijo@ryu.titech.ac.jp
[2] Faculty of Liberal Arts, Tohoku Gakuin University, Sendai, Japan
t-suzuki@izcc.tohoku-gakuin.ac.jp
[3] Graduate School of Science and Technology,
Tokyo University of Science, Chiba, Japan
j7410703@ed.tus.ac.jp
[4] Institute for the Advancement of Higher Education,
Hokkaido University, Sapporo, Japan
ssn@costep.hucc.hokudai.ac.jp

Abstract. This study compares the process of inter-agency collaboration among multi-disciplinary agencies within Japan as they work to provide well-coordinated care for the elderly through a Community Care Access Center (CCAC). Using the KJ method, also known as an "affinity diagram", in two group meetings (before and after CCAC establishment) with practitioners and administrators from 6 agencies in the city of Kakegawa, Japan. 521 comments by agencies were coded into 37 categories. In comparing the comments from the two meetings, the portion of negative comments regarding organization management decreased, while comments on the shared problems of the CCAC increased. A multiple correspondence analysis indicated that the 6 agencies shared a greater awareness of issues after the establishment of the CCAC, but the problems pointed out by the visiting nurses agency differed from those of the other agencies.

Keywords: Community Care Access Center · Multi-disciplinary and Inter-agency collaboration · Elderly care · KJ method · COFOR

1 Introduction

The aging society is a society in which elderly people account for a large proportion of the population. This is a trend we are seeing around the world, but in Japan it is happening more rapidly and in significantly larger numbers than elsewhere. By 2025, Japan will have 36 million people aged 65 and older. This means that the elderly will account for 30 % of the total population. We need an effective health care system for this large cohort of aging population within the demographic onus structure.

© Springer-Verlag Berlin Heidelberg 2015
A. Fred et al. (Eds.): IC3K 2013, CCIS 454, pp. 357–369, 2015.
DOI: 10.1007/978-3-662-46549-3_23

In order to cope with this tendency, the Japanese government changed the system for elderly care from institutional health care to community care. This community care provides the elderly with in-home nursing and medical care through a community general support center (CGSC) system launched in 2008 [1]. However, Japan's CGSCs do not provide the kind of coordinated nursing and medical care that is provided by such agencies as the Community Care Access Centers (CCACs) of Ontario, Canada [2]. A CCAC requires multi-disciplinary and inter-agency collaboration among medical, nursing-care, and welfare practitioners, but for practitioners in different fields to work together effectively, trust is necessary, and this relationship of trust needs to be established at an early stage [3, 4]. There is little research, however, that is based on the analysis of real-world examples of individuals in different professions and organizations cooperating with each other [5–7].

This study elucidates the process of multi-disciplinary and inter-agency collaboration by making a case study of Fukushia, a Japanese-style CCAC health care system in Kakegawa, Shizuoka prefecture, and analyzing the comments shared in group meetings of the participants held just before and after (2010, 2012) the launching of the CCAC.

2   Literature Review

2.1   Inter-agency Collaboration

In the UK, there has been an awareness since the 1970s of the need for multi-disciplinary and inter-agency collaboration in child and adolescent mental health services [8], and many studies have been made of the topic [5, 6, 9, 10]. These studies focus on how practitioners from several different agencies cooperate in the area of public health for youth, but many of their conclusions can be equally applied to the topic of general community care for the elderly. Okamoto, 2001, for example, examines how individuals with different professions in different organizations work together to address the issue of mental health among gangs of young people who are at high risk of becoming criminals.

The elements of successful multi-disciplinary and inter-agency collaboration are communication and cooperation [5, 6, 9, 10]. McKnight et al. 1998, emphasizes the role of communication in forging initial relationships of trust among inter-agency and cross-functional team members, and makes the following propositions: in initial relationships, highly trusting intentions are likely to be robust when (1) the parties interact face to face, frequently and in positive ways, or (2) the trusted party has a widely known good reputation. Still, these studies do not examine the methodology for achieving good communication among prospective collaborators nor explain how their mutual reputations are forged.

2.2   Common Frame of Reference (COFOR)

In a multi-disciplinary and inter-agency team, each member perceives the goals and problems differently depending on their knowledge and interests. In order to reach consensus, social processes focused on an understanding of each partner's real beliefs and motives are necessary [11]. This is precisely why it is important that all participants are aware of their respective perceptions, convictions and motivations. Individuals with different fields of specialty may interpret what they see differently even when they are looking at the same thing. It is necessary, therefore, that they share a common frame of reference for interpreting and integrating the information they communicate among themselves [11, 12].

In order to evaluate and to design assistance in domains of human-machine cooperation, such as air traffic control or aircraft piloting, Hoc, 2001 reviews the literature of cognitive approaches to cooperation, and notes that there is cooperation when at least some agents perform more than their private activities to become involved in cooperative activities. COFOR is at the core of cooperation and may integrate not only shared representations of environment, but also representations of one's own or other's plans, intentions, goals, resources, risks, etc. In this sense, the examination of the evolution of COFOR components over time is very useful in evaluating cooperation [13].

Common frame of reference (COFOR) is a mental structure that it is only accessible to the observer by means of external signals, such as input and output (communication between agencies), or external representations in common media (e.g., a duty roster) [12]. The duty roster is a tangible COFOR component for daily work in one agency, but how do we identify the COFOR components in multi-disciplinary and inter-agency cooperation?

A common frame of reference is usually acquired by having a series of meetings in which participants describe their view of a problem [11]. Given this, by visualizing the affinity of mutual views among team members over time, we can elucidate the evolution of COFOR components.

2.3   KJ Method

A common frame of reference is an informal mental structure, albeit with a societal aspect, that participants need to build together. At the same time, this kind of informal structure can be difficult to recognize and is hard to make transparent. One solution to the problem of achieving COFOR transparency within the context of a multi-disciplinary and inter-agency CCAC is the application of the KJ method in group meetings and the creation of diagrams and charts showing the output from those meetings.

Devised by a Japanese anthropologist named Kawakita Jiro, the KJ method is a generalized brain storming technique—what he called an "idea-generating" methodology—to gather qualitative data [14]. The KJ method has been widely adopted in business circles, not so much for generating new ideas, but for its effectiveness in consensus making [15]. The KJ method is a theory generating methodology like the grounded theory methodology of Strauss and Corbin, 1990 [16]. In group discussions using the KJ method, individuals write their opinions as short phrases on slips of sticky notes or labels. There are four essential steps in the process: (1) label making, (2) label grouping, (3) chart-making, and (4) written or verbal explanation [14]. Everyone in the group participates in the step 1 process of label-making. After that, trained facilitators carry out steps 2 through 4, intuitively sorting the labels into groups and creating a diagram linking the groups with lines (A chart). This diagram, the so-called A chart, will help to show the connections and open the way for new interpretations, and this is the distinguishing feature of the KJ method [17].

Participants in a CCAC who are trying to achieve multi-disciplinary collaboration could apply the KJ method to create a COFOR component for solving the issues that confront them. A comparison of the diagrams created before and after the launching of the CCAC will show how their perceptions of the issues have changed and should help in clarifying the collaboration process. In the KJ method, a trained facilitator creates an A chart giving an overview of the issues, grouping the problems on the basis of experience and intuition. This, of course, means that the diagram will be slanted by the facilitator's personal perceptions and assumptions. For the purposes of this study, the labels generated in the group meetings were sorted according to the similarity of the issues they addressed. We did not attempt to examine the effectiveness of the group meetings in achieving cooperation, but instead used the labels as output of the group meetings to define the process of CCAC collaboration.

3 Case Study Methodology

3.1 Background

Multi-disciplinary and inter-agency collaboration is essential for community-based care of the elderly. Take, for example, the case of an old man who is released from a hospital after suffering a mild stroke. He is unable to walk and shows dementia-like symptoms, but everyone in the family works, and during the day the old man is left at home alone. Even in a large city like Tokyo, there is no facility where an individual like this can be immediately admitted, and in any case the cost is much too high for the family. If this man is to get in-home care so that he will not become totally bed-ridden, he needs the coordinated support of the following: A hospital community coordinator who can decide what kind of support and guidance the man will need after being discharged; a senior nursing care manager who can make arrangements for the home renovations that will be needed for in-home care; the public health care nurses and visiting nurses assigned to the area where the old man lives.

Japanese local administrations are often criticized for being overly compartmentalized, but for effective community-based care of the elderly, this kind of tendency needs to be overcome. On the premise that multi-disciplinary and inter-agency case-level collaboration is best achieved when all parties concerned are housed in the same building, the city of Kakegawa launched a new Japanese-style CCAC called Fukushia in 2011 with plans to build a total of five such facilities throughout the city by 2015. Each Fukushia is staffed by personnel from four different agencies including city hall, the local social welfare council, the community general support center (CGSC), and a visiting nurses agency, with two additional agencies, the Kakegawa senior care manager liaison association, and the local city hospital, cooperating in providing social welfare services.

In 2010, prior to the launching of the new facility, the authors were asked by Kakegawa city to interview the staff of all six agencies. All of the staff interviewed expressed misgivings of the organizational management of Fukushia, including their own agency management: they worried about how they could work effectively with their counterparts in such different organizations. It was evident that collaboration would be difficult even with a new organizational structure and facility. It was therefore decided to hold group meetings in which the KJ method would be applied. This paper examines the results of two group meetings sharing the same protocol that were held before (2010) and after (2012) the Fukushia launching.

3.2   Method

Our research question was, "What is the process of multi-disciplinary and inter-agency collaboration between administration staff and practitioners within a highly differentiated and complex system of care for the elderly?" Our approach to finding an answer was to carry out a quantitative analysis of the KJ method label output from the two group meetings. The labels bore comments made by the staff of the six agencies about each other.

In our analysis, we looked first to see what kinds of comments increased or decreased in relation to the awareness of problems. This was done by comparing the number of comments made at the two meetings before and after the launching CCAC, and recording the difference. Our next objective was to see if there was any change in the affinity of awareness of problems among the meeting participants in the two meetings, and this was done through multiple correspondence analysis of the comments made at the two meetings.

The two meetings were attended each time by 29 practitioners and administrative staffs from the six agencies comprising Fukushia. The first meeting participants were: 8 from city hall; 10 from the CGSC; 3 from the local social welfare council; 3 from the visiting nurses agency; 2 from the local hospital, and 3 care managers. For the second meeting: 10 from city hall; 8 from the CGSC; 4 from the local social welfare council; 2 from the visiting nurses agency; 4 from the local hospital, and 1 care manager. At each meeting, the 29 participants were divided into 5 groups and given sticky labels on which to write their comments; blue labels for comments about their own agency and red labels for comments about the other agencies. On the red labels, participants were asked to write their own agency name and the name of the agency they were commenting about. The meetings were chaired by the lead author of this paper.

The red and blue labels where pasted onto a white board so that everyone could see what kind of comments were being made and which agencies were making the comments. The first meeting produced 220 comments and the second, 314 for a total of 534 comments. After excluding 13 illegible comments, 521 comments were then coded into 37 categories according to the issue or problem they referred to. This task was carried out individually by three researchers, and where the results did not correspond, a final decision was made through discussion among the three. Finally, the comments were sorted in a cross-tabulation table for multiple correspondence analysis. Figure 1 shows the flow of the data gathering and analyses.

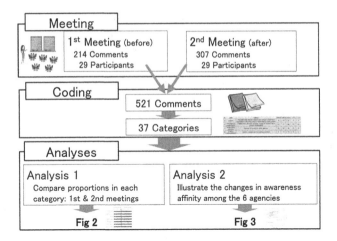

Fig. 1. Flow of the data gathering and analyses

4 Findings

4.1 Changes in Comment Proportions

For a better grasp of the trends, a comparison if the change in number of comments between the first and second meetings was made in categories that had 10 or more comments in total from the two meetings. Figure 2 shows the change in proportion between the comments from the first and second meetings, starting with those showing the greatest increase in the second meeting at the top of the chart.

The comments that showed the greatest increase in the second meeting were those related to specific shared issues of the CCAC. These comments were classified into the categories of "difficult cases", "user support", "regional collaboration", "in-home care", and "patients". There was also a notable increase in the number of comments related to work procedures, in the categories of "effectiveness", "information sharing", and "complicated procedures".

There was little change in the number of comments made at the two meetings in the categories of "inconsistency", "lack of doctors", "insufficient human resources", those related to problems of organization structure and procedures. Likewise, little change was seen in the number of comments related to inter-agency and intra-agency collaboration. To be more precise, there was an increase in the actual number of comments, but little change in the proportionate share of these comments within the designated categories. A decrease was evident in the number of comments related to the organization as such. These were comments on "agency management", "compartmentalization", "developing human resources" and "insufficient publicity". Figure 2 indicates the proportion of comments from two group meetings.

These results indicate that while the six agencies had many critical comments related to the organization management at the time of the launching of Fukushia, after the facility was set up their comments focused more on such factors as the quality of general community care services and specific shared issues of concern, rather than on criticisms of organizational structure or attitudes.

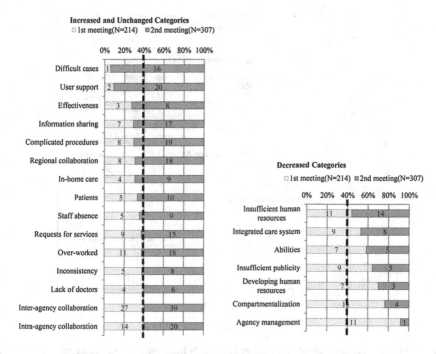

Fig. 2. Proportion of comments from the two group meetings.

4.2   Changes in Awareness Affinity Among the Six Agencies

Figure 3 shows the result of multiple correspondence analysis of the comments from the first and second group meetings.

At the first meeting, the six agencies expressed concern for problems peculiar to the own organizations, such as agency management for the Social Welfare Council, in-home care for the hospital, and compartmentalization for city hall. However, at the second meeting only the visiting nurses agency expressed concern for its own peculiar problems, and this agency did not share any concerns with the other agencies. By the second meeting all but the visiting nurses agency shared similar concerns.

The meeting participants from the visiting nurses agency only raised issues within their own organization. The issues they raised included such topics as—"With only 3 fulltime staff, there is considerable after-hours burden", "it is difficult to establish an effective visiting program plan", "there are citizens and care managers who are unaware of the visiting nursing program" and "financial difficulties in management"—all issues that are difficult for the visiting nurses agency to solve on its own. The fact that the visiting nurses agency is the only private business participating in Fukushia is probably a contributing factor to the problems the visiting nurses appear to have in communicating with the other agencies, but it should also be noted that the issues raised by the nurses tend to be introverted. The services provided by the visiting nurses are crucial to Fukushia and there is a critical need to address the issue of how the other agencies may provide better support to the visiting nurses agency.

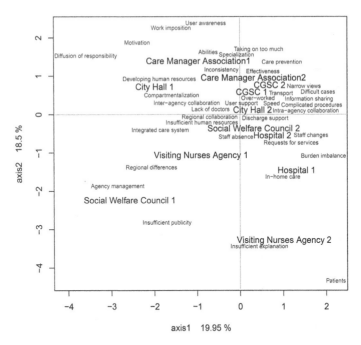

Fig. 3. Multiple correspondence analysis of comments from the two group meetings. Phrases in black indicate agencies. The numbers 1 and 2 indicate the first or second meeting, and phrases in grey indicate categories.

4.3 Positive Comments

In the second meeting only, participants asked if they could write positive comments. The facilitator agreed to this and instructed the participants to mark positive comments with a star. The final total of 76 positive comments includes only those comments that were so marked.

Table 1 shows the number of positive comments with those making the comments listed on the left and those receiving the comments listed at the top of the chart. The total shown is 69, because only comments whose sender and recipient could be identified were counted.

Figure 4 shows the number of comments by category. Here it can be seen that the positive comments are concentrated on inter-agency collaboration, regional collaboration, information sharing, and user support, while Table 1 shows that the CGSC (community general support center) and Social Welfare Council are highly evaluated by the other agencies. The primary task of the CGSC is to advise users on the services they are entitled to through long-term care insurance, and this advice, of course, ties in with the services provided by the six Fukushia agencies. The primary function of the Social Welfare Council is to support local volunteer activities and promote local safety measures, making it the linchpin for regional cooperation. The fact that these two agencies were highly rated can be interpreted to mean that the other agencies felt they were fulfilling their respective roles in furthering interagency and regional collaboration.

Table 1. Positive comments (shaded blocks indicate self-evaluations).

From \ To	Care manager association	CGSC	City hall	Visiting nurses agency	Hospital	Social Welfare Council	Total
Care manager association	0	0	0	0	0	0	0
CGSC	2	3	4	1	1	4	15
City hall	2	10	10	4	1	8	35
Visiting nurses agency	0	1	0	0	0	0	1
Hospital	1	3	0	1	1	0	6
Social Welfare Council	0	4	4	2	0	2	12
Total	5	21	18	8	3	14	69
from the other agencies	5	18	8	8	2	12	53
from their own agency	0	3	10	0	1	2	16

City hall stands out for its own high self-evaluation, based, in part, on the fact that it was city hall that established Fukushia. At the same time, it is almost equally highly evaluated by the other agencies, suggesting that there is a certain balance here. In Fig. 3, it can be seen that in the second meeting, the comments on city hall have brought it closer to the center of the graph. This means that by the second meeting, the

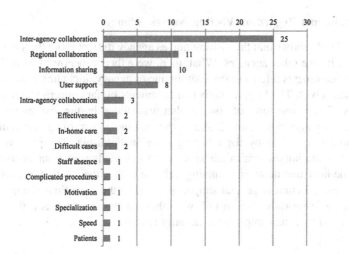

Fig. 4. Positive comments (category).

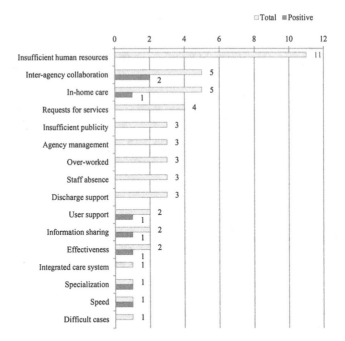

Fig. 5. Comments regarding visiting nurses agency (Total /Positive).

city hall staff had developed an awareness of the issues that was more in line with those of the other agencies. It can be surmised from this that in the one year since the establishment of Fukushia, city hall, which was highly evaluated, both by itself and by the other agencies, and was now playing a central role within Fukushia.

4.4   Comments Regarding Visiting Nurses Agency

Figure 3 clearly shows that the visiting nurses agency did not share the same awareness of issues with the other agencies. What, then, were their concerns? Figure 5 shows the number of comments related to the visiting nurses agency, including those made by the nurses themselves. The largest number of comments by far concern insufficient human resources. This is one topic of concern that was shared by the visiting nurses agency with the other agencies. Table 2 shows the comments regarding the visiting nurses agency that were made by the other agencies in Fukushia. The positive comments indicate that the nurses, with their specialized knowledge, play an important role in linking medical treatment and nursing care, and that the visiting nurses agency is pivotal to the community general support system. At the same time, the nurses were so busy that we were only able to talk with them at the meetings and they lacked the representation that their importance actually required.

Table 2. Comments regarding visiting nurses agency (P indicates positive comments).

From	Code	P	Comments
CGSC	Requests for services		Would like them to also care for those certified as needing support
			Would like the agency to act as the local town health clinic
	Information sharing		Please talk to us!
	Speed	P	Speedy response
	Difficult cases		Grateful that they take on difficult cases
City hall	Inter-agency collaboration	P	Smooth relaying of information on similar cases
	Requests for services		More services and more nurses
			Would like them to prepare care plans for those certified as needing support
	Agency management		Isn't it unsettling to make home visits alone?
	Insufficient human resources		Regular agency members bear a heavy (light) burden
			Not enough staff
			Hard to secure staff
			Not enough nurses
			Very little regular, stable employment
	Over-worked		Too busy!
	Specialization	P	Give professional advice
	Effectiveness	P	Important core of regional medical care
	In-home care	P	Makes it possible to die peacefully at home
Social welfare council	Integrated care system		Fee-based services seem expensive
	Inter-agency collaboration	P	Easy to collaborate because we're on the same floor
	Information sharing	P	In our meetings provide clear explanations of different cases
	Over-worked		Rely on them but feel sorry that they seem so busy
	Staff absence		Only meet the nurses at meetings

5 Discussion

The analysis of the comments made at the two group meetings held before and after the launching of Fukushia show that there was a change from criticism of organizational management to a shared focus on specific issues confronting Fukushia as a CCAC. It is evident that the six agencies had come closer to a common awareness of the issues before them. Clearly, the six agencies had overcome their mutual fear to forge a stronger awareness of their shared role as a public provider of general community care services. At the same time, however, it was evident that the private visiting nurses agency did not share this general awareness.

Only two or three individuals from the visiting nurses agency attended the group meetings and they can hardly be said to be representative of their organization. If general community care is to evolve from a mere concept to a truly multi-disciplinary and inter-agency undertaking to provide specific community services, and if it is to include private enterprise, strategies will be needed to tackle the issues that have arisen since the launching of Fukushia, issues which are represented by the keywords of "regional collaboration", "lack of doctors" and "discharge support". The next step is to decide what kind of communication among the six agencies is needed to achieve this.

A qualitative analysis of the multiple correspondence analysis and the positive comments would seem to indicate that in the one year since the establishment of Fukushia, the city hall has come to share a common frame of reference with the other agencies and is now playing a central role within Fukushia.

In this study, we also proposed a method to clarify the COFOR in problem awareness among the Fukushia members. It was found that a degree of objectivity could be achieved by applying multiple correspondence analysis to the awareness affinity diagram created by the meeting facilitators based on their subjective observations in previous studies. This led us to the conclusion that it may be possible to objectively externalize the latent potential for a multi-disciplinary and inter-agency collaboration COFOR, using the group meetings and the analysis of the comments. Also, as previous literature on this kind of research has indicated, the examination of the evolution of the COFOR components over time is very useful in evaluating cooperation [13]. The multiple correspondence analysis diagram of Fig. 3 shows the evolution of the COFOR components. The significance of our study, we believe, is that we have presented a method of measuring multi-disciplinary and inter-agency collaboration that makes it possible to see how COFOR can evolve.

Better collaboration between the visiting nurses agency and the other agencies is essential if the coordination of the services provided by Fukushia is to be improved. Simple improving communications, however, will not solve the basis problem which is lack of manpower. As the nurses themselves pointed out, often, even hospitals and care managers are unaware that the visiting nurses agency even exists. Something needs to be done to resolve this lack of awareness, One possible solution is have a duty roster, one of the COFOR components cited in Hoc, Carlier, 2002, that can be accessed for viewing from various locations. Care will be needed, however, to ensure that the application of the duty roster as a COFOR component does not interfere with the work of the visiting nurses agency, which is, after all, a private company. It is also our opinion that sharing the results of the analyses made in this paper with the Fukushia members through interviews with them may also be an effective way to further clarify the changes in their awareness.

Acknowledgements. We would like to express our deepest appreciation to the staff of the Kakegawa City Hall senior citizens support section and the staff of the East Fukushia facility who helped us with the case study presented in this paper. We would also like to thank Dr. Hiroko Otsuka, Dr. Hiroshi Takeda, and Mr. Taku Hirano for acting as meeting facilitators in cooperation with the authors.

References

1. Ministry of Health, Labour and Welfare: Act for Partial Revision of the Long-Term Care Insurance Act, Etc., in Order to Strengthen Long-Term Care Service Infrastructure (2011). http://www.mhlw.go.jp/english/policy/care-welfare/care-welfare-elderly/dl/en_tp01.pdf
2. Ontario Association of Community Care Access Centres. 2009/2010 CCAC Quality Report. http://www.ccac-ont.ca/uploads/201106-CCAC_Quality_Report/CCAC_Quality_Report_EN/index.htm
3. Bromiley, P., Cummings, L.L.: Transactions costs in organizations with trust. Res. Negot. Organ. **5**, 219–250 (1995)
4. McKnight, D.H., Cummings, L.L., Chervany, N.L.: Initial trust formation in new organizational relationships. Acad. Manag. Rev. **23**(3), 473–490 (1998)
5. Okamoto, S.K.: Interagency collaboration with high risk gang youth. Child Adolesc. Soc. Work J. **18**(1), 5–19 (2001)
6. Salmon, G.: Multi-agency collaboration: the challenges for CAMHS. Child Adolesc. Mental Health **9**(4), 156–161 (2004)
7. Paletz, S.B., Schunn, C.D., Kim, K.H.: The interplay of conflict and analogy in multidisciplinary teams. Cognition **126**(1), 1–19 (2013)
8. Department of Health: NHS planning and priorities guidance 1997/98. HMSO, London (1997)
9. Robinson, M., Cottrell, D.: Health professionals in multi-disciplinary and multi-agency teams: changing professional practice. J. Interprof. Care **19**(6), 547–560 (2005)
10. Salmon, G., Faris, J.: Multi-agency collaboration, multiple levels of meaning: social constructionism and the CMM model as tools to further our understanding. J. Fam. Ther. **28**(3), 272–292 (2006)
11. Marmolin, H., Sundblad, Y., Pehrson, B.: An analysis of design and collaboration in a distributed environment. In: Proceedings of the Second Conference on European Conference on Computer-Supported Cooperative Work, pp. 147–162. Kluwer Academic Publishers (1991)
12. Hoc, J.M., Carlier, X.: Role of a common frame of reference in cognitive cooperation: sharing tasks between agents in air traffic control. Cogn. Technol. Work **4**(1), 37–47 (2002)
13. Hoc, J.M.: Towards a cognitive approach to human–machine cooperation in dynamic situations. Int. J. Hum.-Comput. Stud. **54**(4), 509–540 (2001)
14. Scupin, R.: The KJ method: a technique for analyzing data derived from Japanese ethnology. Hum. Organ. **56**(2), 233–237 (1997)
15. Takeda, N., Shiomi, A., Kawai, K., Ohiwa, H.: Requirement analysis by the KJ editor. In: Proceedings of IEEE International Symposium on Requirements Engineering, pp. 98–101 (1993)
16. Strauss, A.L., Corbin, J.: Basics of Qualitative Research, vol. 15. Sage Publications, Newbury Park (1990)
17. Kawakita, J., Matsuzawa, T., Yamada, Y.: Emergence and essence of the KJ method: an interview with Jiro Kawakita (in Japanese). Jpn. J. Qual. Psychol. **2**(2), 6–28 (2003)

Semiotics in Interoperation for Information Systems Working Collaboratively

Weizi Li[1(✉)], Kecheng Liu[1,2], and Shixiong Liu[1]

[1] Informatics Research Centre, University of Reading, Reading, UK
{weizi.li,k.liu}@henley.ac.uk,
shixiong.liu@pgr.reading.ac.uk
[2] School of Information Management and Engineering,
Shanghai University of Finance and Economics, Shanghai, China

Abstract. The interoperability for information systems to work collaboratively has long been seen as a critical issue in achieving organisations' objectives. However, the lack of scientific foundation and effective approach for interoperation has always been recognised by researchers and practitioners. There is very little published in the nature of information systems' behaviour in interoperation. Organisational semiotics provides a theoretical foundation for systems interoperability. In this paper, we have synthesised current knowledge on interoperability paradigm which covers requirement articulation, knowledge foundation, analysis and measurement as well as artefact development to enhance systems interoperability. A notion of 'semiotic interoperability' is proposed in this paper as a paradigm, guiding systems interoperation and measuring degree of interoperability, covering aspects from physical properties, transmission structure of signs, placing emphasis on communicating meaning, intention to social consequence of information. Furthermore, a conceptual research framework on semiotics interoperability paradigm is proposed to guide further research in assessing, analysing, explaining and predicting as well as designing systems' behaviour by covering broad semiotic issues of interoperability.

Keywords: Organisational semiotics · Semiotic framework · Information systems integration · Semiotic interoperability · Communication

1 Introduction

Information systems interoperability is becoming critical as organisations seek to optimise their IT investment in order to support information sharing to become more agile and competitive [16]. The drivers for system interoperation often include both organisational and technical factors, such as the extended use of existing legacy systems often generates needs for integration with new systems; the need to consolidate and globalise especially in the circumstances of mergers and acquisitions where many legacy mission-critical systems need to be integrated to enhance information exchange; the seeking for productivity increase as well as cost cut through integrating business processes, transactions and applications [56].

© Springer-Verlag Berlin Heidelberg 2015
A. Fred et al. (Eds.): IC3K 2013, CCIS 454, pp. 370–386, 2015.
DOI: 10.1007/978-3-662-46549-3_24

Therefore systems interoperability is not just the capability of connection among IT systems. It is the issues of the whole organisation where both social and technical aspects need to be considered. The information systems interoperation is defined as the process that ensures the interaction between information systems necessary to achieve organisational objectives. The ideal interoperation of information systems should be organic and seamless communication among not only technical systems but also process, norms, people in the context of certain cultures as well as organisational strategies.

The nature of information systems interoperation lies in the successful signs communication among different systems. Semiotics, as the study of signs that examines the nature and properties of all kinds of signs [47, 57], provides the theoretical foundation on how signs can be successfully communicated among systems. Organisational semiotics is one of the branches of semiotics particularly related to business and organisations [42]. Stamper has developed a semiotic framework which guides us in examining all the aspects of the signs and studying how signs are used for communication and coordination in an organisational context. Therefore organisational semiotics provides a solid theoretical foundation to guide the information systems integration. This paper will introduce a new concept of semiotics interoperability which is defined into different level of interoperability. Semiotic interoperability provides a solid conceptual framework by explaining how signs can be successfully communicated in different level. This can be further developed to guide the design of seamless interoperation as well as to assess organisation's interoperability level to identify organisation's requirement towards comprehensive interoperation.

2 Interoperability Paradigm

Interoperability as one of the critical factor for information systems working together is a key factor to be encompassed throughout the activities and processes of information systems research. Therefore the research methodology of interoperability should be a problem-solving paradigm which needs to cover the whole iterative processes of information systems research, throughout the analysis, design, implementation, management as well as the use of interoperable information systems. Specifically the paradigm should address the following questions:

- What is the theoretical underpinning and knowledge foundation of interoperability that helps to understand the nature of interoperation and how they can be used to support interoperability research?
- What are the requirements of interoperability for information systems working collaboratively and how they can be identified and elicited?
- How can an interoperability approach be modelled, designed and implemented to remove interoperation barriers and meet the requirements?
- How can information systems' interoperability be assessed, evaluated, and measured with respect to organisation's requirements for interoperation?

Currently only few attempts to address the above questions from a paradigmatic perspective. For example, a structured approach has been proposed by Chen and Daclin [11], where main phases such as requirement definition, existing system analysis, solution design and implementation has been identified in a sequential way [12]. More specifically, research methodology for interoperability at the pragmatic level has been proposed by Asuncion [6] aiming to address the above questions through a regulative research cycle. But they are not able to address these questions in a paradigmatic way as there is a lack of a uniform theoretical underpinning throughout the research.

In order to provide an overview on what current research has done to understand, improve and evaluate systems' interoperability, we adopt the information systems research framework [32] to synthesise the current research on interoperability, by

Table 1. Research aspects on interoperability.

Research aspects	Specific topics	Sources
Requirement Articulation	*Driven by identifying barriers and opportunities* • Barriers at four dimensions (business, knowledge, applications, and communications) • Enterprise interoperability (models interoperability, process interoperability, business information integration)	[20; 55; 54]
	Driven by measuring interoperability • interoperability assessment process (i.e. requirements, standards, data elements, node connectivity, protocols, information flow, latency, interpretation, and information utilisation) • Enterprise interoperability methodology measuring gap between desired interoperability goal and actual status of the system	[19; 39]
	General discussion on requirements • Social and organisational level: resources reallocation, political issue, privacy and security, people issues, culture change, behavioural capability, etc. • Pragmatic level: information flow, process and norms, cross-functional integration, policy and procedure, privacy and security • Semantic level: use of language and terminologies, information overload, cross-referencing, terminology translation, semantic heterogeneity, modelling language,	[3; 5; 8; 9; 25; 28; 30; 38; 51; 58; 69; 72]
Artefact development	*Knowledge based approach* • Knowledge sharing and spanning boundaries of interoperation • Knowledge reuse and recombination	[46; 49]
	Process based approach • Process coupling approach • Operation integration approach • Flow based integration (i.e. information flow, physical flow, and financial flow) • Information based integration (product information, process information and rules)	[33; 58; 59; 60]

(Continued)

Table 1. (*Continued*)

	Standard based approach • Standard-based modelling language (e.g. XML, UML, Roset- taNet, BPML, IDEF3) • Ontology-based integration methods (e.g. OWL-S, METEOR- S, WSMO) • Ontology matching approaches • Data standards (e.g. HL7, EIC, PHIN) for establishing com- mon semantics	[15; 14; 36; 37; 70]
Evaluation/ measurement	Software and systems engineering domain • Levels of Information Systems Interoperability (LISI) • Levels of Conceptual Interoperability Model (LCIM) • Layers of Coalition Interoperability (LCI) • Capability Maturity Model Integration (CMMI)	[10; 17; 53; 66]
	Evaluation methodology • Interoperability Assessment Methodology (IAM) • Enterprise Interoperability Methodology (EIM)	[18; 19; 39]
	Industry evaluation framework • National defence: Organisational Interoperability Maturity (OIM) • Government service: European Interoperability Framework (EIF); Government Interoperability Maturity Matrix (GIMM) • Healthcare: NEHTA • Collaboration Engineering Maturing Model (CEMM); Supply Chain Management Maturity Model	[1; 17; 50; 52]
	Qualitative evaluation methods: • Non-technical Interoperability (NTI) • Organisational Interoperability Agility Model (OIAM) • Enterprise Interoperability Framework (EIF)	[17; 23; 65; 68; 66]
	Quantitative evaluation methods: • System of Systems Interoperability (SoSI) • Stoplight • Interoperability Score	[26; 29; 39; 48]
Knowledge foundation	Software engineering and IT context • Physical and network interconnectivity, compatible communi- cation, data exchange/link, application interaction and process automation	[34; 35; 45]
	Enterprise context • Data exchange, common meaning, business applications coor- dination, business processes automation, process monitoring, integrated decision support	[4; 13;27; 56]
	Organisational context • Organisation purpose, principle, attitudes, functions, schedule, economical factor, people-oriented perspectives	[33; 71]
	Merge and Acquisition context • A dynamic process from physical consolidation to coordina- tion of functions, integral/organic combination of organisa- tional assets, people, process, and technology	[31; 44; 61]
	Human communication context • Language and meaning. Process of sign-based communication	[22; 41]

categorising them into requirements identification, knowledge foundation, interoperation artefact and interoperability measurement. Table 1 illustrates the types of interoperability research with examples drawn from the IS literature.

2.1 Requirements Articulation

Requirements articulation defines the problem space in which the requirements for interoperability are contextualised. Specifically it includes the identification of goals, tasks, problems, and opportunities that defines interoperation requirement in the context of organisation [62, 63].

Interoperability challenges and issues are the most discussed topic in current research. Barriers at the social level and organisational level are most widely highlighted, such as resources reallocation, political issue, privacy and security, people issues, culture change, and behavioural patterns, etc. [3, 25, 28, 38, 69]. There are also discussions on the pragmatic level and semantic level, such as information flow, process and norms, cross-functional integration, use of language and terminologies [3, 5, 8, 51, 58].

Most interoperability requirements are articulated to overcome the interoperability barriers and realise the opportunities in organisations. Panetto and Molina [55] analyse and characterise several research challenges for Enterprise Integration and Interoperability. Their results are elaborated by a more intensive summary and contributions highlighted [54]. The challenges are classified from four dimensions (business, knowledge, applications and communications) where challenges of interoperability in enterprise are identified to include model interoperability, process interoperability and business information integration, etc. Therefore interoperability requirement can also be identified by combining conceptual, organisational and technological barriers with business, process and data concerns [20].

On the other hand, requirement articulation can also be driven by measuring interoperability among existing information systems in the organisation. For example, the interoperability assessment process defined by Leite [39] goes through key processes (i.e. requirements, standards, data elements, node connectivity, protocols, information flow, latency, interpretation, and information utilisation) in sequence, and for each process in which a calculation is made for identification of the degree of interoperability. Chen and Daclin have also purposed a methodology [12, 20] that defines steps for conducting interoperability measurement of existing systems in the enterprise. In this sense, interoperation is also seen as a methodological process to periodically measuring the gap between desired interoperability goal and actual status of the system, and to adjust both the goal and interoperation actions if necessary.

2.2 Artefact Development

Artefact development involves the creation of models, methods and techniques to remove barriers and span boundaries of interoperations among information systems and enhance the interoperability to meet organisation's requirements.

Interoperability improvements can be achieved by approaches enabling knowledge sharing and spanning boundaries of interoperation. Boundaries at different levels such as syntactic, semantic and pragmatic can be facilitated by developing common language and meaning as well as concrete functional interests [49]. Furthermore, approaches to facilitate boundaries in social interaction among individuals and organisations are also studied with the perspective of knowledge reuse and recombination [46].

Process oriented approaches provide a dynamic perspective to span boundaries for information systems working collaboratively. Joint activity approaches such as process coupling [60] and operational integration approach [59] are proposed that indicates activities sharing in order to span boundaries in processes. The process integration approach is another way to facilitate interoperation including flow based and information based integration. For example a flow-based integration is used in supply chain [58] by considering information flow, physical flow, and financial flow. Information based integration involves product information, process information and rules [33]. Moreover, concerns such as legal interoperability, business procedures, and policies are also related to process integration.

Standard based approaches are often studied to enable syntactic and semantic data exchange, especially in private and public sector. Approaches to improving syntactic data exchange include standard-based techniques (e.g. XML, UML, RosettaNet, BPML, IDEF3), middleware, enterprise application integration and service oriented architecture [36]. The most comment approaches for data interpolation at semantic level are ontology based approaches such as ontology-based integration methods (e.g. OWL-S, METEOR-S, WSMO) to specify ontology architecture [70] and ontology matching approaches [37]. Moreover, public and private sectors have developed a number of data standards (e.g. HL7, EIC, PHIN) for establishing a common semantic understanding among information systems in emergency related industry as well as healthcare industry [14, 15].

2.3   Analysis and Measurement

Analysis and measurement involves activities that analyse and assess the status of interoperability of existing information systems in the organisation, with the purpose of identifying interoperability gaps for further improvement.

The interoperability evaluation framework is regarded as the most popular means to assess the capability of interoperation. It is usually highly industry specific as most of them tend to solve not only the technical issues, but also concerns at higher levels such as management, process, and organisation goals [41]. Therefore, the interoperability evaluation frameworks are either developed for specific domain or able to adjust itself for fitting targeted industry. Its development initiates in the domain of software and systems engineering for guiding the analysis and measurement of interoperability spanning from technical [10], environmental and organisational [17] and conceptual [66] aspects. Extension and adaptation of these frameworks are made for different purposes and regions, and have been widely used, e.g. for developing European e-government systems [21]. The strengths of this type of interoperability evaluation frameworks are the separation of interoperability levels, which can clearly identify interoperability status and are easy to be adjusted in order to meet specific requirements.

However, another viewpoint on interoperability from a process perspective is proposed. It develops a whole process that evaluates interoperability from methodology design to questionnaire design [18, 19, 39]. Its strength is its tailored solutions for organisations, but it would be more difficult to be applied in other domain without significant changes.

In industry, the development of interoperability evaluation framework put more efforts on collaboration at the levels of process, organisational missions, business objectives, and human factors [52]. The researchers start to move their attention from syntactic to pragmatic and even up to social level of interoperability. However, the contribution is limited at the moment and requires further development [43]. Interoperability evaluation framework has also been widely used for government projects. In terms of national defence, government service, digital government, and e-Health, the interoperability evaluation frameworks mainly aid to identify the interoperability status. They extend the previous frameworks to reach information level, semantic level, and organizational level in order to measure the specific industrial factors [17, 50].

Besides, several generic interoperability evaluation frameworks have been developed for consolidating an evaluation methodology including both qualitative and quantitative techniques. The qualitative techniques include questionnaire, interview, and observations which are used for identifying interoperability status [17, 23, 65, 68]. The quantitative techniques are such as mathematical equation, matrix, and integral calculus for scoring the interoperability status. Key measurements include quality, time and cost [26, 29, 39, 48].

2.4   Knowledge Foundation

The knowledge foundation for interoperability study involves theories, concepts, definitions, constructs, models that reveal the nature of interoperability and support requirement articulation, approach development and measurement of interoperability. Current research on knowledge foundation mainly involves the identification of content themes in interoperability that incorporates various viewpoints, scopes, levels and layers [36]. Information systems researchers have incorporated a wide range of content themes in their studies and they differ with the way of characterising the concepts in different context. In the context of software engineering and information technology, interoperability can be featured into physical and network environment interconnectivity, compatible communication, data exchange/link, application interaction and process automation [34, 35, 45].

Interoperability content theme in enterprise context is more business and organisation oriented. It involves data exchange, common meaning, business applications coordination, business processes automation/monitoring and integrated decision support [4, 13, 27, 56]. Organisational factors such as purpose, principle, attitudes, functions and schedule [71] as well as people-oriented, economical, and organisational perspectives [33] are also essential interoperability knowledge elements. When it comes to merge and acquisition, interoperability knowledge tends to be a dynamic process from physical consolidation to coordination of functions [44, 61], in order to achieve an integral/organic combination of organisational assets, people, process, and technology [31, 61]. In the context of human communication, on the other hand, interoperability is more about language and meaning [22] as well as the process of sign-communication [41].

The above synthesis provides a portrait of current research in interoperability paradigm. However, notwithstanding the contributions of research to date, our knowledge of interoperability in information systems remains limited and fragmented in addressing the questions of interoperability paradigm. This knowledge fragmentation is amplified by the fact that conceptualizations of interoperability differ among researchers and areas [26]. There is a lack of scientific foundation [7] for the whole interoperability paradigm, which has led to an inability in assisting the development of interoperable information systems. The call for a unified paradigm with a sound theoretical underpinning of interoperability has risen to focus on broad socio-technical context and dimensions.

Semiotics, a long-established theory of signs, information and human communication, seems to be a suitable candidate for such a theory. There are some researches investigating interoperability issues from semiotic perspective. For example, a model that has a semiotic focus on human understanding and combines social and technical factors proposed to analyse semantic interoperability in an organization [2]. The semiotic dynamics focusing on emergent semantics in information exchange is investigated to address semantic interoperability. Although current semiotic research on interoperability mainly focuses on the semantic aspect, the studies in modelling and simulation (MandS) standardization take a step forward with a semiotic view incorporating syntactic, semantic and pragmatic perspectives into MandS systems interoperability [67]. However, all the aforementioned researches only focus on certain semiotic levels and therefore are not able to address the fundamental questions of interoperability paradigm. We argue that rather than focusing on certain aspects, interoperability should include all aspects of sign-based activities in the process of communications.

3   Semiotic Interoperability

Information systems interoperation can be seen as the process of sign-based communication not only among technical systems but also interactions in the social environment surround the systems. The effective and comprehensive information systems integration should fully consider and depends on the successful use of sign at different levels of communications, interactions and social activities. Organisational semiotics provides a sound theoretical foundation on how signs are used in the communications to support information systems integration in organisation. A new concept of semiotic interoperability is introduced in this paper which defines interoperability from a broader perspective of signs with the purpose of giving a theoretical guidance for effective information systems integration. Semiotic interoperability will look at information systems integration by explaining not only how signs are structured and used in language, how signs are organised and transmitted, what physical properties signs have, but also how signs function in communicating meanings and intentions, and what the social consequences are of the use of signs.

3.1   Organisational Semiotics and Semiotics Framework

Organisational semiotics, as a sub-branch of semiotics particularly related to organisations, is the study of organisations using the concepts and methods of semiotics. Organisational semiotics provides a holistic view about signs, information, systems and organisations. An organisation can be seen as an information system where information is created, stored, and processed for communication and coordination and for achieving the organisational objectives.

Stamper has developed a semiotic framework (Fig. 1) that guides us in examining all the aspects of the signs and studying how signs are used for communication and coordination in an organisational context. Organisations have both a technical and a social dimension and their performance relies heavily on their ability to integrate both of these dimensions. From this semiotic perspective, information systems integration is about how signs are used for successful communications in IT platforms serving technical business operations as well as human information functions supporting social dimension of business activities.

The physical world, as a separate layer concerned with the physical aspects of signs gives a handle to deal with the factors governing the economics of signs, which has become important in business contexts. Empirics, has been defined as another branch to study the statistical properties of signs when different physical media and devices are used. Syntactic, semantics and pragmatics respectively deal with the structures, meanings and usage of signs. Social world has been defined as the effects of the use of signs in human affairs.

The three upper layers in the framework are concerned with the use of signs, includes functions of signs in communicating meanings and intentions, and the social consequences while using signs. The three lower layers in the framework aim to answer questions on how signs are structured and used in language, how signs are organised, what physical properties signs have, etc. This work has been widely used in analysing business organisation and information systems design.

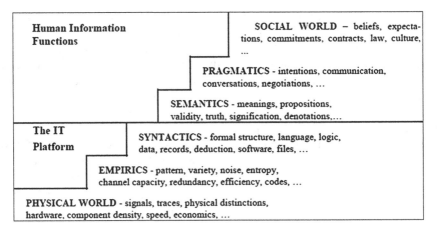

Fig. 1. Semiotic framework [64].

3.2   Semiotic Interoperability

It is often conceivable to see information systems integration as a process of sign communication. The semiotic framework that explains all aspects of how signs can be used and communicated for successful communication, determines the level of interoperability of information systems integration. Therefore we say systems are integrated in a certain level of interoperability if signs among systems are successfully communicated in a certain semiotic framework level.

Our previous research [40] proposes a set of theoretical and methodological methods for information systems integration in digital hospital. The theory of organisational semiotics has been used for developing a list of methods (e.g. modelling methods, implementation methods) and models (e.g. ontology model, activity model). In addition, a theoretical and conceptual architecture has been proposed based on the semiotic framework. This architecture identifies the requirements for information systems integration at several levels (i.e. organisational goal level, business services level, business activities level, system functions level, system communication level, and infrastructure level), which support the communication among information systems. Li's work elaborates the application of organisational semiotics theories, and reveals that the semiotic framework is able to illustrate the operation of organisation as an information system, and to guide successful communication between information systems. Hence we focus on interoperability and communication in this paper and propose the concept of semiotic interoperability to identify the level of system integration based on semiotics framework [24, 42, 64] and previous work on how organisational semiotics supporting information systems integration [40]. This new coined notion and framework will provide a sound theoretical and knowledge foundation for interoperability paradigm.

Semiotic Interoperability allows information systems to work together through communication with insight into the physical properties, transmission structure of signs, placing emphasis on communicating meaning, intention and social consequence of information. Semiotic interoperability can be further explained by the semiotic interoperability framework (Fig. 2):

Social Interoperability	*The resultant interoperable systems should be coherent with social commitment, obligations and norms in order to support organisation's strategy, vision and objectives*
Pragmatic Interoperability	*Processes supported by the systems in individual contexts can be aggregated to achieve the overall intended purpose*
Semantic Interoperability	*Ability of interpreting and converting information into equivalent meaning to allow information sharing between systems*
Syntactic Interoperability	*Data exchange between systems through compatible formats and structures*
Empiric Interoperability	*Sign transmission through compatible channels and protocols between systems*
Physical Interoperability	*Connectivity between networks and hardware and devices*

Fig. 2. Semiotic interoperability framework [41].

- Physical Interoperability. The interoperability at physical level indicates that connectivity between networks, hardware and devices. This level is concerned with the physical connection and transmission channel in sign communication. Signs are modelled by physical signals (varying in time) and marks (static in time), their sources and destinations and the routes over which they are transmitted. Physical interoperability is achieved when a chain of physical tokens, transmitted along a route, is received at the other end, by the receiver, conserving the same physical properties. For example, the hardware devices of the systems must be interconnected in order to support the data transmission. However, although the physical level has been achieved, the data transmission cannot be succeeded without proper communication channel and protocols. These issues are concerned at empiric interoperability.

- Empirics Interoperability. Empirics interoperability ensures sign transmission through compatible channels and protocols between systems. This level is concerned with the matching of coding and decoding between sign sender and receiver based on statistical properties of information. In such communication, coding is done at the sending end and decoding at the receiving end. Interoperability at this level is achieved when the receiver can reconstitute the same sequence of symbols that were sent by the sender, irrespective of any problems at the physical level. Research at this level involves the study of communication devices that are well matched to the statistical characteristics of the media. The word meaning/matching in this level means the equivalence of codes. The empiric level ensures the capacity of communication channels and protocols of different information systems are matched. For example, specific bandwidth and proper communication protocols such as IEEE 108.11g have to be matched for both systems in order to successfully transmit the data. However, the information exchange still may fail if the structures of the data for information systems are incompatible. The issue of data structure is concerned at the syntactic level.

- Syntactic Interoperability. Syntactic interoperability ensures data exchange between systems through compatible formats and structures. Syntactics is concerned with rules of composing complex signs from simple ones. Information can be coded following a certain structure; a complex sign, a word, a mathematical expression, or a sentence can be composed of some more basic parts according to the rules. Communication at syntactic level presumes the existence of devices able to process symbolic expressions. Communication is successful if the devices are able to identify and internally rebuild each other's symbols and expressions, irrespective of the symbols that are used. The syntactic interoperability is achieved when the expression of information, or language, or formula can be recognised by different information systems. The data structures and format of file and message have to be readable to both ends of communication. For example, a program wrote in JAVA cannot be recognised by other non-JAVA supported information systems. However, although the syntactic level is achieved, the communication still may fail if the message cannot be understood by other information systems. These issues are concerned at the semantic level.

- Semantic Interoperability. Interoperability at semantic level indicates the ability of interpreting and converting information into equivalent meaning to allow information sharing between systems. Semantic level is concerned with the meaning of signs.

Meaning or semantics of a sign is normally considered as a relationship between a sign and what it refers to. Under the definition of meaning, there has to be a 'reality' assumed, so that signs can be mapped onto objects in the 'reality'. At the semantic level of use of language, meaning acts as the operational link between signs and practical affairs. People use signs or a language in communication. To enable one person to understand another, there must be some ways/processes/principles governing the use of signs which are established and shared in a language community. At the semantics level, communication is successful if signs are interpreted for both sender and receiver according to same principles. If an utterance may be interpreted in several different ways this means that the interpretation selected by the speaker would be the same one selected by the hearer. The semantic interoperability requires that the content's meaning and governing norms can be shared for different information systems. The semantic level addresses the issue of semantic interoperability, and involves terminology aspects (homonyms, synonyms, scope) as well as human language aspects. The semantic interoperability can be achieved to give the same meaning to exchanged information between information systems, and it requires a conceptual model which describes what information is exchanged in terms of concepts, properties, and relationships between these concepts.

- Pragmatics Interoperability. Pragmatics interoperability ensures that processes supposed by the systems in individual contexts can be aggregated to achieve the overall intended purpose. Pragmatics is the level of semiotics concerned with the relationship between signs and the potential behaviour/intention of responsible agents, in a social context. Within a social community, there exist common knowledge and shared assumptions. These basic assumptions serve as a minimum basis for communication. Therefore, successful communication at this level is achieved if the hearer understands the speaker's intentions, irrespective of the semantic interpretation of the communicative act. The communication may be successful even if the hearer does not do what the speaker wishes him to do, as long as the hearer correctly interprets the intention expressed by the speaker

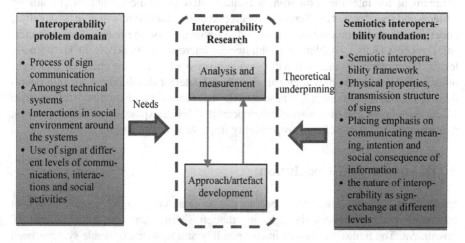

Fig. 3. Semiotic interoperability conceptual research framework.

Interoperability is achieved at this level when processes serving different purposes under different contexts by different information systems can be composed to jointly support a common intention. The emphasis is on the context awareness for processes integration. The following elements can be considered in the context: information system itself, intention, purpose, theme, time, location etc.

• Social Interoperability. Social interoperability ensures that the resultant interoperable systems should be coherent with the social commitment, obligation and norms in the organisation and support organisation's strategy, vision and objectives. Communication at the social level requires the hearer and the speaker to share social norms. A sign is meaningful if it actually alters those norm structures. Knowledge at the social level is thus essentially defined in terms of norms, i.e. regularities of behaviour, perception, and judgement. The interoperability in social level ensures the intention or purpose of the sender has led to a social consequence to the receiver through the resultant interoperable systems. Social commitments and obligations can often be created or discharged as the result of a conversation. Even if the hearer does not do apparently anything, the communication succeeds at the social level as long as a result of the communicative act, any social commitments are created or modified.

3.3   Semiotic Interoperability Paradigm

In order to provide a unified paradigm to address questions in interoperability paradigm, we have developed the semiotic interoperability paradigm overview (as shown in Fig. 3) based on the knowledge foundation of semiotic interoperability notion and framework as well as the information system research framework [32].

Figure 3 presents our conceptual research framework of interoperability for understanding, analysing, designing and assessing interoperability, combing organisational semiotics and information systems research paradigms. This framework demonstrates how semiotic interoperability paradigm address questions (see Sect. 2) in interoperability research. The problem domain refers to the organisation in which reside the requirement for interoperation from a boarder view of sign-communications among systems and social activities. Semiotics interoperability foundation will give a theoretical underpinning and comprehensive knowledge foundation on the nature, key characteristics, success factors, enablers and inhibitors of interoperability across the six semiotic levels. Interoperability research will provide process of generating methods and techniques aiding analysis, measurement, design and artefact development for interoperations at various semiotic levels. These models will construct a methodological roadmap where assessment and diagnosis of interoperability level, explanation of interoperation requirements and design for interoperability improvement can be conducted

4   Discussion and Conclusion

Information systems integration becomes a critical concern for organisations in enhancing competitiveness through information sharing and maximising their IT investment. The issues on systems interoperability and how interoperable systems need to be as a strong support for organisational competitiveness have drawn many attentions of researcher. The scientific foundation of information systems' behaviour in interoperation is very much an understudied area, despite its importance in the context of organisations. The main outcome of our future work will fill this gap with the development of the new notion of "semiotic interoperability" - a uniform scientific paradigm covering aspects of physical properties, signal transmission and structure of signs to communicating meaning, conveying intention and achieving intended social consequences. Although the concept and framework of semiotic interoperability is new, some of the constituent claims have been studied in our previous researches on how semiotics supports information systems integration. The research results on semiotic interoperability paradigm will bring significant contribution to the theory of organisational semiotics and information systems, particularly in understanding the nature of information systems' behaviour in interoperation.

References

1. ADatP-34: NATO C3 Technical Architecture (NC3TA). NATO Allied Data Publication 34 (2003)
2. Ahlin, K., Saariko, T.: A semiotic perspective on semantics interoperability (2012)
3. Allen, D.K., Karanasios, S., Norman, A.: Information sharing and interoperability: the case of major incident management. Eur. J. Inf. Syst. **22**(1), 1–15 (2013). doi:10.1057/ejis.2013.8
4. Amice, C.: CIMOSA: CIM Open System Architecture. Springer, Heidelberg (1993)
5. Amrani, R., Rowe, F., Geffroy-Maronnat, B.: The effects of enterprise resource planning implementation strategy on cross-functionality. Inf. Syst. J. **16**(1), 79–104 (2006). doi:10.1111/j.1365-2575.2006.00206.x
6. Asuncion, C.: Pragmatic interoperability in the enterprise: a research agenda. Ph.D thesis (2011)
7. Berre, A., Elvesæter, B., Figay, N., Guglielmina, C., Johnsen, S., Karlsen, D., Lippe, S.: The ATHENA interoperability framework. In: Gonçalves, R.J., Müller, J.P., Mertins, K., Zelm, M. (eds.) Enterprise Interoperability II, pp. 569–580. Springer, London (2007)
8. Bidan, M., Rowe, F., Truex, D.: An empirical study of IS architectures in French SMEs: integration approaches†. Eur. J. Inf. Syst. **21**(3), 287–302 (2012). doi:10.1057/ejis.2012.12
9. Bishr, Y.: Overcoming the semantic and other barriers to GIS interoperability. Int. J. Geogr. Inf. Sci. **12**, 299–314 (1998)
10. C4ISR Interoperability Working Group: Levels of information systems interoperability (LISI). Work report (1998)
11. Chen, D., Daclin, N.: Barriers driven methodology for enterprise interoperability. In: Camarinha-Matos, L.M., Afsarmanesh, H., Novais, P., Analide, C. (eds.) Establishing the Foundation of Collaborative Networks. IFIP, vol. 243, pp. 453–460. Springer, Boston (2007)
12. Chen, D., Vallespir, B., Daclin, N.: An approach for enterprise interoperability measurement. In: Proceedings of MoDISE-EUS, pp. 1–12 (2008)
13. Chen, D., Vernadat, F.: Standards on enterprise integration and engineering—state of the art. Int. J. Comput. Integr. Manuf. **17**(3), 235–253 (2004). doi:10.1080/09511920310001607087
14. Chen, R., Sharman, R.: Emergency response information system interoperability : development of chemical incident response data model. J. Assoc. Inf. Syst. **9**(3), 200–230 (2008)

15. Chen, R., Sharman, R., Rao, H., Upadhyaya, S.: Data model development for fire related extreme events: an activity theory approach. MIS Q. **37**(1), 125–147 (2013)
16. Clabby, J.: Web Services Explained: Solutions and Applications for the Real World. Prentice Hall PTR, Upper Saddle River (2003)
17. Clark, T., Jones, R.: Organisational interoperability maturity model for C2. In: Proceedings of the 1999 Command and Control (1999)
18. Cornu, C., Chapurlat, V., Quiot, J., Irigoin, F.: Customizable interoperability assessment methodology to support technical processes deployment in large companies. Ann. Rev. Control **36**(4), 300–308 (2012). doi:10.1016/j.arcontrol.2012.09.011
19. Daclin, N., Chen, D., Vallespir, B.: Enterprise interoperability measurement-basic concepts. In: Proceedings of the EMOI, vol. 1, pp. 1–5 (2006)
20. Daclin, N., Chen, D., Vallespir, B.: Methodology for enterprise interoperability. In: Proceedings of the 17th World Congress: The International Federation of Automatic Control Seoul, Korea, pp. 12873–12878 (2008)
21. EIF: European interoperability framework for pan-European egovernment services. IDA Working Document (2004)
22. Euzenat, J.: Towards a principled approach to semantic interoperability. In: Proceedings of the IJCAI 2001 Workshop on Ontology and Information Sharing, Seattle (WA US) (2001)
23. Fewell, S., Richer, W., Clark, T., Warne, L., Kingston, G.: Evaluation of Organisational Interoperabiity in a Network Centric Warfare Environment (2004)
24. Filipe, J.B.L.: Normative organisational modelling using intelligent multi-agent systems, Ph.D. thesis. Staffordshire University (2000)
25. Fink, L.: How do IT capabilities create strategic value? Toward greater integration of insights from reductionistic and holistic approaches. Eur. J. Inf. Syst. **20**(1), 16–33 (2010). doi:10.1057/ejis.2010.53
26. Ford, T., Colombi, J.: The interoperability score. In: Proceedings of the Fifth Conference on Systems Engineering Research (2007)
27. Goodchild, M., Egenhofer, M.: Interoperating GISs: report of a specialist meeting held under the auspices of the VARENIUS project (1997)
28. Hamilton, D.: Linking strategic information systems concepts to practice: systems integration at the portfolio level. J. Inf. Technol. **14**(1), 69–82 (1999)
29. Hamilton, J., Rosen, J., Summers, P.: An Interoperability Road Map for C4ISR Legacy Systems (2002)
30. Hammond, W.: EU eHealth Interoperability Roadmap. Studies in Health Technology and Informatics, vol. 134, pp. 245–253 (2010)
31. Henningsson, S., Carlsson, S.: The DySIIM model for managing IS integration in mergers and acquisitions. Inf. Syst. J. **21**(5), 441–476 (2011). doi:10.1111/j.1365-2575.2011.00374.x
32. Hevner, A., March, S., Park, J., Ram, S.: Design science in information systems research. MIS Q. **28**(1), 75–105 (2004)
33. Huat Lim, S., Juster, N., de Pennington, A.: Enterprise modelling and integration: a taxonomy of seven key aspects. Comput. Ind. **34**(3), 339–359 (1997). doi:10.1016/S0166-3615(97)00069-9
34. IEC: IEC-65-290-DC – TC65: Industrial process measurement and control. International Electrotechnical Commission (2000)
35. ISO/IEC: 7498-1. Information Technology - Open Systems Interconnection - Basic Reference Model: The Basic Model. Work report (1996)
36. Izza, S.: Integration of industrial information systems: from syntactic to semantic integration approaches. Int. J. Enterp. Inf. Syst. **3**(1), 1–58 (2009)
37. Kalfoglou, Y., Schorlemmer, M.: Ontology mapping: the stage of the art. Knowl. Eng. Rev. **18**(1), 1–31 (2003)

38. Khoumbati, K., Themistocleous, M., Irani, Z.: Evaluating the adoption of enterprise application integration in health-care organizations. J. Manage. Inf. Syst. **22**(4), 69–108 (2006). doi:10.2753/MIS0742-1222220404
39. Leite, M.: Interoperability Assessment. PRC Inc., Arlington, VA (1998)
40. Li, W.: The architecture and implementation of digital hospital - information system integration for seamless business process, Ph.D thesis. Univerity of Reading (2010)
41. Li, W., Liu, K., Liu, S.: Semiotic interoperability - a critical step towards systems integration. In: The 5th International Conference on Knowledge Management and Information Sharing, Vilamoura, Portugal, pp. 508–513 (2013)
42. Liu, K.: Semiotics in Information Systems Engineering. Cambridge University Press, Cambridge (2000). doi:10.1017/CBO9780511543364
43. Liu, S., Li, W., Liu, K., Han, J.: Evaluation frameworks for information systems integration: from a semiotic lens. In: The 3rd International Conference on Logistics, Informatics and Service Science, Reading, UK, pp. 559–568 (2013)
44. Mehta, M., Hirschheim, R.: Strategic alignment in mergers and acquisitions : theorizing is integration decision making. J. Assoc. Inf. Syst. **8**(3), 143–174 (2007)
45. Mitche, J.: Manufacturing modelling and integration. In: Presentation at a Meeting of the Computer Department at CETIM (slides) (1997)
46. Mitchell, V.: Knowledge integration and information technology project performance. MIS Q. **30**(4), 919–939 (2006)
47. Morris, C.W.: Signs, Language and Behaviour. Braziller, New York (1946)
48. Morris, E., Levine, L., Meyers, C., Place, P., Plakosh, D.: System of systems interoperability (SOSI): final report. Carnegie Mellon Software Engineering Institute, Pittsburgh (2004)
49. Nazir, S., Pinsonneault, A.: IT and firm agility : an electronic integration perspective. J. Assoc. Inf. Syst. **13**(3), 150–171 (2012)
50. NEHTA: Interoperability Framework. Work report (2007)
51. Otjacques, B., Hitzelberger, P., Feltz, F.: Interoperability of e-government information systems: issues of identification and data sharing. J. Manage. Inf. Syst. **23**(4), 29–51 (2007). doi:10.2753/MIS0742-1222230403
52. Ouksel, A.M., Sheth, A.: Semantic interoperability in global information systems. ACM SIGMOD Rec. **28**(1), 5–12 (1999). doi:10.1145/309844.309849
53. Palomares, N., Campos, C., Palomero, S.: How to develop a questionnaire in order to measure interoperability levels in enterprises. In: Popplewell, K., Harding, J., Poler, R., Chalmeta, R. (eds.) Enterprise Interoperability IV, pp. 387–396. Springer, London (2010)
54. Panetto, H.: Enterprise integration and networking: theory and practice. Ann. Rev. Control **36**(2), 284–290 (2012)
55. Panetto, H., Molina, A.: Enterprise integration and interoperability in manufacturing systems: trends and issues. Comput. Ind. **59**(7), 641–646 (2008). doi:10.1016/j.compind.2007.12.010
56. Panian, Z.: Why enterprise system integration is inevitable? WSEAS Trans. Bus. Econ. **2006**, 590–595 (2006)
57. Peirce, C.S.: Collected Papers of C. S. Peirce. Harvard University Press, Cambridge (1931)
58. Rai, A., Patnayakuni, R., Seth, N.: Firm performance impacts of digitally enabled supply chain integration capabilities. MIS Q. **30**(2), 225–246 (2006)
59. Robicheaux, R., Coleman, J.: The structure of marketing channel relationships. J. Acad. Mark. Sci. **22**(1), 38–51 (1994)
60. Saraf, N., Langdon, C.S., Gosain, S.: IS application capabilities and relational value in interfirm partnerships. Inf. Syst. Res. **18**(3), 320–339 (2007). doi:10.1287/isre.1070.0133
61. Schweiger, D.: MandA Integration: A Framework for Executives and Managers. McGraw-Hill, New York (2002)

62. Silver, M., Markus, M., Beath, C.: The information technology interaction model: a foundation for the MBA core course. MIS Q. **19**(3), 361–390 (1995)
63. Simon, H.: The Sciences of the Artificial, 3rd edn. MIT Press, Cambridge (1996)
64. Stamper, R.: Information in Business and Administrative Systems. Batsford, London (1973)
65. Stewart, K., Cremin, D., Mills, M., Phipps, D.: Non-technical interoperability: the challenge of command leadership in multinational operations. QINETIQ LTD Farnborough (UK) Centre for Human Sciences (2004)
66. Tolk, A.: Beyond technical interoperability-introducing a reference model for measures of merit for coalition interoperability. In: 8th International Command and Control Research and Technology Symposium (2003)
67. Tolk, A., Diallo, S., King, R., Turnitsa, C.: A layered approach to composition and interoperation in complex systems. Complex Syst. Knowl. Based Environ. Theory Models Appl. **168**, 41–74 (2009)
68. Tolk, A., Muguira, J.A.: The levels of conceptual interoperability model. In: Fall Simulation Interoperability Workshop, Orlando, Florida (2003)
69. Turban, E., Watkins, P.: Integrating expert systems and decision support systems. MIS Q. **10** (2), 121–137 (1986)
70. Wache, H., Voegele, T., Visser, U., Stuckenschmidt, G., Schuster, G., Neumann, H., Hubner, S.: Ontology-based integration of information a survey of existing approaches. In: IJCAI 2001 Workshop: Ontologies and Information Sharing, Seattle, pp. 108–117 (2001)
71. Waring, T., Wainwright, D.: Interpreting integration with respect to information systems in organizations – image, theory and reality. J. Inf. Technol. **15**(2), 131–147 (2000). doi:10. 1080/026839600344320
72. Yahia, E., Aubry, A., Panetto, H.: Formal measures for semantic interoperability assessment in cooperative enterprise information systems. Comput. Ind. **63**(5), 443–457 (2012). doi:10. 1016/j.compind.2012.01.010

Challenges to Construct Regional Knowledge Maps for Territories' Sustainable Development

Montserrat Garcia-Alsina[1(✉)], Christian Wartena[2],
and Sönke Lieberam-Schmidt[2]

[1] Information and Communication Sciences Department, Universitat Oberta
de Catalunya, Rambla Del Poblenou, 156, 08018 Barcelona, Spain
mgarciaals@uoc.edu
[2] Department of Media Information and Design, University of Applied Science
and Arts Hannover, Expo-Plaza 12, 30539 Hannover, Germany
{christian.wartena,soenke.lieberam-schmidt}
@hs-hannover.de

Abstract. Knowledge management in organizations is a practice progressively implemented during the last decades. Their value is recognized as an asset to generate competitive advantage. This article translates the knowledge management's potentials to a geographic territory, and explores how to apply to a territory the framework developed by knowledge management discipline to be applied in companies. The main point is how to build regional knowledge maps to satisfy needs expressed by some actors involved in regional policy and innovation in a territory. This work reports a research in progress, whose objective is to define a methodology to efficiently design territorial knowledge maps. It presents the theoretical background offered by other disciplines that could support knowledge management discipline. These disciplines are on the one hand, regional innovation systems and competitive - territorial intelligence, to explore components involved in a territory, and on the other hand text mining to extract information of big volumes of data contained in diverse sources of information related to a region. It shows also the theoretical framework designed to deal with this research, that identify which components should collect a regional knowledge map (structures, actors, infrastructures, resources and social capital), and how to study the contexts where knowledge is created to innovate and contribute to the regional development. Finally this paper summarizes the potentials and the challenges to construct regional knowledge maps, which constitutes sub lines of research.

Keywords: Knowledge maps · Regional development · Text mining · Territorial intelligence · Regional policy

1 Introduction

In this work we argue the role of regional knowledge maps for regional innovation management, their structure and the challenges faced to construct them. It takes as context the value given to knowledge management (KM) in organizations and investigates its value if it is applied to geographic regions.

© Springer-Verlag Berlin Heidelberg 2015
A. Fred et al. (Eds.): IC3K 2013, CCIS 454, pp. 387–399, 2015.
DOI: 10.1007/978-3-662-46549-3_25

We are installed in a new knowledge-driven economy and we are moving towards a knowledge-based society [1, 2]. Hence, knowledge is an important asset that should be managed in organizations to achieve organizational objectives, generate competitive advantages and create richness [3, 4]. Besides, knowledge is a basis to innovate [5, 6].

Applied to geographic regions, knowledge management could be also of great relevance because it enables to identify how the knowledge cycle works in a territory to create value and competitive advantage. Knowledge is a basis to innovate, and innovation is important in an economy based on knowledge and innovation [2]. Consequently, managing the knowledge of a territory could contribute to regional innovation and therefore their sustainable development [6–8]. In this sense, different disciplines such as Territorial Intelligence (TI) and Regional Innovation Systems (RIS) point out frameworks and instruments to identify knowledge, regional actors, and structures to recognize motors of innovations in a specific territory. Besides, some authors have recently pointed out knowledge maps as an instrument to manage knowledge in a territory and to promote its development [9–11]. Despite of this recognized relevance of knowledge maps, literature about how to build them is disperse and scarce [12–18].

Hence, building territorial knowledge maps to support innovation and regional development represents a major step forward, due to the complexity of aspects that these maps should collect: which processes take place in a territory, which knowledge is created or needed in the different processes, how and where can knowledge be created and stored, which structures and infrastructures exist in a region related to knowledge and innovation, and the information fluxes among its actors.

Therefore, we argue that more research is needed in this area in order to collect, structure and make efficient use of the vast amount of mostly textual and unstructured information that is available for almost every region in industrial countries.

This work reports our research in progress, which aim is the design of a methodology to develop territorial knowledge maps as a tool for regional development. We take into account the theoretical corpus of knowledge management and other disciplines related to region studies and text mining. The remainder of this article is structured in four parts. Firstly, we specify the objectives and the research questions. Secondly, we expose shortly the theoretical backgrounds used as basis to develop our theoretical framework, which is presented in this second section. Thirdly, we describe the challenges faced to build knowledge maps and the research line faced to overcome them. Lastly we present the conclusions.

2   Objectives and Research Questions

Our objective is defining a methodology to efficiently design territorial knowledge maps, by extracting information of big volumes of data contained in diverse sources of information related to the territory. The methodology, of course, should be independent of the specific.

Hence, the work implies facing several challenges, which we have concretized in the following objective: (a) drawing the structure and the components of the knowledge maps; (b) knowing how we can extract knowledge from a wide amount of sources related to one region and (c) visualizing it to get a complete picture of the region's innovation potential. To conceptualize how to do so is the case for the project in process presented in this paper, which turns on the following research questions:

1. Which information sources help to find this knowledge?
2. Which are the territory's actors involved in innovation?
3. How can we identify the knowledge available in a certain region?
4. How can we study the social relationships, conventions, norms and rules that influence innovation?
5. How can we mine the territorial data to create knowledge maps that enable stakeholders to apply them for innovation and regional development?
6. How can we identify innovation potentials that remain hidden and unobserved until now?
7. How can we identify obstacles hindering innovation?

The first five questions are directly related to the project's objective, because they point out to which methodology is adequate to build knowledge maps. The last two are oriented to detect which knowledge is important to be incorporated in a regional knowledge map to assure the focus on innovation.

3 Theoretical Background

Taking into account the aims mentioned above, our research is based on four perspectives, each of which has its own corpus of knowledge. These perspectives are: knowledge management, competitive and territorial intelligence, regional innovation systems, and text mining. These four perspectives will be used to construct our theoretical background.

3.1 Knowledge Management

Knowledge Management (KM) is a methodology that integrates the activities embedded in the organizational processes to obtain organizational aims and manage intellectual capital [4, 19–22]. KM applied to private or public organizations and to the territory contributes to generate competitive advantages [23], because knowledge plays an important role in the innovative processes. Owing to advances in technologies and greater flows of information, knowledge is more and more viewed as a central driver of economic growth and innovation [5].

Knowledge management practices in organizations have been progressively incorporated in the last 30 years, and intellectual foundations of knowledge management are in construction, as we can see in the remarkable increase in articles, books conferences and job titles [24]. Nowadays there are a considerable theoretical corpus and models about how knowledge could be managed in organizations [25]: identifying which knowledge is created or needed in the different organizational processes, and how it can be created and stored to share it to generate value in an organization [4, 19]. Therefore, valuable instruments has been developed such as knowledge audit [26–28] and knowledge maps [12–18].

Knowledge audit enables the identification and collection of information and knowledge resources inside organizations, information and knowledge requirements, organizational processes and structures through which knowledge could be managed [27, 29].

Knowledge maps are a way to represent the intellectual capital in an organization [12, 14–16, 18]. Which content should have these organizational maps depends on where they must be oriented. Besides, there are different ways to represent this knowledge and also different software tools to create maps and support its visualization [13, 14, 16]. To elaborate knowledge maps, the processes in which knowledge is needed and created are an important topic [4, 16].

More concretely, regional knowledge maps could collect: (a) the territorial knowledge created by the different actors involved in a region (authorities, clusters, companies, universities, NGO, etc.), (b) territory's structures and (c) infrastructures. They could also collect the information's fluxes between those actors. Their analysis contributes to identify the strengths and weaknesses at national or regional level, to produce new insight about territory's capabilities for regional stakeholders (industry, academia and civil society) to identify new areas of applied research, how to promote industrial leadership in a region, and how to coordinate and integrate research agendas and actions. Consequently, they represent a powerful tool for the strategic management of information for regional policy, and could help in a local way the decision-making needed for a consistent and coordinate implementation of regional policy in order to optimize resources.

Although KM as discipline has pointed out the role and value of knowledge maps, it offers little and disperse research about how to do knowledge maps [12, 14–18]. On the other hand, although much research about territorial knowledge and innovation systems exists, the body of research how to build regional knowledge maps is – as said before – not only disperse but scarce. Hence, it is a research field that needs more empirical works.

3.2   Competitive and Territorial Intelligence

Competitive Intelligence (CI) until the present has developed a framework to manage strategic information to generate intelligence in the organizations scale. This framework is based on a cycle, which considers the following phases: identification of information's needs, its acquisition, its organization, its analysis, and knowledge/intelligence creation [30].

In the past 20 years some authors have translated competitive intelligence to the territory, and a new research field has emerged named *territorial intelligence (TI)*. Nevertheless studies in the territory scale are incipient [31].

This research field explores the territory possibilities by the collection of information and its treatment in order to detect strengths, opportunities, weakness and threats to the territorial development [32, 33]. To collect these insights and watch the territory some authors have developed tools such as Catalyse focused in three topics: qualitative and quantitative diagnosis to identify people's needs, resources available on the territory and territorial indicators to measure the impact of the services upon the territory and the territorial community [32, 34]. This tool could act as the instrument to audit knowledge in a territory. Besides, territorial knowledge map could be the tool to collect and present the knowledge existing o needing in a territory.

3.3 Text Mining and Analytics

Text Mining and *analytics* offers the tools to exploit the wide quantity of unstructured data to extract information, which once analyzed becomes information and knowledge. Text mining enables the automatic analysis of large amounts of text from the internet (companies, institutes, governments, etc.) [35, 36]. Using techniques like *named entity recognition* [37] and topic detection [38] it is possible to find the main (innovative) products an industry in a region is working on, and not only its existing relations, but the missing relations, too.

Text mining is used for business intelligence, for KM within companies and for studying customer behavior and satisfaction [39], and to monitor new developments in a certain field of technology [40]. Text mining is also used to find experts and networks of experts [41].

The application of text mining to regional intelligence to systematically mine the expertise and innovative potentials of all companies in a region is new and will have to deal with a host of challenges concerning the interaction of crawling and analyzing web content [42].

3.4 National and Regional Innovation Systems

National or *Regional Innovation Systems* (*NIS* or *RIS*) has been defined in different ways but all have in common the study of which components in a region yield innovation. A group of actors emphasized the importance of the regional institutions, and others focalized in knowledge [2]. Nelson [43] defines NIS as a set of institutions whose interactions determine the innovative performance of national firms. To Lundvall [8] NIS is constituted by elements and relationships, which interact in the production, diffusion and use of new, and economically useful, knowledge. David and Foray [44] agree that an efficient system of distribution and access to knowledge contribute to increase the amount of innovative opportunities. Consequently, regional factors can influence the innovative capacity of firms [45], and a process of learning influences innovation. In this process it is important the knowledge's flow between actors, structures and infrastructures involved in a region [7, 46, 47].

Innovation Systems (national, regional o sectorial) gather actors that are working in the same area to create and use knowledge with economic aims [8, 43, 48–50]. This organizational formula enables joint efforts, optimizes resources and defines innovation politic. These systems take different forms such as technologic parks or clusters [46, 47, 49, 50, 51].

Hence, NIS and RIS offer a framework to identify which kind of actors, resources, social capital, structures, institutions and organizations are involved in the diffusion of new technologies and infrastructures in a geographical area to support innovation and regional development [7, 46–50, 52–54]. Consequently, it is a relevant part of our theoretical framework.

3.5 Theoretical Framework

Taking into account the above-mentioned backgrounds, we have constructed our theoretical framework to cope with the aim of this research: design a methodology to build knowledge maps. This framework deals with the topics that can lead or promote innovation and development in a region. These topics are grouped in two blocs: (1) components involved in a regional innovation systems that should be translated into a knowledge map; (2) context in which interactions of these components happen to generate knowledge and innovation, including the context offered by knowledge cycle (Fig. 1).

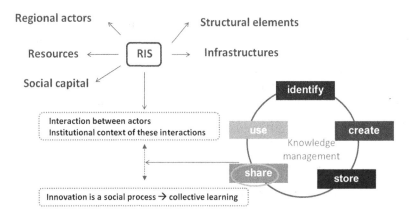

Fig. 1. Parts of a regional innovation system.

Concerning to the first bloc, we consider the following components: regional actors, resources, structural elements, infrastructures and social capital.

- Firstly, we must identify which are the *regional actors* in a region, who have been identified in previous studies: public and private sectors such as, teachers and secondary or higher education institutions, researchers and their research centers, employees and their companies (large or SME) and social spheres [46, 48, 55–57].
- The second element to study is the *structural elements* related to: (a) economy (kind of sectors or firms' size, clusters), (b) knowledge (disciplines, locations), (c) politic and administration (parties, parliament, local government, ministries and associations such as trade unions or Chambers of Commerce) [56, 57].
- Thirdly we focus on *infrastructures* related to different areas: (a) technique (water, wastewater, energy and transport), (b) social (health, education, and culture system), (c) finance (banking sector, venture capital and government aid), (d) knowledge infrastructures that support regional innovation such as research and higher education institutions, (e) policies of government and local authorities, and (f) science and innovation systems [50, 55, 57].

- The next element to be considered is the *resources* existing in a territory that are related to: (a) Nature (geology, water, air, climate, accumulation capacity), (b) raw materials (minerals, vegetarian, animal), (c) demography (inhabitants, structure), (d) regional innovative capabilities and intangible assets (intellectual capital that include human capital, structural capital and relational capital) [47, 57].
- Lastly, as innovation is a social process through which organizations learn [47], we incorporate the *social capital* as a component to be considered in the knowledge map. These components act as a link between the two blocs of the framework above mentioned. Social capital refers to which social relationships and in which context takes place to generate knowledge as basis to innovation. More concretely, this framework gives insights about which are the institutional collaborations (universities – government – industry) [58], and the different dimensions that could influence the social capital, and then the knowledge creation. These dimensions are: *structural* dimension (networks ties, network configuration, and appropriable organization); *cognitive* dimension (shared codes and languages and shared narratives); and *relational* dimensions (trust, norms, obligations identification) [59].

The second bloc of our framework refers to the context in which knowledge is created, and innovation could emerge in the interactions between actors. We take into account that innovation is a social process where collective learning increases knowledge. To study this context we consider the three dimensions of social capital above-mentioned and the knowledge cycle. On one side, the identification and analysis of the structural, cognitive and relational dimensions of the social capital can influence the innovation potentials in a region and the obstacles hindering innovation. On the other side, the knowledge cycle give detailed insights about the context (actors, structures, infrastructures and social capital) in which knowledge is identified, created, stored, shared and used.

4 Challenges

When trying to build a regional knowledge map we face a number of difficulties. The most important problems are (1) the lack of clear definition of regional knowledge maps and a methodology for their development, and (2) the wide variety of topics to be considered, which require identifying the adequate information sources, dealing with different kinds of formats and structures, and different methods to design the data collection, sampling, and data analysis. Other challenges are consequence of these central problems.

More specifically, we find challenges in the following areas:

- methodology to build knowledge maps: how audit the knowledge of an organization to identify the components and which structure employ,
- identifying elements from a RIS (actors, structures, infrastructures and resources) and their information sources, taking into account the specificities of each region and the different languages of these sources, specially in bilingual territories,
- discovering social capital and social networks in the region between actors, and
- data analysis process.

4.1   Methodology

The few works dedicated to knowledge maps have developed techniques oriented to: (a) capture explicit and tacit knowledge, (b) analyze knowledge areas in organizations, (c) identify through which organization's information sources the organizational knowledge can be captured, (d) illustrate how knowledge flows throughout an organization [16], (e) which structure and formats are more convenient to represent knowledge maps [13, 60]; and (f) which software could help to construct knowledge maps [13] and regional knowledge maps [11].

Nevertheless, the typology of the sources to consider, and the actors and processes developed in an organization are different in front of those elements studied in regional studies. Besides, the methods used to collect and extract information to draw knowledge maps often are qualitative.

Thus, new ways to develop knowledge maps must be explored. At this moment we have established four working lines:

- Identify the functions in which the processes and activities of a territory take place and that are de basis to their management. Examples of these functions are: education, health, or commodities production. This framework can help to identify the components of a knowledge map above mentioned.
- Find the information sources of these functions, processes, and maps' components, exploring the functionalities of Catalyse.
- Explore methodologies of social networks analysis to visualize interactions between actors, in order to detect for instance underuse of knowledge, lacks of knowledge, innovation potentials, enablers or inhibitors.
- Planning the application of text mining tools, keywords extraction and which controlled vocabularies are more convenient to mine the information sources located, according to their characteristics.

4.2   Identifying Elements from an RIS

Identifying the elements included in an RIS and their sources is another challenge to this project. On the one hand side, the vast amount of actors in a reasonable region that can easily exceed the number of 100,000, and the diversity of the information sources, make it to a real challenge to achieve completeness and an equable granularity of information. Especially those actors that are either very small, very new or not well connected in networks and/or the Web will be difficult to identify. The same happens with structures and institutions, because they are different between regions, and also they could lack presence on Internet. Moreover, there are hardly any lists available of all actors in a region. Since websites of companies not necessarily link to other companies in the same region, crawling the web by following links is not an efficient option. Thus information hubs for a region have to be found that link to, or mention the most important actors in a region. These information hubs might serve as a starting point for crawling information. Potential information hubs include chambers of commerce, business associations, business networks etc. Also printed lists from such organizations and from local governments constitute useful sources of information.

Despite the utility of these sources, a previous treatment is required, in order to test information's accuracy and reliability.

More fundamentally, we have to define what the actors in a region are that can contribute to innovate. According to the theoretical framework above described, these actors come both from the public and private area. So actors could be administrative departments and politic structures but also companies. These last kind of organizations are structured across borders of regions. Companies might be present in a region but only with a small part of their activities. It can be very hard to identify the role a company plays in a region. Public available sources of information are not designed to reveal these structures, but are usually consumer oriented and give only addresses of headquarters and sales offices. Finally, each region has a lot of actors that might be uninteresting for a knowledge map for regional innovation. Each village will have a bakery, a plumber etc. that are uninteresting for the final picture of the innovative potentials of a region. Nevertheless, also a bakery or a plumber can be an innovative company and advance to an important regional or supra regional actor. Thus the decision, which actors to include in a regional knowledge map and which not, is a further challenge. It could be useful establish an operative definition of what innovation is and their indicators, adapted to each territory in accordance to their environment, strengths, opportunities, threatens and weakness. In this sense, we can take as reference the frameworks offered by internationals organizations such as European Union and the Organization for Economic Co-operation and Development (OECD), and more specifically the innovation system to which a territory is attached, if that exists.

The identification of social capital and social networks between actors and their influence in innovation process is another challenge. However, this discipline in the last years has advanced and methodologies and software tools for social network analysis have become available. Hence, we should explore the potentials of this discipline to develop knowledge maps.

4.3   Analysis

Finally, during the analyzing process the text and link mining procedures [61] have as challenge to find the right tradeoff between manually performed work and machine based automatisms. While manual work may achieve higher quality of results, only automated text mining methods will be able to process the vast amount of information sources. Using information crawled form websites of companies, preliminary results show that it is feasible to find addresses, phone numbers, etc. by *named entity detection* on the crawled texts. For this purpose regular expressions for these entities were defined and the number of occurrences of entities on the web pages is counted. The most frequently found address and phone number usually indeed present the company's main contact information. Also, we have very encouraging results with respect to the classification of the main activity of a company using their Web presence. For this purpose we used the main economic sectors of the STW Thesaurus for Economics [62, 63]. Others controlled vocabularies to explore their potentials are: NACE of the European Union (Classification of Economic Activities in the European Community), ISIC (International Standard Industrial Classification of all economic activities) of the

United Nations, CPA (PA Statistical Classification of Products by Activity) of the Eurostat, IPC (International Patent Classification of the WIPO (World Intellectual Property Organization), or ICB (Industry Classification Benchmark) of the FTSE Group (http://www.icbenchmark.com/).

On the other hand it turns out to be much more challenging to find relations between companies or to find products or services made by a company. The main problem is the diversity of the resources and of the type of actors. Again a clear definition of actors and a missing correspondence between actors and information resources poses a major problem. Large companies and institutions, like universities, cannot be seen as single actors that have one main activity. Rather we have a complex and partly obscure hierarchy of actors for which it is already difficult to identify at which level in the hierarchy the activities should be classified.

5   Conclusions

The design of regional knowledge maps emerges as an important topic of research because their content could enable the strategic planning and decision-making process in regional policy. It could give detailed insights about the knowledge state of a territory, for instance which of knowledge lacks or which knowledge is produced but underused, or which knowledge could be exported. Regional knowledge maps could also enable the relations between actors and the collective learning as a basis to the innovation.

Their design requires capturing all the information about the regional competencies composed by infrastructures, structures, resources, actors, knowledge and social capital. The collection and the representation of these elements present a number of challenges that must be resolved.

The challenges could be faced with the conjunction of different disciplines to define a methodology. The theoretical corpus of knowledge management as discipline offers a basis to develop this methodology, so in the future this topic could be one more to be considered with the alliance of other areas of research: text mining, regional innovation systems and competitive and territorial intelligence.

This work has the will to help starting this new line of research. The challenges presented are the result of the first phase of an exploratory research project. On the one hand the project focuses on the identification of information sources related to a region, from which we extract information as a first approach to the objective. On the other hand we identify which tools facilitate extracting information of these sources, according to the theoretical framework designed. During this phase we delimit the search to some specific elements of the RIS in a small geographical area, concretely the region of Hanover (Germany). Finally, we will validate this methodology applying that to other regions.

Acknowledgements. This research is funded by the Spanish Ministry of Education, Culture and Sport (Ref. CAS 12/00155).

References

1. Lisbon European Council. Presidency conclusions. European Parliament (2000). http://www.europarl.europa.eu/summits/lis1_en.htm
2. Godin, B.: The knowledge-based economy: conceptual framework or buzzword? J. Technol. Transf. **31**(1), 17–30 (2006)
3. Diakoulakis, I.E., Georgopoulos, N.B., Koulouriotis, D.E., Emiris, D.M.: Towards a holistic knowledge management model. J. Knowl. Manage. **8**(1), 32–46 (2005)
4. CEN (European Committee for Standardization – Comité Européen de Normalisation – Europäisches Komitee für Normung). European Guide to good Practice in Knowledge Management - Part 1: Knowledge Management Framework. CWA 14924-1:2004. Brussels (2004)
5. Godin, B.: The linear model of innovation: the historicalconstruction of an analytical framework. Sci. Technol. Human Values **31**(6), 639–667 (2006)
6. OECD (1996). *The Knowledge-Based Economy.* OECD, Paris.5 OECD (1996)
7. Asheim, B., Coenen, L.: Knowledge bases and regional innovation systems: comparing Nordic clusters. Res. Policy **34**(8), 1173–1190 (2005)
8. Lundvall, B.-A. (ed.): National Systems of Innovation: Towards a Theory of Innovation and Interactive Learning. Pinter, London (1992)
9. Barinani, A., Agard, B., Beaudry, C.: Competence maps using agglomerative hierarchical clustering. J. Intell. Manuf. **24**(2), 373–384 (2013)
10. España (2011). Proposición no de Ley presentada por el Grupo Parlamentario Popular en el Congreso, relativa al desarrollo de un mapa de conocimiento. *Boletín Oficial de las Cortes Generales. Congreso de los Diputados.* IX Legislatura, Serie D: General, 29 de abril de 2011, no. 563, p. 12 (2011)
11. Plazas Tenorio, A. (Coord.) Desarrollo de un mapa de conocimiento como aporte a la consolidación del sistema regional de innovación del Cauca. Informe final. Colciencia. CREPIC N° 496 (2004)
12. Watthananon, J., Mingkhwan, A.: Optimizing knowledge management using knowledge map. Procedia Eng. **32**, 1169–1177 (2012)
13. Eppler, M.J., Burkhard, R.M.: Visual representations in knowledge management: framework and cases. J. Knowl. Manage. **11**(4), 112–122 (2007)
14. Driessen, S., Huijsen, W.O., Grootveld, M.: A framework for evaluating knowledge-mapping tools. J. Knowl. Manage. **11**(2), 109–117 (2007)
15. Huijsen, W., Van Vliet, H., Plessius, H.: Picture this: mapping knowledge in higher education organizations. In: Proceedings EISTA 2004, pp. 429–434, Orlando, FL (2004)
16. Kim, S., Suh, E., Hwang, H.: Building the knowledge map: an industrial case study. J. Knowl. Manage. **7**(2), 34–45 (2003)
17. Eppler, M.J.: Making knowledge visible through intranet knowledge maps: concepts, elements, cases. In: Proceedings of the 34th Hawaii International Conference on System Sciences, vol. 4, pp. 4030 (2001)
18. Wexler, M.N.: The who, what and why of knowledge mapping. J. Knowl. Manage. **5**(3), 249–263 (2001)
19. Raghu, T.S., Vinze, A.: A business process context for knowledge management. Decis. Support Syst. **43**(3), 1062–1079 (2007)
20. Ergazakis, K., Karnezis, K., Metaxiotos, K., Psarras, I.: Knowledge management in enterprises: a research agenda. Intell. Syst. Account. Finance Manage. **13**(1), 17–26 (2005)
21. Bollinger, A.S., Smith, R.D.: Managing organizational knowledge as a strategic asset. J. Knowl. Manage. **5**(1), 8–18 (2001)

22. Beijerse, R.P.: Questions in knowledge management: defining and conceptualising a phenomenon. J. Knowl. Manage. **3**(2), 94–110 (1999)

23. Danskin, P., Englis, B.G., Solomon, M.R., Goldsmith, M., Davey, J.: Knowledge management as competitive advantage: lessons from the textile and apparel value chain. J. Knowl. Manage. **9**(2), 91–102 (2005)

24. Serneko, A., Bontis, N., Booker, L., Sadeddin, K., Hardie, T.: A scientometric analysis of knowledge management and intellectual capital academic literature (1994–2008). J. Knowl. Manage. **14**(1), 3–23 (2010)

25. Echevarria, L., Garcia-Alsina, M., Vélez, J., Barrios, C.: Contribución de la tecnología en la gestión del conocimiento entre los grupos de investigación del área de informática. Puente **6** (2), 21–28 (2012)

26. Drus, S.M., Shariff, S.S.: Analysis of knowledge audit models via life cycle approach. In: International Conference on Information Communication and Management IPCSIT, vol, 16, pp. 176–180 (2011)

27. Levantakis, T., Helms, R., Spruit, M.R.: Developing a reference method for knowledge auditing. In: Yamaguchi, T. (ed.) PAKM 2008. LNCS (LNAI), vol. 5345, pp. 147–159. Springer, Heidelberg (2008)

28. Perez-Soltero, A., Barcelo-Valenzuela, M., Sanchez-Schmitz, G., Martin-Rubio, F., Palma-Mendez, J.T., Vanti, A.A.: A model and methodology to knowledge auditing considering core processes. ICFAI J. Knowl. Manage. **5**(1), 7–23 (2007)

29. Liebowitz, J., Rubenstein-Montano, B., McCaw, D., Buchwalter, J., Browing, C.: The knowledge audit. Knowl. Process Manage. **7**(1), 3–10 (2000)

30. Garcia-Alsina, M., Ortoll, E.: La Inteligencia Competitiva: evolución histórica y fundamentos teóricos. Trea, Gijón (2012)

31. CAENTI (2012). Territorial Intelligence portal. http://www.territorial-intelligence.eu/. Consulted: 16 May 2013

32. Girardot, J.-J.: Evolution of the concept of territorial intelligence within the coordination action of the european network of territorial intelligence. Ricerca e Sviluppo per le politiche Sociali **1**(1–2), 11–29 (2008)

33. Herbaux, P.: Tools for territorial intelligence and generic scientific methods. In: International Annual Conference on Territorial Intelligence, Besançon, 16 October 2008

34. Masselot, C.: Information territoriale: une construction collective nécessaire. In: Les Cahiers de la SFSIC, pp. 71–76, Bordeaux (2011)

35. Kosala, R., Blockeel, H.: Web mining research: a survey. ACM SIGKDD Explor. Newsl. **2** (1), 1–15 (2000)

36. Moens, M.-F. (ed.): Information Extraction: Algorithms and Prospects in a Retrieval Context. Springer, Dordrecht (2006)

37. Nadeau, D., Sekine, S.: A survey of named entity recognition and classification. Lingvisticae Investig. **30**(1), 3–26 (2007)

38. Wartena, C., Brussee, R.: Topic detection by clustering keywords. In: 19th International Workshop on. IEEE Database and Expert Systems Application, DEXA 2008, pp. 54–58 (2008)

39. Saggion, H., Funk, A., Maynard, D., Bontcheva, K.: Ontology-based information extraction for business applications. In: Proceedings of the 6th International Semantic Web Conference (ISWC 2007), Busan, Korea, November 2007

40. Färber, M., Rettinger, A.: A semantic wiki for novelty search on documents. In: DIR, Delft, 26 April (2013)

41. Ehrlich, K., Lin, C. Y., Griffiths-Fisher, V.: Searching for experts in the enterprise: combining text and social network analysis. In: Proceedings of the 2007 international ACM conference on Supporting group work, pp. 117–126 (2007)

42. Wartena, C., Garcia-Alsina, M.: Challenges and potentials for keyword extraction from company websites for the development of regional knowledge maps. In: Fifth International Conference on Knowledge Management and Information Sharing (KMIS 2013) (2013)
43. Nelson, R.R. (ed.): National Innovation Systems: A Comparative Study. Oxford University Press, Oxford (1993)
44. David, P., Foray, D.: Assessing and expanding the science and technology knowledge base. STI Rev. **14**, 13–68 (1995)
45. de Oslo, M.: Guía para la recogida e interpretación de datos sobre innovación. OECD, Francia (2005)
46. Jimenez, F., Fernández, I., Menéndez, A.: Los Sistemas Regionales de Innovación: revisión conceptual e implicaciones en América Latina. In: Listerry, J.J., Pietrobelli, C. (eds.) Los Sistemas Regionales de Innovación en América Latina. Banco Interamericano de Desarrollo, Washington (2011)
47. Doloreux, D., Parto, S.: Regional innovation systems: current discourse and unresolved issues. Technol. Soc. **27**, 133–153 (2005)
48. Sharif, N.: Emergence and development of the national innovation systems concept. Res. Policy **35**, 745–766 (2006)
49. Cooke, P., Gómez, M., Etxebarria, G.: Regional innovation systems: Institutional and organisational dimensions. Res. Policy **26**(4–5), 475–491 (1997)
50. Edquist, C. (ed.): Systems of Innovation: Technologies, Institutions and Organizations. Pinter, London (1997)
51. Asheim, B.: Next generation regional innovation policy: how to combine science and user driven approaches in regional innovation systems. EKonomiaz **70**(1), 26–43 (2009)
52. Andersson, G.: Rethinking Regional Innovation. Syst. Pract. Action Res. **26**(1), 99–111 (2013)
53. Chen, K., Guan, J.: Mapping the functionality of China's regional innovation system: a structural approach. China Econ. Rev. **22**(1), 11–27 (2011)
54. Cooke, P.: Regional innovation systems, clusters and te knowledge economy. Ind. Corp. Change **10**, 945–974 (2001)
55. Normann, R.: Regional Leadership: A Systemic View. Syst. Pract. Action Res. **26**(1), 23–38 (2013)
56. Fröhlich, K.: Innovationssysteme der TV: Unterhaltungsproduktion: Komparative Analyse Deutschlands und Großbritanniens. VS Verlag für Sozialwissenschaften, Wiesbaden (2009)
57. Voß, R. (ed.): Regionale Innovationssysteme. News & Media, Berlin (2002)
58. Etzkowitz, H., Leydesdorff, L.: The triple Helix: University-industry-government relations: a laboratory for knowledge based economic development. EASST Rev. **14**, 14–19 (1995)
59. Nahapiet, J., Ghoshal, S.: Social capital, intellectual capital, and the organizational advantage. Acad. Manage. Rev. **23**(2), 242–266 (1998)
60. Eppler, M.J.: A process-based classification of knowledge maps and application examples. Knowl. Process Manage. **15**(1), 59–71 (2008)
61. Lieberam-Schmidt, S.: Analyzing and Influencing Search Engine Results. Springer, Wiesbaden (2010)
62. Gastmeyer, M. (ed.): Standard-Thesaurus Wirtschaft. Deutsche Zentralbibliothek für Wirtschaftswissenschaften, Kiel (1998)
63. Neubert, J.: Bringing the "Thesaurus for Economics" on to the web of linked data. In: Proceedings of the Linked Data on the Web Workshop (LDOW2009) (2009)

Exposition of Internal Factors Enhancing Creativity and Knowledge Creation

Lina Girdauskiene[✉] and Asta Savaneviciene

Department of Management, Kaunas University of Technology,
Kaunas, Lithuania
{lina.girdauskiene,asta.savaneviciene}@ktu.lt

Abstract. A creative organization is distinguished among traditional organizations. Creative organizations are trying to strike a balance between business and creativity when creating a proper environment for knowledge management, at the same time, encourages employees to develop creative work and produce new products that meet market needs. Thus, the research purpose is defined as follows: what features of internal factors do influence creativity and knowledge creation in a creative organization? A qualitative research method, based on a scientific analysis and identification of the key factors, allowed reveal the features of a task, group and time influencing creativity and knowledge creation in a creative organization. The research results show that different features of tasks, group and time make an impact on different employee groups, types of creativity and knowledge in a creative organization.

Keywords: Creativity · Knowledge creation · Task · Group · Time · Creative organization

1 Introduction

Creativity is a core factor for each organization seeking to maintain a competitive advantage and successful development. A creative organization is distinguished among traditional organizations. It has unique projects, which leads to high staff turnover, especially among creators. It is characterized by individual artistic creativity, which is transformed into production and products. Therefore, creative organizations are trying to strike a balance between business and creativity when creating a creative environment for knowledge, at the same time, encourages employees to develop creative work and produce new products that meet market needs. The creative duality leads to the natural need of organization's specific business management to ensure the two parallel processes of the organization, consistent with each other - individual creativity and empowerment of the creativity.

During the recent decade creative industry and creative organization as a research topic is relevant. Scientists pay a lot of attention to the genesis of creative industry, identification of performances and various management issues [10–12, 18, 20, 22, 23, 26, 28, 33], concept of creativity and formation of creative environment [2, 3, 5–9, 17, 20, 21].

Creative organization in the context of this topic is fragmented, and touches only the concept of knowledge bases [12, 16], artists and administrative staff training issues

© Springer-Verlag Berlin Heidelberg 2015
A. Fred et al. (Eds.): IC3K 2013, CCIS 454, pp. 400–410, 2015.
DOI: 10.1007/978-3-662-46549-3_26

(cultural sociologists), creativity and creative process management techniques and methodologies (educational psychology) and innovation issues [12], but still lacks researches how creative organizations remain creative and innovative [30].

2 Internal Factors, Creativity and Knowledge Creation in a Creative Organization

2.1 The Concept of a Creative Organization

A creative organization reflects the conceptual and individual talent and large production convergence by new media technologies (ICT) in knowledge-based economy. This organization is unique, because it attempts to strike the balance between production and artistic creativity. The project based organizational structure is identified as the most common type in a creative organization [15], which allows to justify creative organization's specificity: to experiment constantly by creating new products and forming new groups.

Two employees' types could be identified in the creative organization-administrators and creators. Administrators mostly belong to permanent employees' group, who are responsible for managerial, administrative and economic issues. Although creators produce artistic products or services, thus adding value to the organization and ensuring a competitive advantage, they often migrate among groups, projects or even external organizations. It could be argued that various experience, rotation and movement from one to other projects extend employees' competence and uphold their creativity.

2.2 Creativity and Knowledge Creation

Creativity is a base for knowledge creation. Usually creativity is defined as the production of novel, useful ideas or problem solutions. Creativity and its resulted knowledge creation keep the key position in a creative organization theory. All components of creative organizations are creative: creative process, products and employees, as well as work environment and work culture, even the first word of the title is directly related to creativity [2, 3, 6–8, 17]. Duality of creativity is expressed through creativity in the creative content of organizations (arts and culture in the traditional sense), and creativity as a competitive economic base.

Competence of creative employees results successful performance of creative organization. It consists of knowledge, abilities, skills, talent and other personal features and can be directly affected by the balance among skills, abilities and complexity of tasks. Amabile [4] determines three main components of creativity: expertise, motivation and creative thinking skills. Other scientists [7, 12, 30], analyze interaction between individual and organization. It is stated that special abilities of creative employees can be developed by learning or by setting proper environmental conditions.

Rahimi et al. [27] state that creativity is a result of the combination of existing knowledge and new knowledge. Scientists [12, 30], define three types of creativity:

- Analyzing;
- Changing;
- Combining.

Very often, during the creative process all types of creativity are assimilated – already known ideas are interconnected in a new context, as well as new context is studied, in which the adaptation of new ideas is applied, or existing system is changed.

Knowledge creation is considered as the four modes of knowledge conversion of model by Nonaka [24, 25]: socialization, externalization, internalization and combination, where these modes of knowledge converse from explicit to tacit. A broad range of factors that can influence the success of knowledge creation has been mentioned in the scientific literature. Wong [32] proposed summarized key factors: management, leadership and support, culture, IT, strategy and purpose, measurement, organisational infrastructure, processes and activities, motivational aids, resources, training and education, HRM. Organizational components as task, group and time are one of the most effectively affecting creativity [1, 3], so it is important to investigate how do they influence creativity in a creative organization.

2.3   Role of the Task, Group and Time in a Creative Organization

Flexibility of the task has received considerable research attention and empirical support as an important situational factor that could influence human creativity [29]. The task is one of the essential dimensions influencing the potential of organization creativity, because it creates the conditions for employee to satisfy their ambitions and self-realize. Correct identification of task specification and characteristics positively impacts an organization and its performance. As creative organizations are innovative and based on each time a new assessment, and often - difficult tasks, standardization becomes a relative concept. However, some degree of standardization of tasks, however, is possible. In this case, these tasks become routine. They can be monitored, regulated and controlled due to its predictable structure: their goals are specific and experience is embedded in the behaviour of employees. These tasks may be controlled of the mid-level managers. Task execution procedure can be transferred (repeated). In such a way the evaluation of a result becomes possible. Creators who perform standardized - routine tasks are more suitable for centralized control, because in this case it is necessary to evaluate the implementation of the objectives and tasks requiring less expertise through self-knowledge and ideas. In addition, the right of decision making should be controlled, and the application of knowledge and ideas should be limited. The opposite situation is with unique or the new tasks. Control of these tasks must be carried out only by top-level managers, because the process is unfamiliar, goals are abstract, creativity is competence based, and there is no experience of executing that task or, at best, not at the organization.

Thus, the standardization of a task is hardly possible in the creative environment. This causes problems of management and coordination. The task specification can directly affect both positively and negatively the organization of creators. Task novelty and complexity results two-fold result of the administrators and creators aspect –it is more complicated for the leaders and more interesting creators to perform this type of task.

Permanent change of tasks, groups, the nature of the tasks (new and complex) leads to limited resources for accomplishment of those tasks. Both creators and administrators, are forced to perform at the same time for a several tasks or they are given too little time to complete the task. This time limitation especially affects creativity. Time as a factor of making creativity-friendly environment becomes very significant and important in order to create a favourable environment for ideas and knowledge creation. Time can influence (positively or negatively) creativity differently: too less time results stress and decrease creativity, on the other hand it concentrates and may increase creativity. Unsworth et al. [31] detected that time demands were positively related to creativity. Wagner (2003) states that characteristics of the working groups, as group size, the degree of harmony and composition, its members' expertise and skill distribution of suitable conditions for the development of creativity and to create and manage knowledge should be considered as a significant criteria, especially how the principles of teamwork improves the microclimate in the organization. Goncalo and Staw [14] state that groups might be more creative than individuals. George [13] suggests that groups composed of diverse members should be more creative than more homogenous groups because they presumably can call upon a greater diversity of knowledge, skills, expertise, and perspectives to generate new and useful ideas [19].

Analyzing characteristics of task encouraging creativity and creation of new knowledge, new and complex tasks creates a potential breeding ground for new ideas and the emergence of knowledge. They become a challenge for creators. Based on the above analysis, the main features of task, group and time are defined:

- Task characteristics: short/long-lasting, clear/uncertain, routine/new, simple/complex.
- Group characteristics: size, integrated/free, group harmony degree, heterogeneous/homogeneous, chemistry of a group, knowledge, skills and composition, approval/objection existing assumptions.
- Time characteristics: the number of different tasks, time properties (a little/a lot of/fragmented/concentrated), job autonomy (full autonomy/narrowly defined objectives).

Hemlin et al. [17] stated, that generally short product lifecycle projects due to constantly changing nature of the task (short/long-lasting, easy/difficult, routine/new, modulated/integrated), the project group composition (size, integrated/free, group harmony degree, heterogeneous/homogenous participants, persons, group harmony, the knowledge, skills and abilities composition consent/objection to existing assumptions), subculture, leadership (transactional, transformational) and the time allocated to the task characteristics (different number of work tasks, time characteristics (few/many, fragmented/concentrated) and work autonomy (full autonomy/narrowly defined objectives) enhace creativity.

The different composition of dimension changes the nature of the task and thus requires different provisions establishing the knowledge creation. The most appropriate strategy for knowledge creation could be implemented through the empowerment and training in routine and non-specific, unrelated tasks aspect. Tasks of administrators often are related, but remain routine and non-specific, so the periodic procedures are proposed. The most appropriate strategy for knowledge creation of creators who work with routine, specific tasks, not connected with each other, would be the balance design of expertise and creativity, when tasks are interrelated - the main provision of the implementation of knowledge creation techniques - through cooperation, informal meetings, practice communities. Then the staff having extensive networking relationships and contacts, use the whole network of knowledge, can faster solve organizational problems and create new knowledge [7, 25].

2.4 Methodology

The qualitative research enabling to reveal the key factors for creativity implementation and knowledge creation was conducted in January of 2012. As a proper source of information for the research TV production organization was selected. 6 respondents, satisfying settled criteria, were tested. The characteristics of respondents are presented in Table 1 below.

Table 1. Characteristics of respondents.

Code	Work position	Work experience	Group
1.	Project manager	20	Administrator
2.	Project manager	9	Administrator
3.	Journalist	17	Creator
4.	Post production director	10	Creator
5.	Director	30	Creator
6.	CEO	22	Administrator

The depth interview as a method of a qualitative research was selected due to organizational issues, uncertainty of the research object and respondents which subject is their responsibility (Table 2).

Table 2. Characteristics of in depth interview.

Code	Interview date	Time	
		Explanatory time, min	Interview time, min
1.	2012 01 09	27	60
2.	2012 01 09	29	120
3.	2012 01 10	24	50
4.	2012 01 10	25	70
5.	2012 01 11	25	100
6.	2012 01 11	20	60

Analyzing the influence of different factors (task, group and time) two questions were raised:

- What types of factor do make an impact on creativity and knowledge creation?
- How do these factors affect creativity and knowledge creation?

Evaluating the impact of the factors on creativity and knowledge creation 3 types of affect were detected: zero (0)- neither negative, nor positive affect; minus (−)- negative affect and positive (+) affect.

3 Results

3.1 Influence of a Task on Creativity and Knowledge Creation

The results of empirical research show the distribution of factors in two groups of employee– administrators and creators. The results were grouped by the criteria of different knowledge types – explicit and tacit, are presented below. It could be stated that tasks are quite similar in two different employee groups – they are routine, simple, clear and additionally complex for creators. The main difference is between creation of tacit and explicit knowledge – tasks usually are new, complex, uncertain and indefinite in a creative organization. It confirms the theoretical insights that tacit knowledge is created executing new uncertain tasks and using their creativity (Table 3).

Table 3. The key types of task for a knowledge creation.

Knowledge type	Administrator	Creator
	Task	
Explicit	Routine, simple, clear	Routine, simple, complex, clear
Tacit	New, complex, uncertain, indefinite	

Table 4 presents the interaction of task and creativity type in a creative organization. When the task is simple, routine, clear and certain, employees usually have to be even more creative and have to find new solutions for the same products or services. But on the other hand it is very convenient for administrators and new explicit knowledge creation. A little bit easier from creativity position is the situation when the task is new, complex, uncertain and indefinite – it is a positive area for creativity and creators. Here tacit knowledge is as usual created. Of course, uncertainty and un-clearness results more stress and tension, it can reduces the creativity or require more time for the same result.

Table 4. Task influence on creativity.

Task type	Creativity type		
	Analyzing	Combining	Changing
New, complex, unclear, indefinite	+	+	+
Routine, simple, clear, certain	+	+	+

Summarizing it could be stated that influence of different types of a task on creativity and knowledge creation is dual – it affects differently two employee groups and their knowledge and creativity in two different ways.

3.2   Group Influence on Creativity and Knowledge Creation

Characteristics of a group are significantly important as to creativity, knowledge creation as to microclimate, teamwork and all results of organization performance. Table 5 presents the key features of a group what influence creativity and knowledge creation. It can be stated that the two groups of employees did not formulate different requirements for tacit and explicit knowledge creation. While creators and administrators expressed the same preference level of the group (small), other characteristics of the group disagreed: administrators wanted to work in homogeneous, with the consent and knowledge, skills and composition, and administrators in heterogeneous conflicts with existing provisions of the existing groups.

Table 5. The key features of group for a knowledge creation.

Knowledge type	Administrator	Creator
	Group	
Explicit Tacit	Small, big, homogeneous, approval, chemistry, composition of knowledge and skills	Small, homogeneous, heterogeneous, approval, conflicts, chemistry

Table 6 presents the affect of different type of a group on creativity and knowledge creation.

Table 6. Group influence on creativity.

Group type	Creativity type		
	Analyzing	Combining	Changing
Small	+	+	+
Big	+	+	+
	−	−	−
Homogenous	−	−	−
	+	+	+
Heterogeneous	+	+	+
Chemistry	+	+	+
Approval	+	+	+
	−	−	−
Conflicts	−	−	−
Composition of knowledge, skill, experience	+	+	+

Big groups and approval of existing ideas or opinions affect creativity and knowledge creation in two ways – it can increase creativity when there are more ideas, experience, skills and knowledge and everybody approve presented items, but on the other hand it can be very difficult to communicate, cooperate and work together. Also, if everybody accept all ideas, there will be no balance of "a true view".

Conflicts affect creativity and knowledge creation negatively, because usually it is destroying process and does not result fruits. Summarizing it could be stated that influence of different types of a group on creativity and knowledge creation is dual – it affects differently two employee groups and their knowledge and creativity in two different ways.

Summarizing it could be stated that influence of different types of a group on creativity and knowledge creation is dual – it affects differently knowledge and creativity in two different ways.

3.3   Time Influence on Creativity and Knowledge Creation

Project based activity results new tasks, which differ in their different durations, level of complexity, clarity and content of tasks simultaneously. The task is one of the strongest catalyst of knowledge creation and creativity. Exciting and challenging tasks leads opportunities to self-realization, the creation of new ideas and results, innovative products or services (Table 7).

Table 7. The key factors of time for a knowledge creation.

Knowledge type	Administrator	Creator
	Time	
Explicit/tacit	Limited, enough, one task at the same moment, special time for a task	

Time is one of the factors that can affect negatively creativity and knowledge creation. If it is enough time the creativity will be increased. But the lack of a time, too many tasks at the same time will decrease analyzing creativity and knowledge creation.

On the other hand enough time is useful for a analyzing creativity, because a lot of researches can be implemented, but it is negatively connected with combining and changing creativity. Too less as well as not limited time make a negative impact.

Summarizing it could be stated that influence of different types of a time on creativity and knowledge creation is dual – it affects differently knowledge and creativity in two different ways (Table 8).

Table 8. Time influence on the creativity.

Time type	Creativity type		
	Analyzing	Combining	Changing
Limited	+	+	+
Non limited	–	–	–
Too less	–	–	–
Enough	+	–	–
One task at the same time	+	+	+
Several tasks at the same time	–	+	+
	–	–

4 Conclusions

Internal factors are essential for a creativity and knowledge creation in a creative organization. The research results show that the task, group and time are the core dimensions what enhance creativity and knowledge creation. They influence these processes accordingly different type of employees, knowledge and creativity. Administrators and creators willing to create tacit knowledge should be given the same tasks, while creators should be provided with routine, simple, complex and clear task in order to create explicit knowledge. Groups' characteristic is the same to for creation explicit and tacit knowledge in two different employee groups. And the time characteristics are the same for both employee groups and knowledge types. Finally assessing the impact of internal factors on a creativity it was detected that the task makes an positive impact on a all types of a creativity while the conflicts in a group usually make negative impact. The time limitation and abundance, several tasks at the same time negatively affects creativity as well.

Also some limitations of the survey could be defined:

- Just the main types of factors (task, group and time) were assessed. No detailed characteristics of each factors were investigated.
- It was a qualitative research. A quantitative research could present deeper insights, relations, connections among different types of factors, employees and creativity.
- The research was conducted in a specific industry – creative industry and in a limited geographical area – Lithuania. The repetition of the survey in different industries and countries could to validate results of this survey as well to reveal new insights.

Future research can be implemented evaluating the impact of the internal factors and the other stages of knowledge processes, assessing individual and organizational creativity and the impact of internal factors on them.

References

1. Adam, R., Clelland, J.: Individual and team-based idea generation within innovation management: organizational and research agendas. Eur. J. Innov. Manage. **5**(2), 86–97 (2002)
2. Afolabi, M.O., et al.: Are we there yet? A review of creativity methodologies. Predicting stock prices using a hybrid Kohonenself Organizingmap,1–8 (2007)
3. Amabile, T.M.: Discovering the unknowable, managing the unmanageable. In: Ford, C.M., Gioia, I.A. (eds.) Creative Actions in Organizations: Ivory Tower Visions & Real World Voices, pp. 77–81. Sage, London (1995)
4. Amabile, T.M.: How to kill creativity (1998). http://gwmoon.knu.ac.kr/Lecture_Library_Upload/HOW_TO_KILL_CREATIVITY.pdf
5. Akhavan, P., Jafari, M., Fathian, M.: Critical success factors of knowledge management systems: a multi-case analysis. Eur. Bus. Rev. **18**(2), 97–113 (2006)
6. Carnero, A.: How does knowledge management influence innovation and competitiveness. J. Knowl. Manage. **4**(2), 87–98 (2000)

7. Csikszentmihalyi, M.: Society, culture and person: a systems view of creativity. In: Sternberg, R.J. (ed.) The Nature of Creativity, pp. 325–339. Cambridge University Press, New York (1988)
8. Cross, R., Parker, A., Prusak, L.: Knowing what we know: Supporting knowledge creation and sharing in social networks. White Paper. IBM institute for knowledge management (2000)
9. Crosick, G.: Knowledge transfer without widgets: the challenge of the creative economy. Goldsmiths, University of London. A lecture to Royal Society of Arts (2006)
10. Evans, S.: Creative Clusters (2008). http://www.creativeclusters.com/
11. Flew, T.: Beyond ad hocery: defining creative industries. cultural cities, cultural theory, cultural policy. In: The 2nd International Conference on Cultural Policy Research, Te Papa, Wellingtn, New Zeland, (2002)
12. Florida, R.: The Rise of the Creative Class: And How It's Transforming Work, Leisure, Community and Everyday Life. Basic Books, New York (2002)
13. George, J.M.: Creativity in Organizations. Acad. Manage. Ann. 1, 439–477 (2007)
14. Goncalo, J.A., Staw, B.M.: Individualism-Collectivism and Group Creativity (2005). http://digitalcommons.ilr.cornell.edu/obpubs
15. Grabner, G.: The project ecology of advertising: task, talents and teams. Reg. Stud. 36(3), 245–262 (2002)
16. Hansen, K.H., Vang, J., Asheim, B.T.: The Creative Class and Regional Growth: Towards a Knowledge Based Approach (2005)
17. Hemlin, S., Allwood, C.A., Martin, B.A.: Creative Knowledge Environments: The Influences on Creativity in Research and Innovation. Edward Elgar, Aldershot and Brookfield, Vermont (2006)
18. Holzl, K.: Creative Industries in Europe and Asia. Definition and potentional (2006). Internetinėprieiga: www.kmuforschung.ac.at
19. Mannix, E., Neale, M.: What differences make a difference? The promise and reality of diverse teams in organizations. Psychol. Sci. Public Interest 6(2), 31–55 (2005). doi:10.1111/j.1529-1006.2005.00022.x
20. Markusen, A., Wassall, G.H., DeNatale, D., Cohen, R.: Defining the creative economy: industry and occupational approaches. Econ. Dev. Q. 22(1), 24–45 (2008)
21. Mayfield, J., Mayfield, M.: The creative environment's influence on intent to turnover. Manage. Res. News. 31(1), 41–56 (2008)
22. Miles, I., Green, L.: Hidden innovation in the creative industry. Research report, London, NESTA (2008)
23. Mueller, K., Rammer, C., Truby, J.: The Role of Creative Industries in Industrial Innovation. Discussion Paper No. 08-109 (2008)
24. Nonaka, I., Toyama, R., Nagata, A.: A Firm as a knowledge creating entity: a new perspective on the theory of the firm. Ind. Corp. Change 9(1), 1–20 (2000)
25. Nonaka, I.: Tacit knowledge and knowlegde conversion: controversy and advancement in organizational knowledge creation theory. Organ. Sci. 20(3), 635–652 (2009)
26. O'Connor, J.: The cultural and creative industries: the review of the literature (2007). Internetinė prieiga: www.creative-partnerships.com/literaturereviews
27. Rahimi, H., Arbabisarjou, A., Allameh, S., Aghababaei, A.: Relationship between knowledge management process and creativity among faculty members in the university. Interdisc. J. Inform. Knowl. Manage. 6, 1–17 (2011)
28. Rickards, T.: Creativity, knowledge production, and innovation studies: a response to Ghassib's, where does creativity fit into productivist industrial model of knowledge production?. Gifted and Talented international. J. World Counc. Gifted Talented Child. 25(1), 99 (2010)

29. Royyon, G.J., Sheenas, I.: Creativity as a matter of choice: prior experience and task instruction as boundary conditions for the positive effect of choice on creativity. J. Creative Behav. **42**, 164–180 (2008)

30. Sternberg, R.J., Grigorenko, E.L.: Guilford's structure of intellect model and model of creativity: contributions and limitations. Creativity Res. J. **3**, 309–316 (2000)

31. Unsworth, K.L., Wall, T.D., Carter, A.: Creative requirement: a neglected construct in the study of employee creativity? Group Org. Manage. **30**(5), 541–560 (2005)

32. Wong, K.Y.: Critical success factors for implementing knowledge management in small and medium enterprises. Ind. Manage. Data Syst. **105**, 261–279 (2005)

33. Wyszomirski, M.: From Public Support for the Arts to Cultural Policy (2004). doi:10.1111/j.1541-1338.2004.00089.x

Cloud Security Assessment: Practical Method for Organization's Assets Migration to the Cloud

Ronivon Costa[1(✉)] and Carlos Serrão[2]

[1] VANTIS, R. Rui Teles Palhinha 6 - 3ºG, 2740-278 Porto Salvo, Portugal
ronivon_costa@iscte.pt
[2] ISCTE-IUL/ADETTI-IUL, Av. das Forças Armadas,
1649-026 Lisbon, Portugal
carlos.serrao@iscte.pt

Abstract. New organizations wanting to surf the Cloud wave face one big challenge, which is how to evaluate how its business will be impacted. Currently, there is no mutually accepted methodology to allow the verification of this information, or to compare security between the organization's systems before and after migrating their resources to a Cloud. In this paper the authors discuss the implications of assessing Cloud security and how to compare two different environment's security in a way to provide enough resources for management to take decisions about migrating or not their systems to a remote datacenter. A practical method is proposed to assess and compare the organization system security before and after migration to a Cloud.

Keywords: Cloud security · Security assessment · OSSTMM 3 · Rav

1 Introduction

The difficulty in evaluating security of complex systems has been an obstacle in the adoption of Cloud by large organizations, since the security properties of such services is dependent of several factors, most of them out of the customer's control. One valid approach for an enterprise to verify how a Cloud provider would satisfy its own security requirements can be based on the execution of actual tests of that Cloud. However, considering that every company wanting to adopt services from a specific Cloud provider would have to run its own tests in the process of evaluation, it is possible to predict huge expenditures both in time and financial terms just to verify viability. This expenditure would even increase more if the process had to be performed for different projects in the same enterprise. Recently, the US Government has launched the FedRAMP project in 2012 to overcome this issue, and will allow participating agencies to jump into Amazon Cloud services with its projects without requiring a new evaluation for every project [1].

Nevertheless, not all the Cloud providers will be willing to allow prospectors to test their own resources against vulnerabilities, since some types of active tests will trigger security alarms that will be difficult to differentiate from actual threats. One way to

© Springer-Verlag Berlin Heidelberg 2015
A. Fred et al. (Eds.): IC3K 2013, CCIS 454, pp. 411–423, 2015.
DOI: 10.1007/978-3-662-46549-3_27

overcome this difficulty is for the Cloud Provider itself to run the security assessments and to make the results available to all prospectors. Using a specific methodology and well accepted metrics, this assessment can be instrumental in helping a company in the decision of migrating resources to a specific Cloud Provider. One such methodology that fulfills this requirement is the Open Source Security Testing Methodology Manual (OSSTMM) and its "rav" concept [2].

Using this "rav" concept, it is possible to compare the actual security of a system related to an optimal state, which can also be compared to other system. In other terms, one can verify the security of the enterprise systems when all resources are hosted in the internal network, and then compare with the security of the same enterprise when some of its resources are located in a Cloud. Although the systems are different, the "rav" will provide a metric that can be related and used for comparison in these very different situations.

In this paper, the authors will explore this characteristic of the OSSTMM methodology and propose an even more direct approach to verify how security is affected when the Cloud is a variable to take into consideration.

2 Problem

The decision to adopt and migrate critical information systems to the Cloud must consider several factors besides the direct financial benefits. Cloud adoption can have different effects in the global security of an enterprise network, and some of these effects will occur due to the assets transfer from the company site to the Cloud and the associated risk transfer [3].

The above concept can be generally expressed with the help of the following metaphor. There are three safes, each one protecting valuable assets: Gold, Silver and Bronze. There is not enough space in the owner's own property to store all three safes, so the owner will have to rent space somewhere to store one of them. The decision about which safe will be stored in the rented space must be based on:

- the importance of the safe contents (assets);
- how well protected the safe is now;
- how well protected will it be in the new place.

The knowledge of a safe existence and place of storage is a risk factor, and therefore should be avoided. For example, if it is brought to thief knowledge the existence of a safe full of valuable assets made of bronze stored in a given place, all of the other two safes will also be at risk of being stolen since they are stored in the same physical location. If the safe with bronze assets is stored in a different location, and someone tries to steal its contents, the other two safes will not be at risk.

The metaphor presented illustrates three characteristics of the concepts used in this paper, which are the following:

- Risk transfer;
- Improved security by limiting the number of visible targets;
- Influence of the surrounding environment over the asset's security.

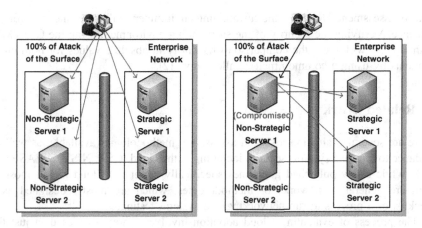

Fig. 1. Global attack surface.

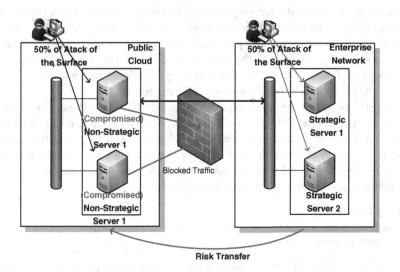

Fig. 2. Split attack surface.

The following image (Fig. 1) illustrates some possible attack vectors to four targets in a simplified view of a network (on the left diagram). If the attacker succeeds using any of the vectors, the compromised internal target can be used as a pivot to attack more important targets in the same network (in the right diagram).

Moving some of the targets to a separate environment will have two effects: the change in the exposure level of the moving targets, and the change in the exposure level of the remaining targets. This change can be for positive or negative, and it will depend on the targets moved, its importance and value, and how it interacts with other enterprise components (Fig. 2).

Every component in a given scope will contribute to the overall security of that environment [4], and this is the main reason why every target must be included in a security assessment. However, the surrounding environment will have an even stronger influence. Verifying the security of the surrounding environment using the OSSTMM methodology will allow the comparison of security in absolute values for both environments by using a common metric called "rav" [2].

3   Related Work

There are several evaluation, risk and assessment methodologies available, as well as guidance to improve systems security including in the cloud (CSA, NIST, OWASP and etc.). While a few published guidelines specifically target the Cloud [5, 6], most of them are general frameworks or methodologies, which are mostly designed as a checklist or a process to run networking and systems testing.

The process of evaluating Cloud adoption involves more variables than just the security properties of the systems and environment. It is necessary to start from the point where all administrative and political decisions have already been dismissed as non-blocking for the cloud adoption it is then necessary to define which resources are more appropriate to move. The technical decisions about what can be moved out can be evaluated against security metrics previously obtained from tests performed on predefined models. The specifics of the tests will not be covered, however, some guidelines should be followed to assure that Cloud specific issues are addressed, such as what to test, the approach, and how to evaluate the results.

OWASP has started a new chapter denominated "Cloud-10 Project" to approach Cloud security risks. The OWASP Top Ten list is important because it helps enterprises to focus on the most serious risks web applications have to face, and the Cloud-10 project is a work in progress that will address this new paradigm in enterprise computing. OWASP Top Ten list is maintained by a community of users and experts in every domain, and are ranked by criteria such as [7]:

- Easily Executable
- Most Damaging
- Incidence Frequency (Known)

The OWASP Cloud-10 project defines the criteria that can guide the security tests, however an appropriate testing methodology is still required. The Open Source Security Testing Methodology or OSSTMM has its focus on operational effectiveness, that is, how it works [8]. OSSTMM3 is an evolution from a penetration testing methodology which evolved to more than just a simple best practices framework by 2005 [2] and finally into a more contemporary security assessment methodology that prioritizes tests (avoiding guesses), concentrates on the interactions and its required protections, and balance between security and operations [8].

OSSTMM has redirected its focus in the earlier releases from testing physical resources such as firewalls and routers to verifying operational security and its related channels, such as Human, Physical, Wireless, Telecommunications, and Data Networks [2] in the latest versions of the methodology. OSSTMM also introduces its own measurement metric, called "rav", which provides a representation of system's states and how the system state changes over time, and are suitable to be used in operational monitoring consoles.

The Cloud Security Alliance (CSA) is a non-profit organization engaged in providing security awareness and tools to adopters. CSA has a specific publications providing guidance to Cloud security [5], which are structured around thirteen domains covering several aspects of Cloud security, including Identity and Access Management. CSA has also started the "Consensus Assessments Initiative" to provide means of documenting existing controls for Cloud services. This initiative is based on a questionnaire available at CSA web site, which can be downloaded, answered and then submitted to the repository of respondents where it can be consulted by customers.

Guidance is also provided by the U.S. Government, and targeted to U.S. Federal Agencies but publicly available. The "Proposed Security Assessment and Authorization for U.S. Government Cloud Computing" has a strong focus on authorization, defines a baseline of security controls and a monitoring process, and also proposes a framework to assess cloud security during vetting of Cloud Service Providers [6].

4   Background Concepts

The following image (Fig. 3) presents our baseline model. Everything inside the enterprise can be seen as a controlled environment, while everything in the outside is beyond its control [9, 10], and therefore, must not be trusted. However this does not means that an intranet is a safe place to run businesses without protection. According to the "2011 Cyber Security Watch Survey - How Bad Is the Insider Threat?" [11] conducted by Carnegie Mellon University to 607 companies, 27 % of all security incidents were caused by insiders in 2010, at the same time that 46 % of all respondents affirm that the internal incidents had caused more damage than the outside attacks.

In Fig. 4 some services have been extended from the internal enterprise network to a Public Cloud, while in Fig. 5 they were further extended to provide employee's access to the organization's resources in the Cloud.

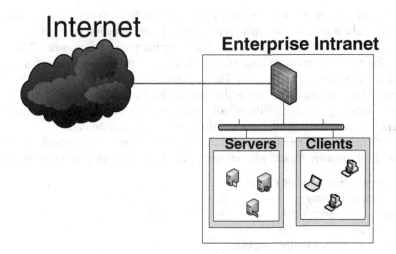

Fig. 3. Baseline.

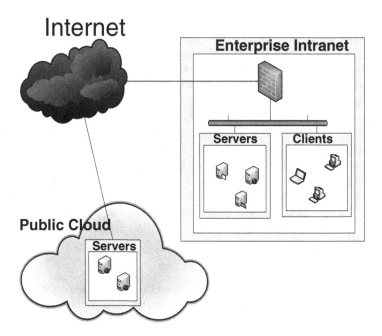

Fig. 4. Enterprise to cloud use Case 1.

Almost any enterprise application can be configured to work in a Public Cloud. However, two important factors must be considered:

- The Cloud is not under the Enterprise's control - therefore, it can be considered an uncontrolled environment [10].
- To work with the applications in the Public Cloud, it is necessary to cross a potential insecure channel: the Internet.

The subtle change in the level of trust in the surrounding environment is more than enough to make it unviable to apply an unprotected authentication protocol for services in a Public Cloud, so there must be a VPN connecting both environments. The risk of exchanging private, important data over the Internet in the above scenario is a show-stopper for most of the companies. The communication between the clients on the company's Intranet and the server in the Public Cloud can be intercepted in a number of ways, including the man-in-the-middle attack and the eavesdropping technique, to intercept and decode authentication information and application data.

Several factors must be considered to choose the resources to be migrated to the Cloud and not all of them are related with security, such as the following provided by [12]:

- The application criticality level
- The data sensitivity
- Functionality over VPNs
- Performance
- Cost to move to another provider
- Bandwidth usage

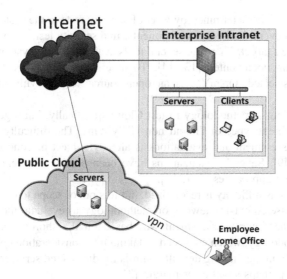

Fig. 5. Enterprise to cloud use Case 2.

Since there is not a specific methodology or framework to assess Cloud security, it was defined to use a generic, more contemporary methodology - OSSTMM - as the basis for the security assessment to be conducted. OSSTMM proposes a methodology to verify and test the Operational Security (OpSec) of systems [2].

Assessing Cloud security is not a trivial activity. For the Cloud, some parts of the OpSec procedures such as, physical security and the internal operational processes can only be assumed to be compliant with the enterprise's policies by means of terms of contracts and SLAs [10] from the perspective of the customer. Other technical security properties can be tested, but the lack of Cloud specific metrics [9] implies that it is necessary to propose a method to quantify security for these use cases, which we do by using the "rav" concept, from OSSTMM [2].

An OSSTMM-based security assessment will result in a numeric value representing the security level of the assessed system – the "rav". When there are several targets in the security assessment scope, the values obtained for all individual targets can be combined to produce a final "rav", representing the actual security for the whole system. The rav calculation can be simplified as [2]:

$$Rav = Controls - (Porosity + Limitations)$$

Where:

Porosity: The number of visible holes in the scope, which means that only what can be detected during the tests is accounted for the "rav".

Controls: The controls in place to protect the targets.

Limitations: Also known as "vulnerabilities", are derived from the porosity and the controls. The higher the porosity, higher will be the limitations. The less controls found, the higher will be the limitations.

The porosity can be determined by a set of security tests in the system, which may be composed by several individual components. In a security test, every component in the system will be a target. There are several tools available to test systems from several attack vectors and documented [13, 14]. To assess the models used in this work, OpenVAS was selected. OpenVas is an open-source security vulnerability scanner software.

OSSTMM is not a methodology to test Cloud specifically, but a generic methodology to test different types of IT and non-IT systems. The difficulty to standardize Cloud security assessment and evaluations is already subject of concerns in the professional sector [9]. Several organizations have been making different contributions, and using different approaches [5–7, 15].

In this paper, this difficulty is recognized, but it is out of scope to develop a specific Cloud security assessment framework or methodology to evaluate the results of our study. Instead, the security assessment was based on the shortcut proposed in the OSSTMM methodology, which consisted of taking into consideration only the Porosity and Limitations found, assigning default controls for discovered services and accepting an uncertain but perhaps small error margin [2].

Using the OSSTMM proposed "rav" metric it will be possible to find a security value for the baseline relative to the scope, and later compare with the values obtained from the assessment of the Cloud model. This comparison between different systems is supported by the methodology by using the concept of "Actual Security", which gives the actual security of any system in terms of "rav" (that can also be seen as a percentage). Using the "rav", it is possible to compare the security of two different systems and actually understand from each one, how much it is prepared for the threats against its attack surface [2].

In order to focus on specific Cloud security issues, the concepts presented in [9] were used. In those concepts the following Cloud-specific vulnerabilities natures were identified:

- is intrinsic to or prevalent in a core cloud computing technology,
- has its root cause in one of NIST's essential cloud characteristics,
- is caused when cloud innovations make tried-and-tested security controls difficult or impossible to implement, or
- is prevalent in established state-of-the-art cloud offerings.

From the above Cloud specific vulnerabilities, a testing process was defined to include:

- Assessment of the Cloud Web Management interface (dashboard);
- Assessment of systems from inside the Cloud;
- Assessment of the systems from inside the Cloud using a separate, hostile Tenant;
- Testing of all knows targets for known Common Vulnerabilities and Exposures (cve.mitre.org);

5 The Method

The method proposed in this paper follows the OSSTMM for assessing security, but the main contribution from that methodology is the "rav" metric. But instead of using the raw value of the "rav" that is obtained from the security assessments, the usage of a delta obtained from the comparison with a baseline model is proposed. Using the actual values computed for every model will give a perception of the security for that specific scenario, when composed by those specific assets with that specific configuration. On the other hand, using a delta will make possible to have an exact perception of the difference of security derived from that environment's influence. The method used in this work will result in a percentage value, which will be the final metric to ultimately take the conclusions about how security is affected by services migration to the Cloud. This percentage can therefore, be applied by different enterprises, or by the same enterprise in different projects when evaluating viability of moving resources to that Cloud. The metric is valid only for that Cloud, but different assessments can be performed for other Cloud infrastructures for comparison.

A security assessment using this methodology should always begin by defining a the baseline model that will be used as a reference for later comparison with other models. For a corporate scenario, the baseline should always be the actual organization private network.

The steps to apply this methodology are:

1. **Identification of Political and Administrative Issues.** Any identified blocking issue should stop the process and no further (operational) tests have to be done. The resource should not be migrated to the Cloud;
2. **Definition of a Security Test Guideline that can be applied to both the Enterprise and to the Cloud.** This test must include:

   - Scope for the enterprise tests. Must include an external security test (from the Internet). The internal test must be also part of the scope, although it may be defined around only those components that interact with the relevant targets.
   - Scope for the Cloud tests. Must include an external security test (from the Internet) and one test from inside the Cloud from a separate (hostile) tenant. A test from inside the tenant itself is optional, and depends on if the systems in the Cloud will have users other than the administrators.
   - Tools to use, including operating system and version where the tools will run, testing tools including version and knowledge base information;

3. **Perform the Cloud Security Test – or Request the Security Assessment Report from the Cloud Provider.** The report must contain metrics according to the OS-STMM methodology, and provide a final Actual Security value.
4. **Perform the Enterprise Security Test and Generate a Report.**

It is recommended to use the "rav" spreadsheet calculator: (http://www.isecom.org/research/ravs.html). In this method, the spreadsheet should be filled with the values obtained in the security tests, but the controls should have assigned default values taken from the field "Total" in the "Porosity" (visibility + access + trust).

The results can be analyzed in two ways:

(a) Compare the Actual Security for the two tests;
(b) Calculate the percentage variation for the Actual Security between the Enterprise security assessment and the Cloud security assessment.

The first method (a) is perfectly allowed by the OSSTMM methodology and provides non-related metrics [2] which can be used to have an overall perception of the security provided by the system.

The second method (b) provides a better perception of the "gain" or "loss" when comparing two systems. This method is our proposed methodology to evaluate the security impact of migrating resources to a Cloud when we have a previous reference. Both methods provide valuable decision information, but the second method will provide a better understanding of how much security will influence future use cases.

6   Case Study

This case study was developed in the context of this work, where the authors have setup labs to apply the method described in this paper. The baseline model (Fig. 3) was composed of one Windows Domain Controller, one internal Windows Web server, one Windows user workstation, one Linux box configured as both a default router and VPN for the entire network, and another Linux box running the public Internet Web Portal. The Internet Web Portal provided an internet presence as found in most enterprises, while the internal Web server is for private, internal use only by the enterprise's employees.

Regarding the physical architecture of the Use Cases, the baseline model was fully implemented in a VirtualBox virtualization environment, while the Enterprise to Cloud Use Cases were implemented using also an OpenStack IaaS Cloud. The Cloud was implemented in a single box, and made accessible from the Internet to allow the management of the services as required by a Public Cloud.

In Figs. 4 and 5, the Internet Web portal and the private internal Web server has been migrated from the organization's internal network to the Public Cloud.

The security assessment of the case study models were based in the shortcut proposed in the OSSTMM methodology, which consist of taking into consideration only the Porosity and Limitations found, assigning default controls for discovered services and accepting an uncertain but perhaps small error margin [2]. Based on that principle, the results presented below were obtained filling the "rav" calculator spreadsheet with the porosity and limitations detected by OpenVAS (Tables 1, 2 and 3 below). Default values were assigned to the controls, what makes it possible to compare different infrastructures using different controls, which is the most probable scenario when comparing an organization's infrastructure with a Cloud infrastructure. After running the security tests using OpenVAS and transporting the metrics to the "rav" calculator, it is possible to compare the results from the Baseline with results from two models (Use Case 1 and Use Case 2), and draw the conclusion that the security is not heavily impacted in the Cloud use cases, with a loss of 2,4601 % in the Use Case 1 and 2.8272 % in the Use Case 2 relative to the Baseline model (Table 1).

Applying the concepts proposed in our method, the variance obtained can be used as a reference value when the enterprise will require the evaluation of other migrations to the same Cloud, thus eliminating the necessity of running further security assessment to determine viability. This process can be used repeatedly by the enterprise which will consequently provide several benefits such as short project life cycles and cost reduction.

Table 1. Case study results.

	Baseline	Use Case 1	Use Case 2
Actual Security (rav)	77.5333	75,6259	75.3413
Variance	N/A	-2,4601 %	-2.8272 %

The above results were obtained from the vulnerability assessment which produced the following data:

Table 2. Porosities found in the case studies.

	Baseline	Use Case 1	Use Case 2
Visibility	101	148	159
Access	2	4	5
Trust	3	3	4

Table 3. Limitations found in the case studies.

	Baseline	Use Case 1	Use Case 2
Vulnerabilities	13	24	24
Weaknesses	28	32	32
Concerns	53	76	86
Exposures	15	328	341
Anomalies	0	0	0

The mapping between the Security Assessment and the inputs to the "rav" calculator is as follows (Table 4):

Table 4. Mapping porosity.

Item	How to get
Visibility	Number of servers in the use case + number of unique open ports for all servers
Access	Interactions points between the servers and the outside world
Trust	Interactions that do not require authentication

The Limitations in the rav calculator were mapped directly from the OpenVAS results as seen on the next table (Table 5).

Table 5. Mapping limmitations.

OSSTMM	OpenVAS
Vulnerability	High Severity
Weaknesses	Medium Severity
Concerns	Low Severity
Exposures	Log
Anomalies	False Positives

7 Conclusions and Future Work

Using the method proposed by this paper will make possible the existence of a practical and objective view of the security provided by a specific Cloud provider. Using the metrics from a security assessment over that Cloud, one can estimate how much the security of its assets will be influenced even before migrating them to that Cloud. Although there will be considerable resistance from the Cloud provider to allow for any customer to perform a security testing, the barrier can be overcome by standardizing the tests and defining the criteria to allow the Cloud provider to perform the security assessment and making the results publicly available.

Regarding to the method presented in this paper, the authors understand that some gaps must be addressed, such as the depth of the security assessment. In our case study, the tests were performed using the OpenVAS security vulnerability scanner software, which does not verify some Cloud specific issues such as VM isolation, memory sharing, storage sharing and reuse, among others. Therefore, a more in depth assessment should include a set of testing tools to verify for example virtualization robustness and isolation effectiveness between tenants.

Another improvement can be introduced in the verification of the Security Controls. In our study case, default controls for all the interaction points and vulnerabilities as supported by the OSSTMM methodology have been used. The Cloud providers can add much more value to their security assessment reports if they decide to actually verify and report all the controls implemented, thus reducing the error margin in the results. These controls may come from an existing source such as from the "Consensus Assessments Initiative" questionnaire [15].

References

1. Reuters: Amazon wins key cloud security clearance from government. http://www.reuters.com/article/2013/05/21/us-amazon-cloud-idUSBRE94K06S20130521
2. Herzog, P.: OSSTMM 3 – The Open Source Security Testing Methodology Manual – Contemporary Security Test and Analysis. Institute for Security and Open Methodologies (ISECOM) (2010)
3. European Network and Information Security Agency (ENISA): Cloud: Benefits, risks and recommendations for information security. http://www.enisa.europa.eu

4. Yildiz, M., Abawajy, J., Ercan, T., Bernoth, A.: A layered security approach for cloud computing infrastructure. In: 2009 10th International Symposium on Pervasive Systems Algorithms, and Networks, pp. 763–767. IEEE 978-0-7695-3908-9/09 (2009)

5. Cloud Security Alliance: Security Guidance for Critical Areas of Focus in Cloud Computing V2.1. http://www.cloudsecurityalliance.org/guidance/csaguide.v2.1.pdf

6. U.S. Chief Information Officer: Proposed Security Assessment and Authorization for U.S. Government Cloud Computing. http://educationnewyork.com/files/Proposed-Security-Assessment-and-Authorization-for-Cloud-Computing.pdf

7. OWASP: Cloud Top 10 Security Risks. https://www.owasp.org/index.php/Category: OWASP_Cloud_%E2%80%90_10_Project

8. Herzog, P.: Analyzing the Biggest Bank Robbery in History: Lessons in OSSTMM Analysis. Banking Magazine, 2/2011. http://hakin9.org/analyzing-the-biggest-bank-robbery-in-history-lessons-in-osstmm-analysis

9. Grobauer, B., Walloschek, T., Stöcker, E.: Understanding cloud computing vulnerabilities. IEEE Secur. Priv. 9(2), 50–57 (2011). doi:10.1109/MSP.2010.115

10. Hiroyuki, S., Shigeaki, T., Atsushi, K.: Building a security aware cloud by extending internal control to cloud. In: 2011 Tenth International Symposium on Autonomous Decentralized Systems, pp. 323–326. IEEE 978-0-7695-4349-9/11 (2011)

11. CERT: 2011 CyberSecurityWatch Survey - How Bad Is the Insider Threat? Carnegie Mellon University. http://www.cert.org/archive/pdf/CyberSecuritySurvey2011Data.pdf

12. Krutz, R., Vines, R.: Cloud Security: A Comphrehensive Guide to Secure Cloud Computing. Wiley Publishing, Indianápolis (2010)

13. Wilhelm, T.: Professional Penetration Testing. Elsevier Inc, Burlington (2010)

14. MacClure, S., Scambray, J., Kurtz, G.: Hacking Exposed: Network Security Secrets and Solutions. Oxborne, California (1999)

15. Cloud Security Alliance: Consensus Assessments Initiative. https://cloudsecurityalliance.org/research/cai

16. Mell, P., Grance, T.: The NIST Definition of Cloud Computing. NIST Special Publication 800–145. National Institute of Standards and Technology – U.S Department of Commerce. http://csrc.nist.gov/publications/nistpubs/800-145/SP800-145.pdf (2011)

Author Index